Numerical and Analytical Approaches in Magnetohydrodynamic Fluid Flow

Payam Jalili
Department of Mechanical Engineering, Islamic Azad University, North Tehran, Iran

Bahram Jalili
Department of Mechanical Engineering, Islamic Azad University, North Tehran, Iran

Pooya Pasha
Mazandaran University of Science and Technology, Iran

Davood Domiri
Babol Noshirvani University of Technology, Iran

IGI Global
Publishing Tomorrow's Research Today

Published in the United States of America by
IGI Global Scientific Publishing
701 East Chocolate Avenue
Hershey, PA, 17033, USA
Tel: 717-533-8845
Fax: 717-533-8661
E-mail: cust@igi-global.com
Website: https://www.igi-global.com

Library of Congress Cataloging-in-Publication Data

CIP Data Pending
ISBN: 979-8-3693-5335-6
eISBN: 979-8-3693-5337-0

Vice President of Editorial: Melissa Wagner
Managing Editor of Acquisitions: Mikaela Felty
Managing Editor of Book Development: Jocelynn Hessler
Production Manager: Mike Brehm
Cover Design: Phillip Shickler

British Cataloguing in Publication Data
A Cataloguing in Publication record for this book is available from the British Library.

Table of Contents

Preface

INTRODUCTION TO MAGNETOHYDRODYNAMIC FLUID FLOW

Magnetohydrodynamics (MHD) is a profound field that intersects the realms of fluid dynamics and electromagnetism. The study of MHD investigates how magnetic fields interact with electrically conducting fluids, such as plasmas, liquid metals, and electrolytes. This interaction creates complex and fascinating phenomena that are vital to understanding both natural processes and the development of advanced technologies.

The origins of MHD can be traced back to the early 20th century, with the pioneering work of Hannes Alfvén, who introduced the concept of magnetohydrodynamic waves, known today as Alfvén waves. These waves describe the propagation of disturbances in a magnetized plasma and are fundamental to many processes in astrophysics and space science. Alfvén's contributions to the field were so significant that they earned him the Nobel Prize in Physics in 1970, cementing his legacy as one of the founding figures of MHD.

Over the past century, MHD has evolved into a critical field of study with applications that span across numerous scientific disciplines. In astrophysics, for example, MHD plays a crucial role in modeling the behavior of plasmas in stellar and interstellar environments. The Sun, with its dynamic magnetic field and constantly shifting plasma, is a quintessential example of an MHD system. Understanding MHD processes in the Sun is vital for predicting solar flares, coronal mass ejections, and other space weather events that can have significant impacts on Earth.

On Earth, MHD is equally important in various engineering applications. The principles of MHD are employed in the design of electromagnetic pumps, which are used to move liquid metals in metallurgical processes without the need for mechanical parts. This is particularly useful in industries that handle molten metals, where traditional pumps would fail due to the extreme temperatures and corrosive nature of the fluids. Moreover, MHD is central to the development of nuclear fusion reactors, where magnetic fields are used to confine and control the hot plasmas necessary for sustaining fusion reactions. The success of nuclear fusion could revolutionize energy production, providing a nearly limitless and clean energy source, thus highlighting the importance of ongoing MHD research.

Despite its numerous applications, the study of MHD is fraught with challenges. The coupling between fluid dynamics and magnetic fields introduces additional forces—such as the Lorentz force and the magnetic pressure gradient—that complicate the governing equations. These added complexities often make it difficult to find analytical solutions, especially for real-world problems that involve irregular geometries, time-dependent behaviors, or turbulent flows. Consequently, researchers frequently rely on advanced numerical methods to solve the partial differential equations governing MHD systems.

This book, "Numerical and Analytical Approaches in Magnetohydrodynamic Fluid Flow," is designed to address these challenges by offering an integrated approach that combines both numerical and analytical techniques. By focusing on semi-analytical methods and the finite element method (FEM), this book provides researchers, engineers, and students with the tools needed to tackle complex MHD problems. The aim is not only to present the theoretical foundations of MHD but also to demonstrate how these principles can be applied to solve practical problems across various fields.

Purpose and Scope of the Book

The purpose of this book is to serve as a comprehensive resource for the study and application of MHD fluid flows, with a particular emphasis on numerical and analytical methods. While many texts cover the basics of MHD, few offer an integrated perspective that blends theoretical insights with practical solutions. This book fills that gap by providing a detailed treatment of the subject, covering both the mathematical underpinnings and the numerical techniques required for MHD research.

The scope of this book is intentionally broad, encompassing a wide range of topics within the field of MHD. Each chapter is dedicated to a specific aspect of MHD fluid flow, ensuring a thorough exploration of the subject matter. The chapters have been carefully organized to provide a logical progression from fundamental concepts to more advanced applications, making the book accessible to both beginners and experienced researchers.

1. **MHD in Porous Media:** Porous media, which include natural materials like soil and rock as well as engineered materials, are characterized by a network of interconnected pores through which fluids can flow. When these fluids are electrically conducting, the introduction of a magnetic field can significantly alter their behavior, leading to phenomena that are both theoretically intriguing and practically important. This chapter delves into the principles of MHD flow in porous media, focusing on the mathematical modeling and numerical simulation of such systems. The discussion includes applications in geothermal energy extraction, where MHD effects can enhance the efficiency of heat transfer in porous rocks, as well as in oil recovery processes, where magnetic fields can be used to improve the mobility of fluids in porous reservoirs.

The chapter also addresses the complexities involved in modeling MHD in porous media. Unlike simple fluid flows, MHD flows in porous media must account for the heterogeneous and anisotropic nature of the materials, as well as the interactions between the magnetic field and the fluid at the microscale. These challenges require sophisticated numerical techniques and a deep understanding of the underlying physics, making this chapter particularly valuable for those working in fields such as environmental engineering, petroleum engineering, and materials science.

2. **Entropy Analysis in MHD Fluid Flow:** Entropy, a measure of disorder or randomness in a system, is a key concept in thermodynamics and plays a crucial role in determining the efficiency of energy conversion processes. In MHD systems, entropy generation is influenced by various factors, including viscous dissipation, Joule heating, and magnetic field gradients. This chapter provides a comprehensive framework for entropy analysis in MHD fluid flows, offering insights into how these factors contribute to the overall thermodynamic efficiency of a system.

Entropy analysis is particularly important in the design of energy-efficient MHD systems. By understanding the sources of entropy generation, engineers can optimize system parameters to minimize irreversibilities and improve performance. This chapter explores several case studies, including MHD power generators and cooling systems, where entropy analysis has been used to enhance system efficiency. The chapter also introduces advanced techniques for entropy minimization, such as the use of optimized magnetic field configurations and the selection of fluids with favorable properties, making it a valuable resource for researchers focused on sustainable energy technologies.

3. **MHD Nanofluid Flow:** The study of nanofluids—fluids that contain nanoparticles—has opened up new possibilities for enhancing heat transfer in various applications. Nanofluids exhibit significantly higher thermal conductivities compared to conventional fluids, making them ideal for applications that require efficient heat dissipation. When combined with magnetic fields, nanofluids offer even greater potential for controlling and optimizing heat transfer processes. This chapter explores the theoretical and experimental aspects of MHD nanofluid flow, with a focus on both the enhancement of heat transfer and the manipulation of fluid flow using magnetic fields.

The application of MHD nanofluids extends to a wide range of industries, including electronics cooling, where efficient heat dissipation is critical to preventing overheating and ensuring the reliable operation of devices. In the biomedical field, MHD nanofluids are being explored for targeted drug delivery and hyperthermia treatments, where the magnetic field can be used to guide and control the distribution of therapeutic agents or to selectively heat and destroy cancerous tissues. This chapter provides a detailed overview of the current state of research in MHD nanofluids, including the latest developments in nanofluid synthesis, characterization, and application.

4. **Analytical Approach to MHD:** Analytical methods provides valuable insights into the fundamental behavior of MHD systems and serve as benchmarks for validating numerical simulations. This chapter presents a range of analytical techniques for solving the equations governing MHD flows, including similarity transformations, perturbation methods, and homotopy analysis. These techniques are applied to a variety of MHD problems, offering exact or approximate solutions that enhance our understanding of the underlying physics.

The chapter also discusses the scenarios in which analytical methods are most useful and their limitations. While analytical solutions are often more elegant and insightful than numerical solutions, they are typically restricted to simpler systems with linear equations or idealized boundary conditions. Nevertheless, analytical methods play a crucial role in developing intuition and guiding the design of numerical models. This chapter includes several examples of how analytical techniques have been successfully applied to solve MHD problems in the literature, making it an essential reference for researchers interested in the theoretical aspects of MHD.

5. **Unsteady MHD Fluid Flow:** Many practical MHD applications involve unsteady or time-dependent flows, where the fluid velocity, pressure, and magnetic field vary with time. Unsteady MHD flows are encountered in a variety of contexts,

from pulsatile blood flow in biomedical devices to transient heat transfer processes in engineering systems. This chapter provides a thorough examination of the mathematical modeling and numerical simulation of unsteady MHD flows, highlighting the challenges and opportunities associated with time-dependent phenomena.

Unsteady MHD flows are particularly relevant in systems where the magnetic field or fluid flow is deliberately varied over time, such as in electromagnetic propulsion systems or pulsed power applications. In these systems, the ability to accurately model and predict the transient behavior of the fluid and magnetic field is crucial for optimizing performance and ensuring stability. The chapter presents several case studies that illustrate the practical applications of unsteady MHD flows, including the design of MHD generators, cooling systems, and biomedical devices. The discussion also includes advanced numerical techniques for solving the time-dependent equations governing unsteady MHD flows, making this chapter a valuable resource for researchers and engineers working on dynamic systems.

6. **Magneto Ferrofluid Flow:** Ferrofluids, which are colloidal suspensions of magnetic nanoparticles in a carrier fluid, exhibit unique properties that make them highly responsive to external magnetic fields. The behavior of ferrofluids under the influence of magnetic fields is of great interest for a wide range of applications, from seals and dampers to biomedical devices. This chapter delves into the mathematical modeling and numerical simulation of ferrofluid flow, with a focus on the effects of magnetic fields on fluid behavior.

The chapter provides a comprehensive overview of the physical properties of ferrofluids, including their magnetic susceptibility, viscosity, and thermal conductivity. These properties make ferrofluids particularly useful in applications where precise control of fluid flow is required. For example, in magnetic sealing applications, ferrofluids act as a liquid O-ring that can be adjusted by changing the magnetic field, offering a flexible and reliable solution for sealing rotating shafts in high-performance machinery. In biomedical applications, ferrofluids are used for targeted drug delivery and magnetic resonance imaging, where their ability to be manipulated by magnetic fields allows for precise control over the distribution of therapeutic agents or the enhancement of imaging contrast. This chapter also explores the latest developments in ferrofluid synthesis and application, providing a valuable resource for researchers and engineers interested in this rapidly growing field.

7. **MHD in Microchannels:** The miniaturization of fluidic systems has led to the development of microfluidics, a field that deals with the behavior of fluids at the microscale. MHD microfluidics is a rapidly growing area of research with applications in lab-on-a-chip devices, biomedical diagnostics, and chemical processing. This chapter explores the unique challenges and opportunities associated with MHD flows in microchannels, including the effects of scaling on fluid behavior and the design of microfluidic devices.

Microfluidic systems offer several advantages over traditional fluidic systems, including reduced sample volumes, faster processing times, and greater control over fluid flow. MHD microfluidics enhances these advantages by allowing for the precise control of fluid flow using magnetic fields, without the need for moving parts or external pumps. This capability is particularly valuable in biomedical diagnostics, where the ability to manipulate small volumes of fluid is critical for the detection of biomarkers and the delivery of therapeutics. The chapter provides a detailed analysis of MHD microfluidics, including the design and fabrication of microchannels, the implementation of MHD forces, and the optimization of fluid flow. Additionally, the chapter discusses the potential for integrating MHD microfluidics with other microfluidic technologies, such as electrokinetics and acoustofluidics, to create multifunctional lab-on-a-chip devices.

8. **Heat Transfer Improvement in MHD:** Heat transfer is a critical aspect of many MHD applications, particularly in energy systems where efficient thermal management is essential for performance and reliability. This chapter examines the role of MHD in enhancing heat transfer, with a focus on the use of magnetic fields to control and optimize thermal processes. The chapter covers both theoretical models and experimental results, providing a comprehensive overview of the current state of research in this area.

Heat transfer in MHD systems is influenced by several factors, including the magnetic field strength, the fluid properties, and the flow geometry. By carefully controlling these factors, it is possible to enhance heat transfer and improve the efficiency of MHD systems. This is particularly important in applications such as MHD power generation, where the ability to manage and dissipate heat is critical for maintaining system performance and preventing thermal damage. The chapter also explores the use of MHD for cooling in electronics and other high-heat-flux applications, where the ability to precisely control fluid flow and heat transfer is essential for reliable operation. Additionally, the chapter discusses the potential for integrating MHD heat transfer techniques with other thermal management strategies, such as phase-change materials and thermoelectric cooling, to further enhance the performance of high-heat-flux systems.

Integration of Numerical and Semi-Analytical Approaches

One of the key themes of this book is the integration of numerical and semi-analytical methods for solving MHD problems. Numerical methods, such as finite element methods (FEM), are indispensable for solving complex, real-world problems where analytical solutions are not feasible. These methods allow for the precise modeling of fluid flow, magnetic fields, and heat transfer, providing detailed insights into the behavior of MHD systems.

However, numerical methods also have their limitations. They can be computationally expensive, requiring significant time and resources to obtain accurate solutions. Additionally, the complexity of the numerical models can sometimes obscure the underlying physics of the problem, making it difficult to identify key parameters or understand the fundamental mechanisms at play.

Semi-analytical methods, on the other hand, offer a more intuitive approach to solving MHD problems. These methods, which include techniques such as perturbation methods, homotopy analysis, and similarity transformations, provide approximate solutions that capture the essential features of the system while remaining mathematically tractable. By combining numerical and semi-analytical methods, it is possible to achieve a balance between accuracy and computational efficiency, allowing for the solution of complex MHD problems with greater insight and understanding.

This book demonstrates how numerical and semi-analytical methods can be used in tandem to tackle a wide range of MHD problems. Each chapter includes a discussion of the relevant numerical and analytical techniques, along with examples and case studies that illustrate their application. By providing a detailed exposition of both the theoretical developments and practical implementations, this book seeks to equip readers with the tools they need to advance the state of knowledge in MHD.

Relevance and Significance of the Work

The significance of this book lies in its comprehensive treatment of both the theoretical and practical aspects of MHD fluid flow. By integrating numerical and analytical approaches, the book provides a holistic view of the field, offering readers a deeper understanding of the complex interactions between magnetic fields and fluid flows. This understanding is crucial for the design and optimization of a wide range of systems that rely on MHD principles.

One of the key contributions of this book is its focus on finite element methods (FEM) in the context of MHD. FEM has become the method of choice for solving complex partial differential equations in various engineering disciplines, and its application to MHD problems has opened up new avenues for research and innovation. This book provides a detailed introduction to FEM for MHD, including the formulation of the governing equations, the implementation of boundary conditions, and the solution of the resulting algebraic systems. By providing both theoretical insights and practical guidelines, the book aims to help readers develop the skills needed to effectively use FEM in their own work.

Another important contribution of this book is its emphasis on entropy analysis in MHD systems. Entropy generation is a key factor in determining the efficiency of thermodynamic processes, and its minimization is essential for the design of energy-efficient systems. By analyzing the sources of entropy in MHD flows, this book provides a framework for optimizing the performance of MHD devices, with potential applications in areas such as power generation, cooling systems, and process engineering.

The book also addresses the emerging field of MHD nanofluids, which combines the benefits of nanofluid technology with the control provided by magnetic fields. This combination has the potential to revolutionize heat transfer processes, offering significant improvements in efficiency and performance. The chapter on MHD nanofluids provides a comprehensive overview of the current state of research in this area, including both theoretical models and experimental studies.

Furthermore, the book's focus on unsteady MHD flows is particularly relevant given the dynamic nature of many practical applications. Understanding the transient behavior of MHD systems is crucial for the design of systems that must operate under varying conditions, such as pulsatile flows in biomedical devices or time-varying magnetic fields in fusion reactors. By providing both numerical and analytical approaches to unsteady MHD problems, this book equips readers with the tools needed to tackle these challenging problems.

The interdisciplinary nature of MHD research is another important theme of this book. MHD fluid flow is a subject that bridges multiple fields, including physics, engineering, materials science, and applied mathematics. The cross-disciplinary approach taken in this book is intended to foster collaboration among researchers from different backgrounds, encouraging the exchange of ideas and the development of innovative solutions to complex problems. Whether you are a physicist studying the behavior of plasmas in space, an engineer designing advanced cooling systems, or a materials scientist exploring new ferrofluids, this book provides the knowledge and tools needed to advance your work.

In conclusion, this book offers a comprehensive and integrated approach to the study of MHD fluid flow, combining numerical and semi-analytical methods to address both fundamental and applied problems. Whether you are a researcher, engineer, or student, I hope this book will serve as a valuable resource in your exploration of the fascinating world of magnetohydrodynamics. The chapters that follow are designed to guide you through the complexities of MHD, offering insights and tools that will enhance your understanding and application of this critical field.

Payam Jalili
Islamic Azad University, North Tehran, Iran

Bahram Jalili
Islamic Azad University, North Tehran, Iran

Pooya Pasha
Mazandaran University of Science and Technology, Iran

Davood Domiri
Babol Noshirvani University of Technology, Iran

Section 1
Introduction to Magnetohydrodynamics (MHD) in Porous Media

This section covers the fundamental principles of MHD and its interaction with porous media, laying the groundwork for subsequent advanced topics.

Chapter 1
MHD in Porous Media

ABSTRACT

Magnetohydrodynamics (MHD) studies electrically conducting fluids' magnetic properties and behavior. It is also known as magneto-fluid dynamics or hydromagnetics. Plasmas, liquid metals, salt water, and electrolytes are examples of magneto-fluids. The term "magnetohydrodynamics" is derived from magneto-, meaning magnetic field; hydro-, meaning water; and dynamics, meaning motion. Hannes Alfvén pioneered the field of MHD, for which he was awarded the Nobel Prize in Physics in 1970. The basic idea behind MHD is that magnetic fields can induce currents in a moving conductive fluid that polarizes the fluid, changing the magnetic field. MHD is described by a set of equations that combine the Navier-Stokes equations of fluid dynamics and Maxwell's equations of electromagnetism. The main objective of this chapter is to study the magnetohydrodynamic flow in a porous medium using numerical and analytical methods.

NONLINEAR EQUATIONS ARISING IN MHD IN POROUS MEDIA

1.1. Steady-State Two-Dimensional Flow of a Micropolar/ Classical Nanofluid on a Stretching Sheet

A consistent Magnetohydrodynamic nanofluid flow is inspected over a stretching plate. In Figure 1, a steady-state two-dimensional flow of a micropolar/classical nanofluid is assumed on a stretching sheet with seven different baffles in the y > 0 direction. On the surfaces, triangular, rectangular, and chamfer baffles are used. T=67°C water streams from the left side. The water temperature is lower than the melting sheet temperature, which is 112 degrees. The surface pulls at a velocity of 2 m/s from both sides. The density of the surface is 8400. 625 is the specific heat value. The velocity of the stretched plate is ($U_W = ax, a>0$). The ambient temperature is T_{inf}, and the melting temperature is T_m (T_{inf} - bx_2). A vertically applied constant

DOI: 10.4018/979-8-3693-5335-6.ch001

magnetic field. The energy equation has considered the effects of Joule heating, viscous dissipation, and thermal radiation, (Peiravi *et al.*, 2019a) (Peiravi *et al.*, 2019b) (Adamo, 2020) (Yen, 2019) (Ganji, Periavi, & Abbasi, 2015) (Seyyedi, 2020) (Peiravi & Alinejad, 2021) (Neuman & Paul, 1970) (Sindhu, Gireesha, & Ganji, 2020) (Jalili et al., 2024a) (Peiravi & Alinejad, 2019) (Ghadikolaei, 2018).

The fractional differential conditions managing the facts are as follows: The managing half differential equations are:

$$\frac{\partial u}{\partial x} + \frac{\partial v}{\partial y} = 0 \tag{1.1.1}$$

$$\rho_f\left(u\frac{du}{dx} + v\frac{du}{dy}\right) = -\frac{dp}{dx} + \mu\left(\frac{d^2u}{dy^2}\right) - \sigma B_0^2 u - \frac{\mu}{k}u \tag{1.1.2}$$

$$\left(\rho C_P\right)_f\left(u\frac{dT}{dx} + v\frac{dT}{dy}\right)$$

$$= k_f\left(\frac{d^2T}{dy^2}\right) + \sigma B_0^2 u^2 - \frac{dq_r}{dy} + \mu\left(\frac{du}{dy}\right)^2 + \left(\rho C_p\right)_p\left[D_B\left\{\frac{dC}{dy}\cdot\frac{dT}{dy}\right\} + \left(\frac{D_T}{T_\infty}\right)\left\{\left(\frac{dT}{dy}\right)^2\right\}\right] \tag{1.1.3}$$

$$u\frac{dC}{dx} + v\frac{dC}{dy} = D_B\frac{d^2C}{dy^2} + \left(\frac{D_T}{T_\infty}\right)\left\{\frac{d^2C}{dy^2}\right\} \tag{1.1.4}$$

div(1/mu. grad(Az))+ Jz= 0 (1.1.5)

Boundary conditions are:

$$u = u_w = ax,\ v = 0,\ T = T_m,\ C = C_m,\ Jz = 1\ at\ y = 0 \tag{1.1.6}$$

$$T \leftrightarrow T_\infty,\ u \leftrightarrow 0,\ C \leftrightarrow C_\infty,\ at\ y \leftrightarrow \infty \tag{1.1.7}$$

$$k\left(\frac{dT}{dy}\right)\Big|y = 0 = \rho_f\left(L + C_s\left(T_m - T_0\right)\right)v(x,0)$$

(1.1.8)

Where v and u are the velocity parameters along the X and y directions, the non-uniform latent variable L (= L_{0x2}). The heat capacity of the solid sheet is Cs, while the heat capacity of the stream is h. The heat transfer to the melting sheet equals the heat melting plus the sensible heat required to raise the solid temperature T (=T_0–Cx_2) to its melting temperature T_m. The similarity transformation method has been utilized to obtain ordinary differential conditions. The following nondimensional parameters are presented.

$$\eta = \sqrt{\frac{a}{\vartheta}}y,\ u = axf'(\eta),\ v = -\sqrt{a\vartheta}\,f(\eta),\ \theta = \frac{T - T_m}{T_\infty - T_m},\ \varnothing = \frac{C - C_\infty}{C_1 - C_\infty}$$

(1.1.9)

By using the above information, the final equations are shown as follows:

$$f''' - \lambda f' - f'^2 + ff'' - Haf' = 0$$

(1.1.10)

$$\frac{1}{Pr}\left(1 + \frac{4}{3}Rd\right)\theta'' + f\theta' - 2f'\theta + 2f' + Ec\,Haf'^2 + Ecf''^2 + Nb\theta'\varnothing' + Nt\theta'^2 = 0$$

(1.1.11)

$$\varnothing'' + Sc(f\varnothing') + \frac{Nt}{Nb}\theta'' = 0$$

(1.1.12)

Moreover, the boundary conditions are as follows:

$$f'(0) = 1, f'(\infty) = 0, \theta(0) = 0, \theta(\infty) = 1, \varnothing(0) = 1, \varnothing(\infty) = 0, f(0) = -\frac{\delta}{Pr}\theta'(0)$$

(1.1.13)

The finite element strategy is a numerical strategy that must be followed. The program FlexPDE and CFD, which discovered the nonlinear semi-differential and conventional differential constraints, is one of the useful applications for this strategy. The code output of FlexPDE and CFD is also a 'problem-solving environment' that performs the entire capacity run required to solve fractional differential constraints. The following section explains ADM. A general nonlinear equation is represented by:

$L(u)+Ru+Nu=g(r)$ (1.1.14)

Using the inverse operator L^{-1} on both sides of (49) and the following conditions:

$U=f(x)-L^{-1}(Ru)-L^{-1}(Nu)$ (1.1.15)

In the Adomian decomposition method, the nonlinear operator Nu=F(u) is observed for nonlinear differential equations:

$$F(u) = \sum_{m=0}^{\infty} A_m$$

(1.1.16)

The Adomian method defines the solution U(x) by the series:

$$u = \sum_{m=0}^{\infty} u_m$$

(1.1.17)

$$F(u) = F(u_0) + F'(u_0)(u - u_0) + F''(u_0)\frac{(u - u_0)}{2!} + F'''(u_0)\frac{(u - u_0)}{3!} + \dots$$

(1.1.18)

By comparing terms, the primary few Adomian polynomials A_0, A_1, A_2:

$A_0=F(u_0)$

$A_1=u_1F'(u_0)$ $A_2=u_2F'(u_0)+$

$\frac{1}{2!}u_1u_1F''(u_0)$

(1.1.19)

Figure 2 shows water flow velocity between rectangular, triangular, and beveled baffles. The changes in the parameter for the flow velocity between the baffles were studied using the lines resulting from the flow of water on the plate with the different baffles. Since the liquid flow is calm at the beginning of the flow on the surface (where the pressure of the water liquid is equal to the ambient pressure), the boundary layer is formed according to the shear stress of the wall. The thickness of the boundary layer is increased by moving the fluid flow from the beginning to the end of the plate. The maximum thickness of the fluid velocity boundary layer occurs around the rectangular baffles since the largest velocity changes are observed around the plate and the baffles. It has high values corresponding to the velocity lines shown in the contour of the rectangular baffles. In the case of eccentric baffles, the fluid velocity changes are the highest in the first two areas, and the reflux is formed with negative pressure. The temperature of the water fluid varies around the rectangular, triangular, and beveled baffles as shown in Figure 3. The changes in the fluid temperature parameter between the vanes have been investigated based on the lines produced by the water fluid flow on the sheet with different vanes. The thermal boundary layer forms before the velocity boundary layer in this fluid flow as it passes over the flat sheet. The conductive heat transfer formula states that heat is initially transferred from the vanes' and sheet's surfaces to the fluid where the fluid's velocity is zero, while displacement takes place farther away from the heat transfer plate. Figure 3's illustrations show that the water fluid is the heater in places that are near a heated surface and that the intensity of heat transmission from the plate to the fluid diminishes with increasing distance from the surface. By putting these blades on the surface, we can increase the conductive heat transfer between the surface and the fluid because the range of chamfer-shaped blades is where the largest amount of heat transfer from the surface to the fluid happens. The degree of compression and heat transfer is also determined by Figures 2 and 3, which are related to the employment of oblong and triangular vanes. By placing oblong and triangular vanes on the surface, only in the initial phase at the end does the temperature of the fluid change, but the heat transfer from the surface to the fluid is not observed in the middle baffles. The fluid concentration fluctuations around the rectangular, triangular, and chamfer baffles are depicted in Figure (1.1 .4).The highest fluctuations in water fluid concentration are seen around the chamfer-shaped baffles, as shown in the two-dimensional figures below. Given that the temperature of the fluid has a direct correlation with the concentration and that the quantity of heat exchange between the surface and the water fluid in these areas is larger than in other areas, the concentration changes with high values. Additionally, the figure on the screen can lower the fluid concentration by inserting triangle baffles. Using elongated and chamfer-shaped baffles increases the heat transfer from the surface with high temperature to the fluid with low temperature, which is self-explanatory for increasing the concentration of the fluid. Figure (1.1 .5) illustrates how the water flow velocity changes around triangular, rectangular, and chamfered baffles. When using rectangular baffles, the water fluid flow velocity reached its maximum at the location $x = 0.4$ with a value of $u = 2.86$ on the tensile sheet. The heat exchange between the fluid and the melting sheet is also minimal at this point due to the low viscosity of the fluid. The screen's installation of rectangular baffles results in significant variations in the fluid's passage velocity. Table 1 shows that the maximum water fluid velocity occurs in the x=0.45 for rectangular baffles, and the minimum velocity occurs in the x=0 for triangular fins by u=-1.31. The maximum water fluid velocity is u=0.45m/s, and the minimum is u=-1.31m/s. The average water fluid velocity on the rectangular baffles is 24.5% bigger than triangular baffles and 45.34% bigger than chamfer modes. The maximum temperature change exists around the triangular baffles, and the thermal boundary layer in this position is larger than in other cases. The maximum water fluid temperature is T=7.07 degrees, and the minimum fluid temperature is T=-3.23 degrees around the triangular baffles. The average water fluid temperature on the chamfer baffles is 38.5% bigger than rectangular baffles and 75.34% bigger than triangular modes. Table 3 shows that the maximum water fluid concentration occurs in the x=0.01 for chamfer baffles, and the minimum concentration occurs in the x=0.2 for chamfer fins by C=-2.31. The maximum water fluid temperature is C=2.09, and the minimum is C=-2.31. The average water fluid concentration on the chamfer baffles is 36.5% bigger than triangular baffles and 55.24% bigger than chamfer modes.

This study investigates the effect of a homogeneous magnetic field on temperature, velocity, and water fluid concentration through triangular, rectangular, and chamfer baffles. The employment of oblong, triangular, and chamfered vanes on a plate to maximize heat transmission and fluid velocity variations in the melting plate region and compare the different types of vanes constitutes the novelty of this article. Let's move on to the finest option for placement on the screen and discuss how to use blades to boost productivity and velocity in the workplace. For the purpose of resolving governing equations, the finite element method is chosen.

Figure 1. Schematic of the treated stretching sheet

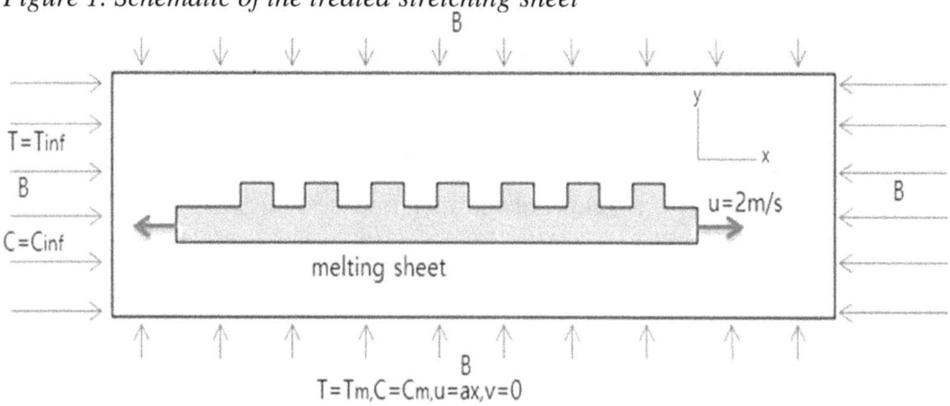

Figure 2. Comparison of velocity changes in the different baffles for stretching sheet by CFD code

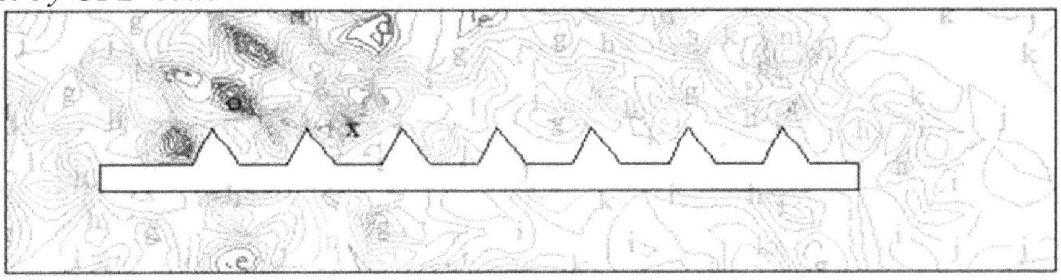

a) Triangular baffles near the top wall

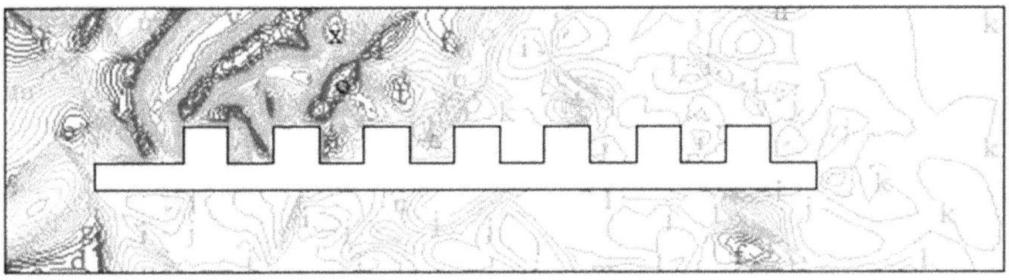

b) Rectangular baffles near the top wall

c) Chamfer baffles near the top wall

Figure 3. Comparison of temperature changes in different baffles for stretching sheet by CFD code

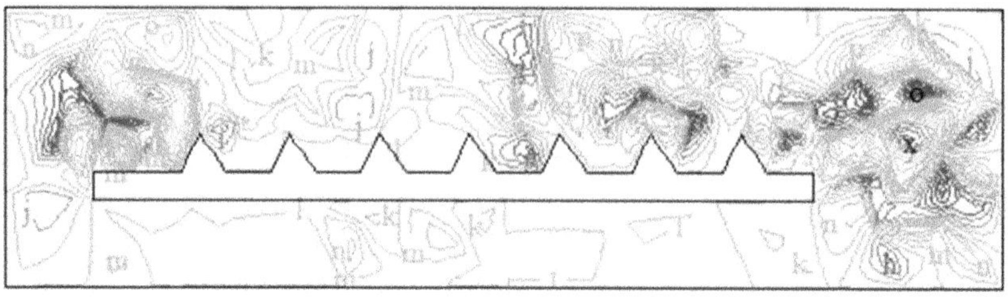

a) Triangular baffles near the top wall

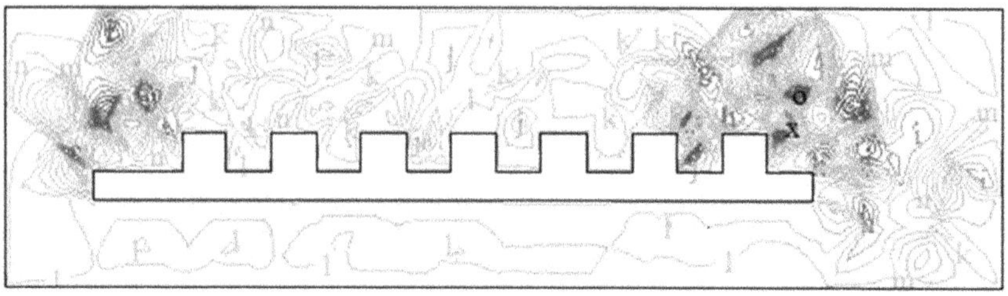

b) Rectangular baffles near the top wall

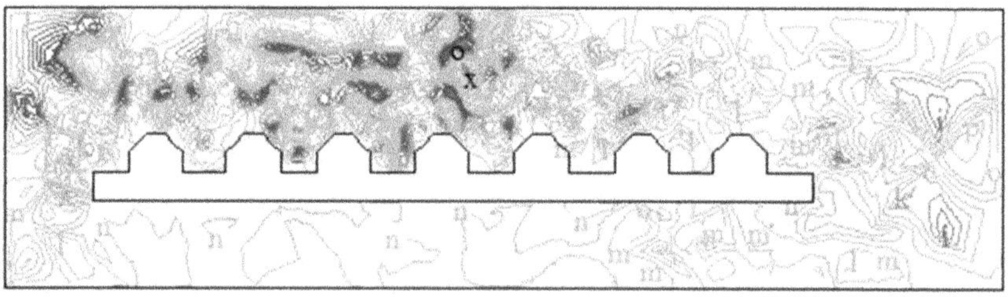

c) Chamfer baffles near the top wall

Figure 4. Comparison of concentration changes in different baffles for stretching sheet by CFD code

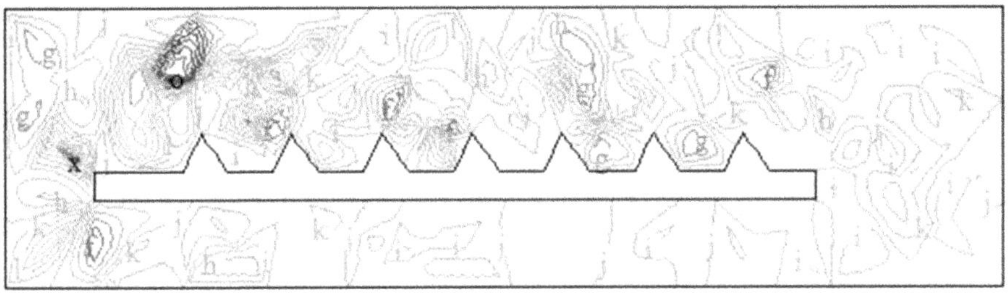

a) Triangular baffles near the top wall

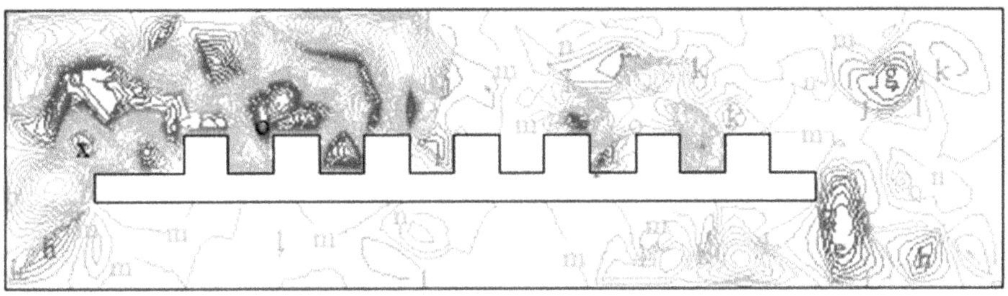

b) Rectangular baffles near the top wall

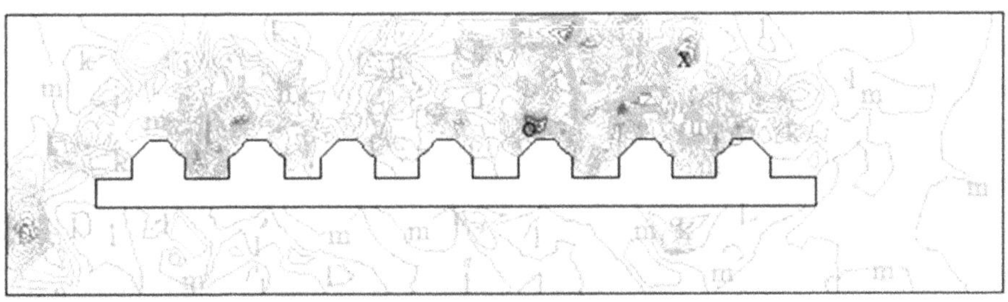

c) Chamfer baffles near the top wall

Figure 5. Comparison of velocity (u) changes in the rectangular, triangular, and chamfer baffles for stretching sheet

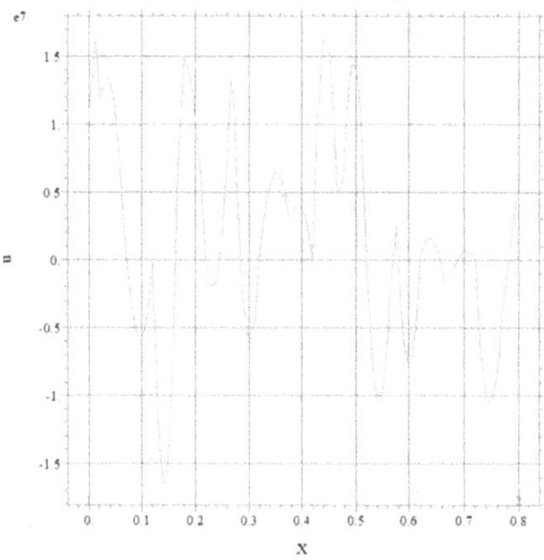

a) Triangular baffles near the top stretching plate

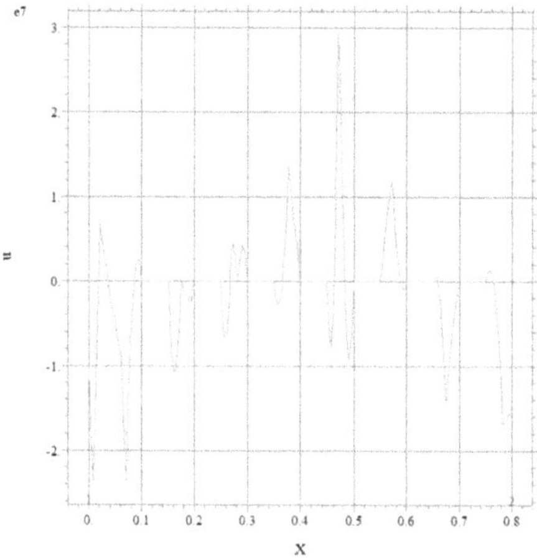

b) Rectangular baffles near the top stretching plate

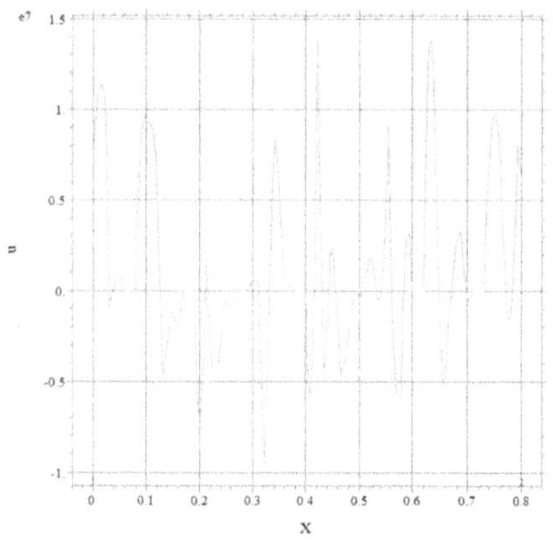

c) Chamfer baffles near the top stretching plate

Table 1. Numerical comparison of water fluid velocity between different baffles

X	$u_{Triangular\ baffle}$	$u_{Rectangular\ baffle}$	$u_{chamfer\ baffle}$
0.0	1.76	-1.90	1.31
0.2	1.04	0	0.18
0.3	-0.09	0.15	0.05
0.5	1.43	2.90	0.06
0.7	0.35	0.02	0.44
0.8	0.5	-1.21	0.77

Table 2. Numerical comparison of water fluid temperature between different baffles

X	$T_{Triangular\ baffle}$	$T_{Rectangular\ baffle}$	$T_{chamfer\ baffle}$
0.0	6.46	0	0.54
0.2	7.07	1.28	0.06
0.3	6.24	0	0.63
0.5	-1.70	0	0.03
0.7	-3.23	0.04	0.75
0.8	0	6.76	0.76

Table 3. Numerical comparison of water fluid concentration between different baffles

X	$C_{Triangular\ baffle}$	$C_{Rectangular\ baffle}$	$C_{chamfer\ baffle}$
0.0	-0.22	1	2.09
0.2	-1.46	-0.04	-2.65
0.3	0.61	0	0.05
0.6	1.82	0.03	0
0.7	0	0.38	2.58
0.8	0.02	0.05	1.09

1.2. Effect of Magnetic Field and Porous Medium on the Stretched Plate

Thread stretching sheet with rectangular, triangular, and chamfer baffles have surveyed the influence of magnetic field and porous medium. The flow, velocity, and heat transfer characteristics are optimized with rectangular, triangular, and chamfer baffles. The boundary conditions for the water flow problem on the stretching sheet and baffles are illustrated in Figure 6. The water flows from the left side with T=170 . The water temperature is more than the melting surface, and the melting surface temperature is 110 degrees. The surface pulls from two sides at a velocity of 2 m/s. The surface density is 8400. The specific heat value is 625. The ambient temperature is T_{inf}, and the melting temperature is $T_m = (T_{inf} - bx_2)$. A vertically applied constant magnetic field. The effects of Joule heating, viscous dissipation, and thermal radiation have been examined in the energy equation, (Jalili et al., 2024b) (Dey & Borah, n.d.) (Goud, 2020) (Bahmani *et al.,* 2024) (Nayak, 2019) (Peiravi, Pasha, & Ganji, 2021) (Gao *et al.,* 2022) (Zar *et al.,* 2024).

The partial differential equations that govern the situation are as follows, the governing partial differential equations are:

$$\frac{\partial u}{\partial x} + \frac{\partial v}{\partial y} = 0 \tag{1.2.1}$$

$$\rho_f\left(u\frac{du}{dx} + v\frac{du}{dy}\right) = -\frac{dp}{dx} + \mu\left(\frac{d^2u}{dy^2}\right) - \sigma B_0^2 u - \frac{\mu}{k}u \tag{1.2.2}$$

$$\left(\rho C_P\right)_f\left(u\frac{dT}{dx} + v\frac{dT}{dy}\right)$$
$$= k_f\left(\frac{d^2T}{dy^2}\right) + \sigma B_0^2 u^2 - \frac{dq_r}{dy} + \mu\left(\frac{du}{dy}\right)^2 + \left(\rho C_p\right)_p\left[D_B\left\{\frac{dC}{dy}\cdot\frac{dT}{dy}\right\} + \left(\frac{D_T}{T_\infty}\right)\left\{\left(\frac{dT}{dy}\right)^2\right\}\right] \tag{1.2.3}$$

$$u\frac{dC}{dx} + v\frac{dC}{dy} = D_B\frac{d^2C}{dy^2} + \left(\frac{D_T}{T_\infty}\right)\left\{\frac{d^2C}{dy^2}\right\} \tag{1.2.4}$$

Where the radiation heat flux q_r is calculated using the Rosseland approximation, and $qr = -\frac{4\sigma_e}{3\beta_R}\frac{dT^4}{dy}$ where σ_e, β_R, are the Stefan-Boltzman constant and the mean absorption coefficient. The temperature variations between the fluid phases in the flow are considered to be modest enough for T_4 to be stated as a linear function of temperature. Boundary conditions are:

$$\{u = u_w = ax, v = 0, T = T_m, C = C_m, at\ y = 0\}$$

$$(1.2.5)$$

$$\{T \leftrightarrow T_{\infty}, u \leftrightarrow 0, C \leftrightarrow C_{\infty}, at\, y \leftrightarrow \infty\}$$

$$(1.2.6)$$

$$k\left(\frac{dT}{dy}\right)\Big|y = 0 = \rho_f(L + C_s(T_m - T_0))v(x,0)$$

$$(1.2.7)$$

Where v and u are the velocity parameters along the X and y directions, the non-uniform latent variable L (= L_{0x2}). The heat capacity of the solid surface is Cs, while the heat capacity of the fluid is h. The heat transfer to the melting surface is equal to the heat melting plus the sensible heat required to raise the solid temperature T (=T_0–c_{x2}) to its melting temperature T_m. The similarity transformation method has been utilized to obtain ordinary differential equations. The following nondimensional parameters are introduced.

$$\eta = \sqrt{\frac{a}{\vartheta}}y, u = axf'(\eta), v = -\sqrt{a\vartheta}f(\eta), \theta = \frac{T - T_m}{T_{\infty} - T_m}, \varnothing = \frac{C - C_{\infty}}{C_1 - C_{\infty}}$$

$$(1.2.8)$$

By using the above transformation, the final equations are obtained as follows:

$$f''' - \lambda f' - f'^2 + ff'' - Haf' = 0$$

$$(1.2.9)$$

$$\frac{1}{Pr}\left(1 + \frac{4}{3}Rd\right)\theta''$$

$$+f\theta' - 2f'\theta + 2f' + Ec\, Ha\, f'^2 + Ecf''^2 + Nb\theta'\varnothing' + Nt\theta'^2 = 0$$

$$(1.2.10)$$

$$\varnothing'' + Sc(f\varnothing') + \frac{Nt}{Nb}\theta'' = 0$$

$$(1.2.11)$$

Moreover, boundary conditions become:

$$f'(0) = 1, f'(\infty) = 0, \theta(0) = 0, \theta(\infty) = 1,$$

$$\varnothing(0) = 1, \varnothing(\infty) = 0, f(0) = -\frac{\delta}{\Pr\theta'(0)}$$

(1.2.12)

Flexpde software is one of the practical applications for the Finite Element Method, which is used to solve nonlinear partial and ordinary differential equations. FlexPDE is also a 'problem-solving environment' that performs the functions required to solve partial differential equation systems.

The current paper describes in detail the last events in this paper. The fluid flow in this paper is water. The temperature of the water is higher than that of a melting surface. The water temperature is 170 degrees Fahrenheit, and the melting surface temperature is 110 degrees Fahrenheit. The surface pulls from both sides at a velocity of 2 m/s. The density of the surface is 8400. 625 is the specific heat value. The images below show the velocity contours in the direction x in all different baffle configurations:

Figure 7 depicts the various states of the shape, triangular, and chamfer baffle surfaces. The fluid velocity near the triangular baffles is less than near the rectangular and chamfer baffles in the images (e-7)). In contrast to the three-baffle model, the fluid flow at five baffles is higher, and the red area in front of the plate is much larger at five baffles. The amount of relative velocity around the surface generally increased by adding more baffles to previous baffles on the surface. The above figures (Figure 8) denote the velocity value of fluid (v) around the plate. The rate of the fluid velocity in five rectangular baffles is more than three and four rectangular baffles on the surface. In Figure. (b-8), by passing fluid flow over the rectangular baffles, the velocity (v) value decreased, and the maximum fluid velocity was observed near the chamfer baffles. The temperature contours of the fluid around the plates with different baffles are shown in Figure 9. The temperature decreased, and the temperature gradient was low for three rectangular baffles. The temperature around the baffles is much lower in the case of four baffles compared to the preceding ones. The temperature of the fluid flow is balanced in the case of five baffles at (g-9), (h-9), and (i-9). The temperature of the fluid flow is low. The fluid temperature is balanced by reducing the number of baffles to three. In the case of four triangular baffles, there is an area at the beginning of the surface at the top of the baffle where the temperature is maximum. Comparing the case of four rectangular baffles with a triangular shape, the fluid flow temperature around the surface is much higher than the liquid temperature in the four rectangular baffle modes. In all cases, the fluid temperature is balanced. Figure 10 depicts the fluid concentration around the baffles. The fluid concentration around the chamfer baffles has less value than other modes. The fluid concentration around the baffles is in equilibrium and has a moderate value. The maximum fluid concentration near the triangular baffles is shown in Figure (e-10). The fluid concentration is high near the surface. The fluid concentration near the chamfer baffles is lower in Figure (h-10) than near the rectangular and triangular baffles. The three baffles at the top of the surface's surface are shown in Figure 11 to show how the velocity of the fluid changed. The fluid flow velocity around the first baffle in Figure (a-11) is the largest and reaches the value u=7. The flow rate then decreases until it reaches a minimum of 0.5 in the third baffle. The rate of velocity changes is substantially higher for the chamfer baffles than it is for the other two options. The chamfer mode's velocity fluctuation around the baffles has a very high maximum value of u=1.7 and a very low minimum value of u=-0.9. The first baffle's maximum velocity lies between x=0.2 and x=0.3. The first baffle (T=19 degrees) in Figure a-12 showed the highest heat and temperature.

The temperature drop is particularly high at x=0.8 with fluid flowing over the top of the baffles, and it will reach T=0.2 in the fourth baffle. The highest temperature is seen near the fourth baffle in triangle-shaped baffles. At x=0.6, the temperature is at its highest point (T=1), while at x=0.8, it is at its lowest point. The greatest temperature change was seen with chamfer baffles. The greatest temperature in these baffles is T=3, while the lowest temperature is T =2.5.

The influence of magnetic field and porous media on thread stretching sheets with rectangular, triangular, and chamfer baffles has been studied. The analysis of fluid flow over a stretching sheet in the presence of a magnetic field uses the finite element model. Additionally, the magnetic parameter, porous parameter, velocity profile, temperature, and fluid concentration were studied in this research.

- The temperature of the fluid flow flowing over the triangle baffles is extremely low and falls between -1 and -2. The temperature rises as the fluid moves through the first and second baffles, peaking in the third blade.
- The fluid temperature vector surrounding triangular baffles is more intense than other modes. The area around it has a higher temperature gradient than before, and the fluid flowing out of the end of the plate is also greater than the rectangular baffles.
- The fluid concentration is higher near the triangle baffles than in other modes. The chamfer baffles on the surface are where the most fluid is concentrated. Their concentration increases from the first baffle to a value of 2.88 and then decreases to 1.4 in the final baffle. The chamfer baffles provide a 55% larger fluid concentration for fluid flow than conventional baffles.

Figure 6. Schematic of hydro magnetic effect on different thread stretching sheets

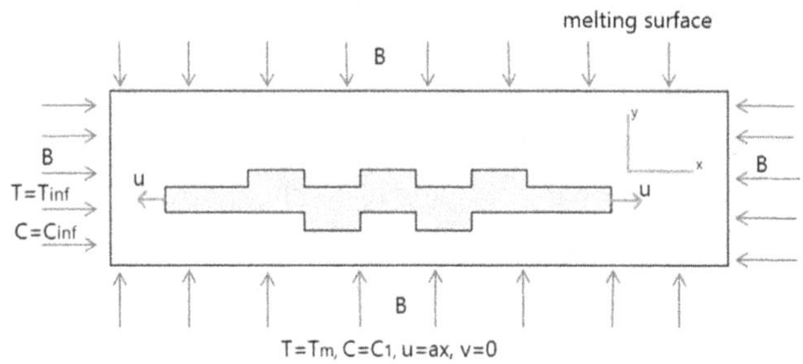

a) **Stretching sheet with rectangular baffles**

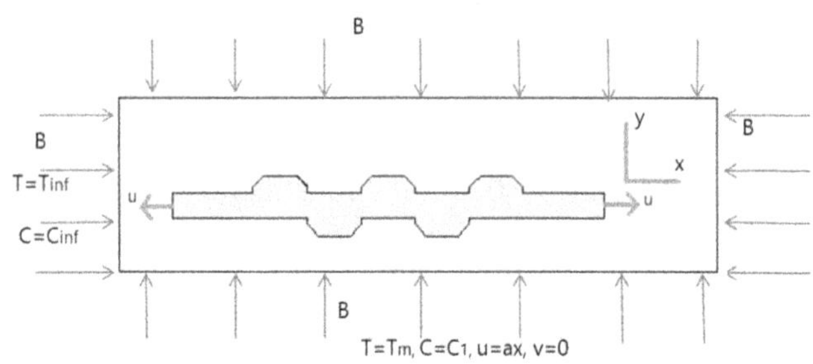

b) **Stretching sheet with chamfer baffles**

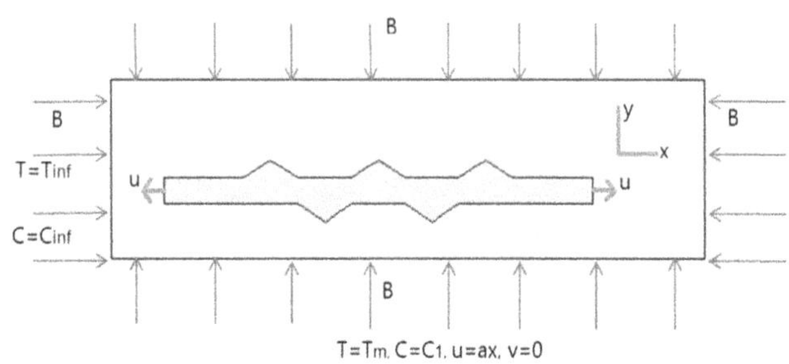

c) **Stretching sheet with triangular baffles**

Figure 7. Comparison of different values of u for rectangular and triangular, and chamfer baffles

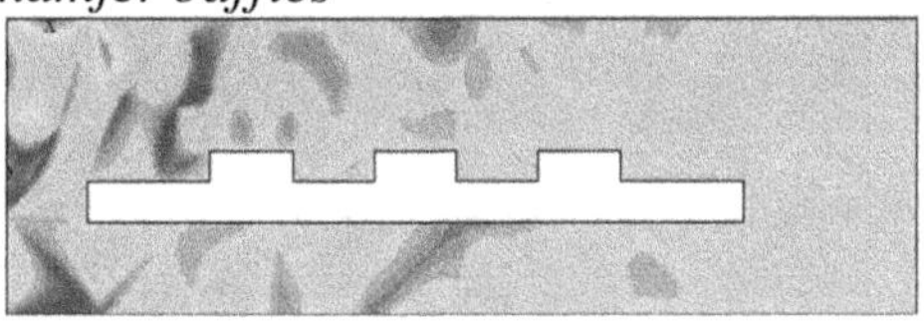

a) Stretching sheet with three rectangular baffles

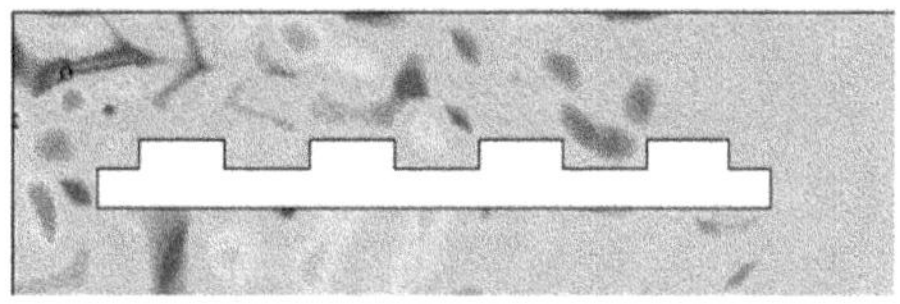

b) Stretching sheet with four rectangular baffles

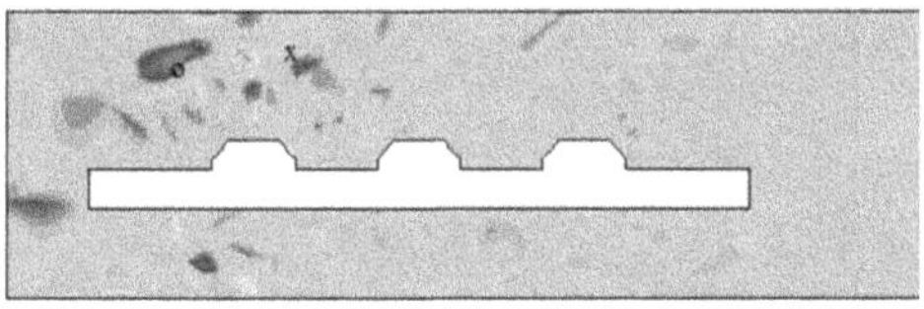

c) Stretching sheet with three chamfer baffles.

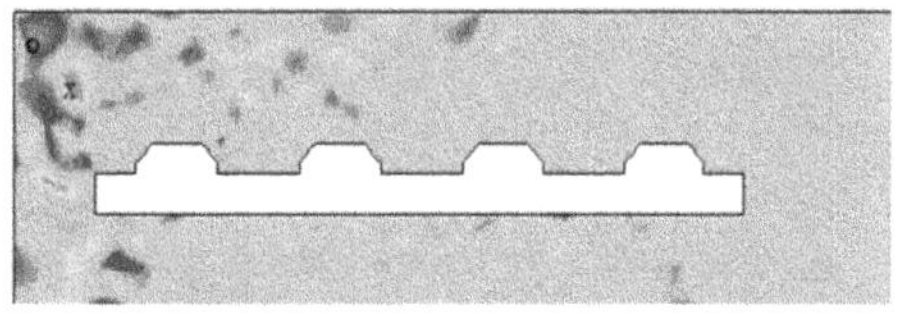

d) Stretching sheet with four chamfer baffles.

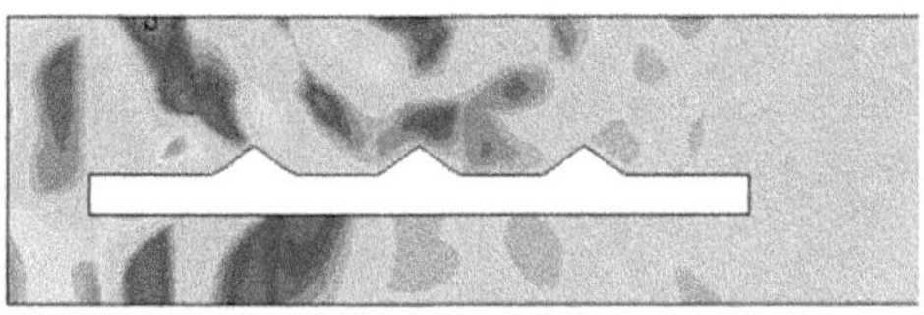

e) Stretching sheet with three triangular baffles.

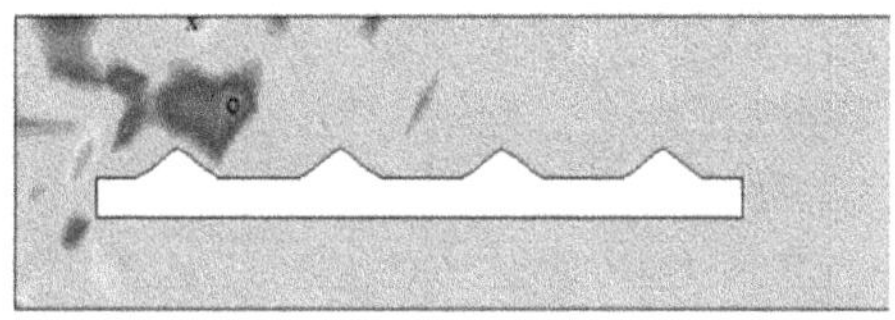

f) Stretching sheet with four triangular baffles.

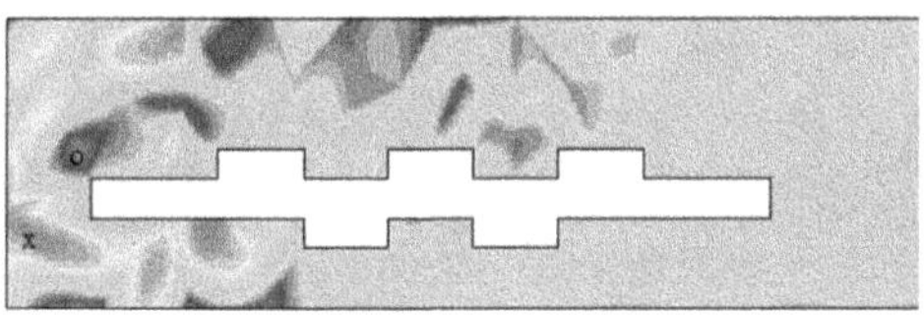

g) Stretching sheet with five rectangular baffles.

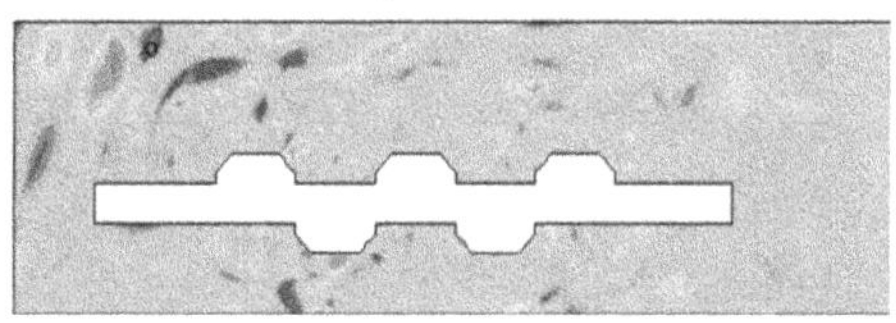

h) Stretching sheet with five chamfer baffles.

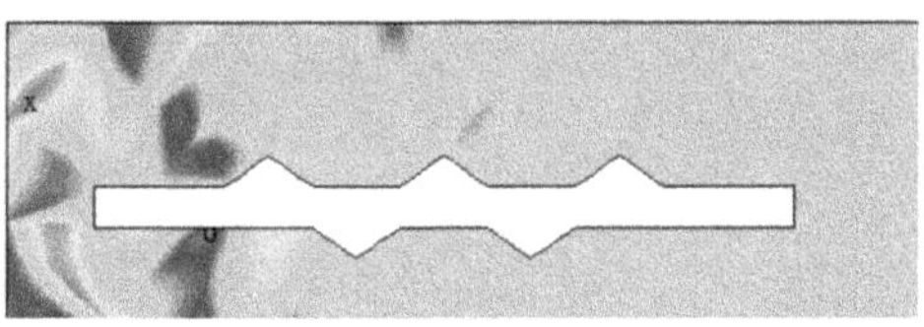

i) Stretching sheet with five triangular baffles.

Figure 8. Comparison of different values of v for rectangular and triangular, and chamfer baffles

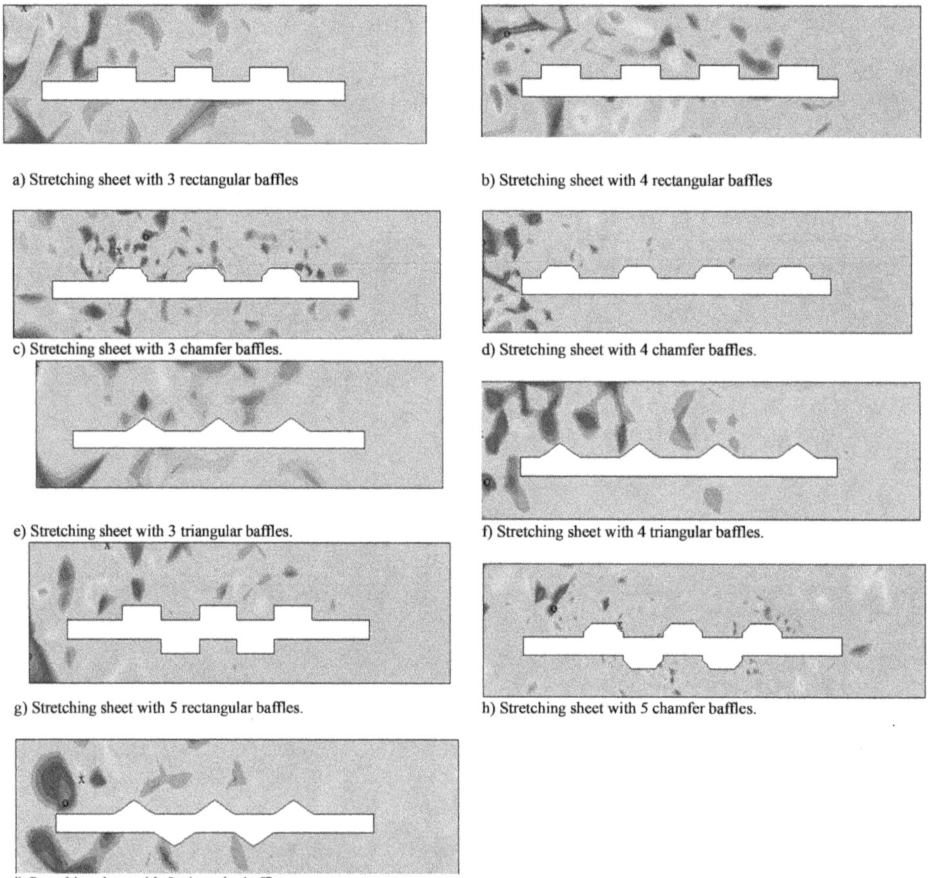

a) Stretching sheet with 3 rectangular baffles

b) Stretching sheet with 4 rectangular baffles

c) Stretching sheet with 3 chamfer baffles.

d) Stretching sheet with 4 chamfer baffles.

e) Stretching sheet with 3 triangular baffles.

f) Stretching sheet with 4 triangular baffles.

g) Stretching sheet with 5 rectangular baffles.

h) Stretching sheet with 5 chamfer baffles.

i) Stretching sheet with 5 triangular baffles.

Figure 9. Comparison of different temperature values for rectangular and triangular and chamfer baffles

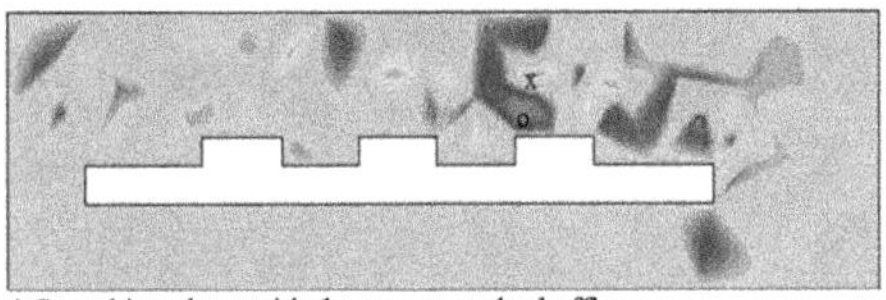

a) Stretching sheet with three rectangular baffles.

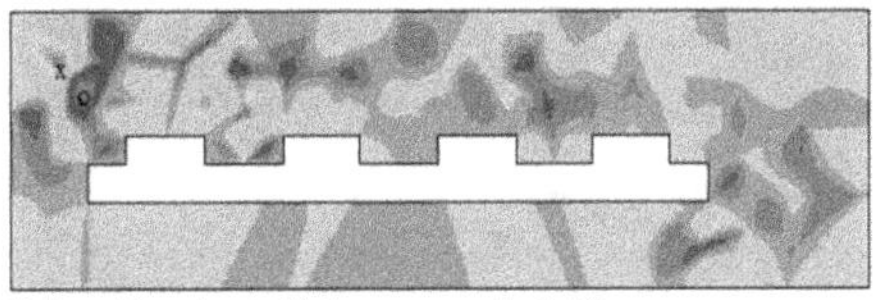

b) Stretching sheet with four rectangular baffles.

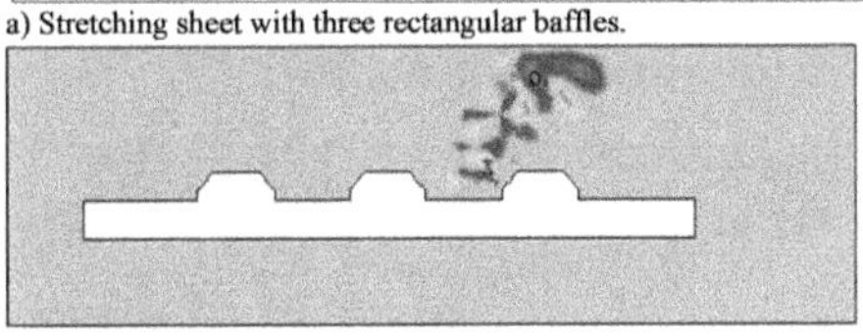

c) Stretching sheet with three chamfer baffles.

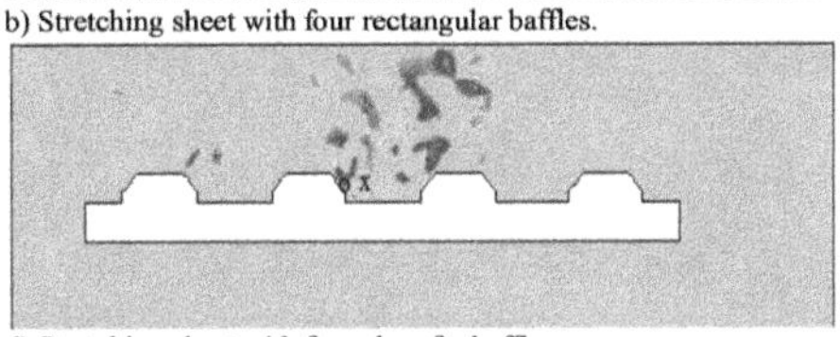

d) Stretching sheet with four chamfer baffles.

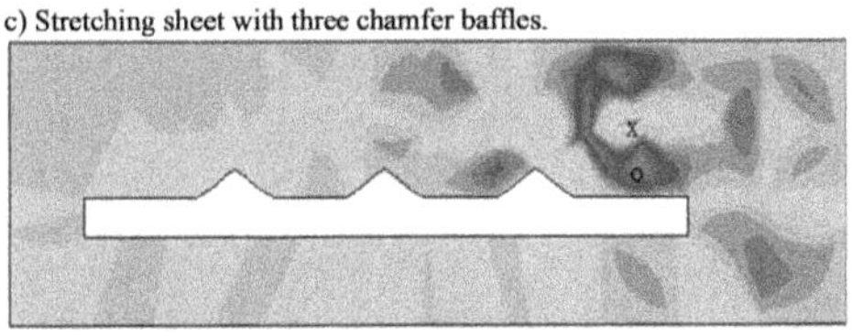

e) Stretching sheet with three triangular baffles.

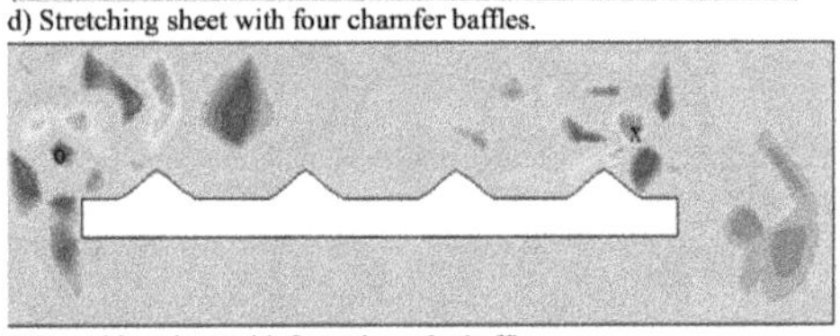

f) Stretching sheet with four triangular baffles.

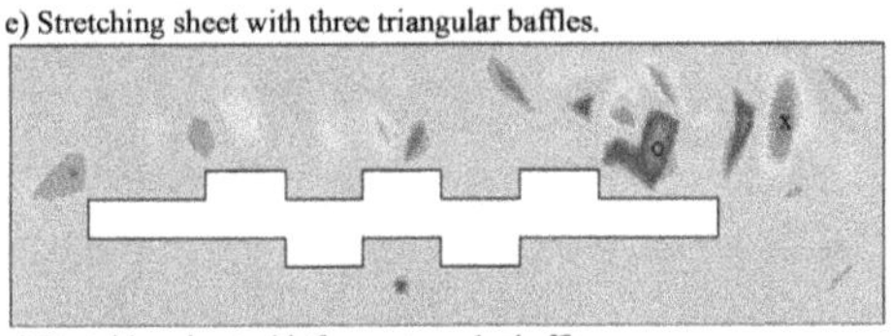

g) Stretching sheet with five rectangular baffles.

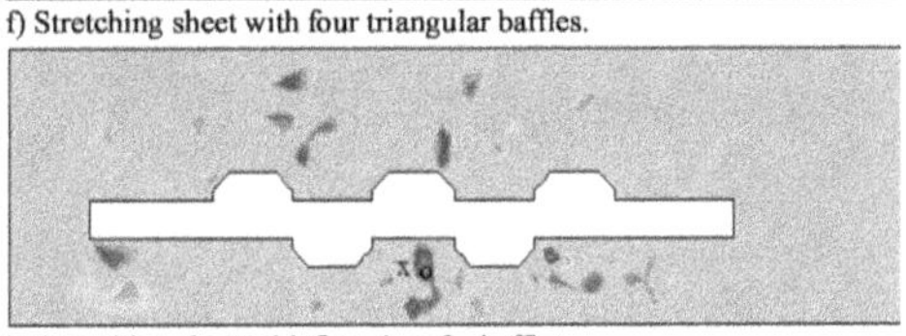

h) Stretching sheet with five chamfer baffles.

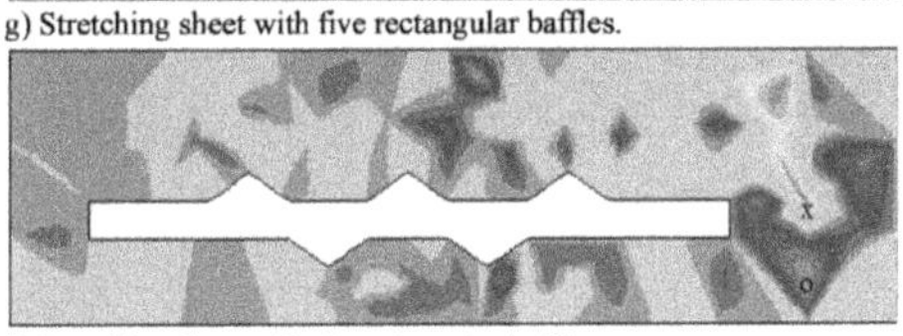

i) Stretching sheet with five triangular baffles.

Figure 10. Comparison of different concentration values for rectangular and triangular, and chamfer baffles

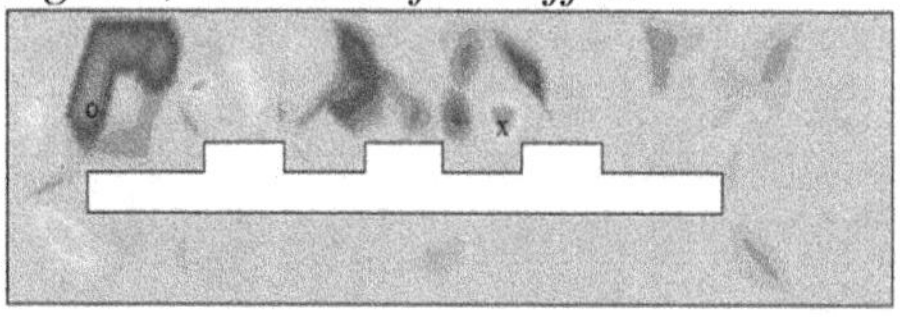

a) Stretching sheet with three rectangular baffles.

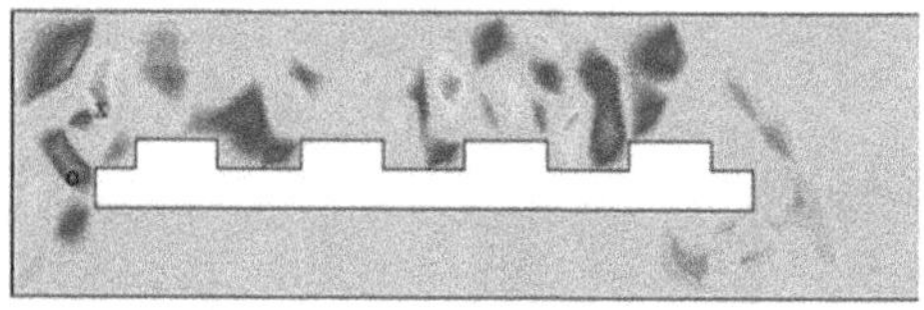

b) Stretching sheet with four rectangular baffles.

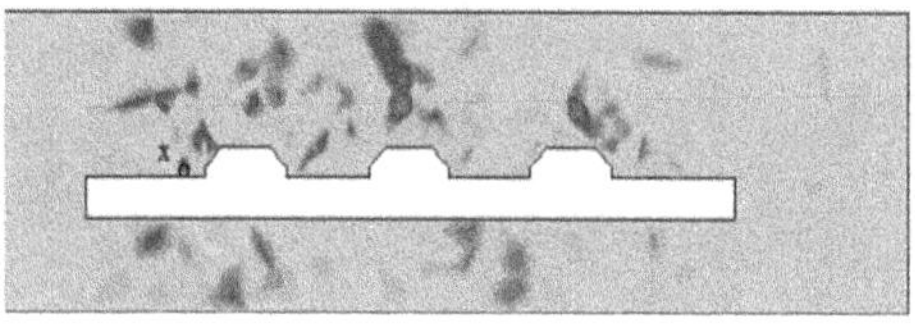

c) Stretching sheet with three chamfer baffles.

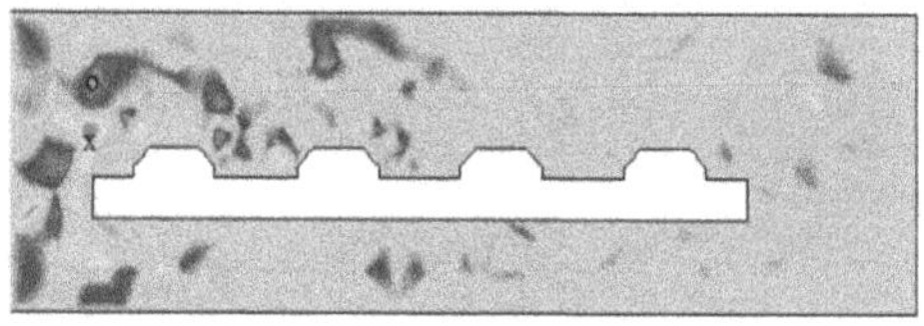

d) Stretching sheet with four chamfer baffles.

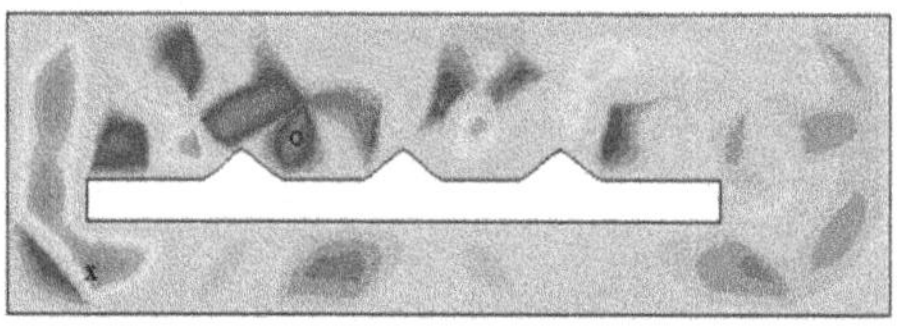

e) Stretching sheet with three triangular baffles.

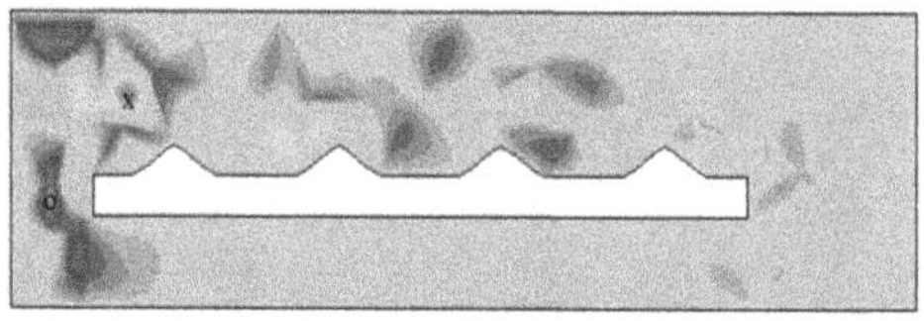

f) Stretching sheet with four triangular baffles.

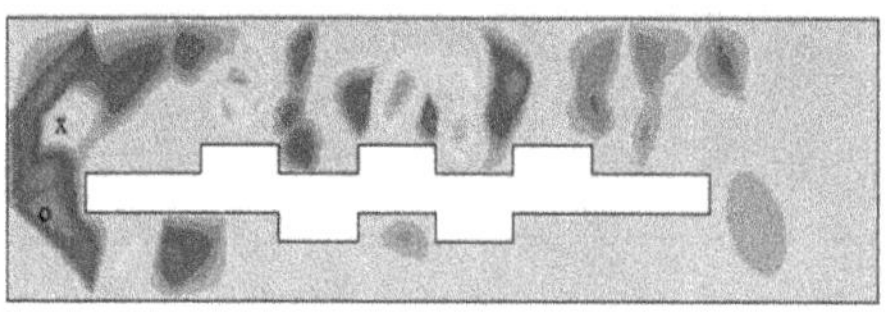

g) Stretching sheet with five rectangular baffles.

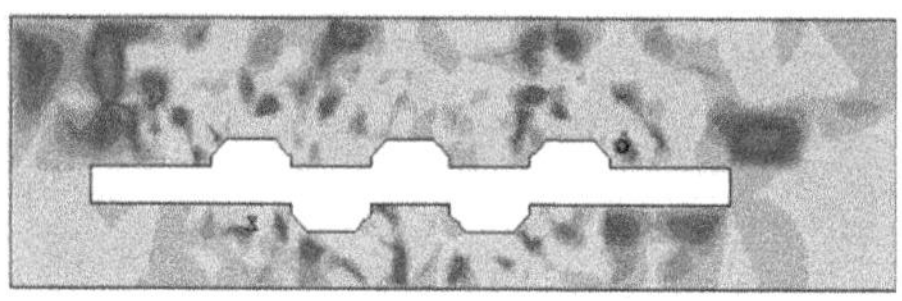

h) Stretching sheet with five chamfer baffles.

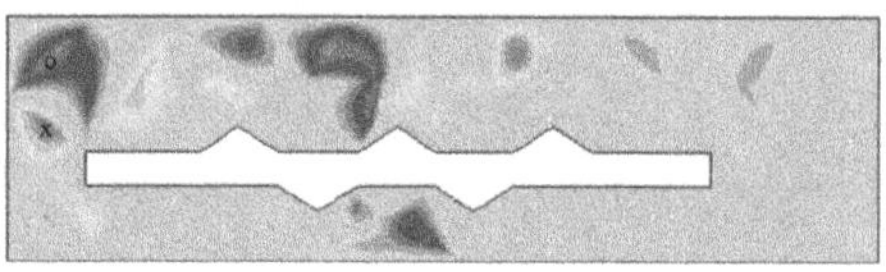

i) Stretching sheet with five triangular baffles.

Figure 11. Comparison of different velocity (u) values for rectangular and triangular, and chamfer baffles

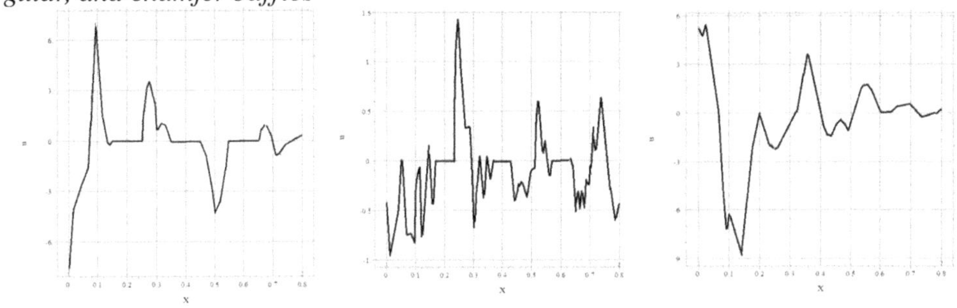

a) Stretching sheet with rectangular baffles. b) Stretching sheet with chamfer baffles. c) Stretching sheet with triangular baffles.

Figure 12. Comparison of different values of temperature for rectangular and triangular, and chamfer baffles

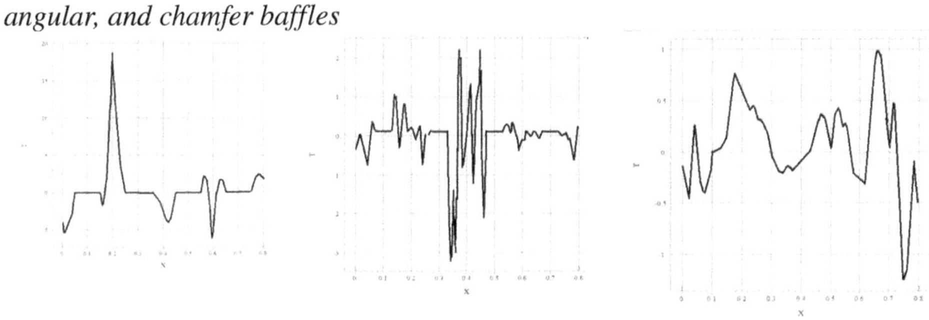

a) Stretching sheet with rectangular baffles b) Stretching sheet with chamfer baffles c) Stretching sheet with triangular baffles

1.3. A Numerical Study of Magnetic Nanofluid Natural Convection in the Porous Enclosure With Brownian Motion

The natural convection of magnetic nanofluids in a porous enclosure is investigated using Brownian motion. Figure 13 depicts this work's sample grid distribution and physical model. The inner and outer circular walls are kept at constant temperatures, T_h and Tc ($T_c < T_h$), respectively, while the other two walls are insulated. In addition, the effect of a constant magnetic field of strength is B_0. The continuity, momentum, and energy are assumed to be constant under the Boussinesq approximation and the assumptions above (Zhang et al., 2022) (Vatani, Ghasemi, & Ganji, 2016) (Alizadeh et al., 2021) (Pasha, Mirzaei, & Zarinfar, 2021) (Gnaneswara et al., 2021) (Muhammad, Hayat, & Alesaedi, 2021) (Waini, Ishak, & Pop, 2021).

Equations for the current work can be defined as:

$$\frac{\partial u}{\partial x} + \frac{\partial v}{\partial y} = 0 \tag{1.3.1}$$

$$u\frac{\partial u}{\partial x} + v\frac{\partial u}{\partial y} = -\frac{1}{\rho_{nf}}\frac{\partial p}{\partial x} + \frac{\mu_{nf}}{\rho_{nf}}\left(\frac{\partial^2 u}{\partial x^2} + \frac{\partial^2 u}{\partial y^2}\right)$$
$$+\frac{\sigma_f \beta_0^2}{\rho_{nf}}\left(- u\sin^2\gamma + v\sin\gamma\cos\gamma\right) - \frac{v_{nf}}{K}u \tag{1.3.2}$$

$$u\frac{\partial v}{\partial x} + v\frac{\partial v}{\partial y} = -\frac{1}{\rho_{nf}}\frac{\partial p}{\partial y} + \frac{\mu_{nf}}{\rho_{nf}}\left(\frac{\partial^2 v}{\partial x^2} + \frac{\partial^2 v}{\partial y^2}\right) + \beta_{nf}g(T - T_c)$$
$$+\frac{\sigma_f \beta_0^2}{\rho_{nf}}\left(- v\cos^2\gamma + u\sin\gamma\cos\gamma\right) - \frac{v_{nf}}{K}v \tag{1.3.3}$$

$$u\frac{\partial T}{\partial x} + v\frac{\partial T}{\partial y} = \frac{k_{nf}}{\left(\rho C_p\right)_{nf}}\left(\frac{\partial^2 T}{\partial x^2} + \frac{\partial^2 T}{\partial y^2}\right) \tag{1.3.4}$$

Here u, T, B_0, K, σ_f, γ, and v depict the velocity in the x direction, temperature, magnetic field, permeability of the porous medium, fluid's electric conductivity, angle of magnetic field parameter and velocity in the y direction, respectively. Further, ρ_{nf} and $(\rho C_p)_{nf}$ depict the effective density and heat capacitance of nanofluid which are defined as follows:

$$\rho_{nf} = (1 - \varnothing)\rho_f + \varnothing\rho_s \tag{1.3.5}$$

$$\left(\rho C_p\right)_{nf} = \left(1 - \varnothing\right)\left(\rho C_p\right)_f + \varnothing\left(\rho C_p\right)_s \tag{1.3.6}$$

The thermal conductivity and viscosity of nanofluid are expressed by Corcione model in which the impacts of particle size, particle volume fraction, and temperature dependence on the thermal conductivity are considered.

$$\frac{K_{nf}}{K_f} = 1 + 4.4 Re_B^{0.4} Pr^{0.66} \left(\frac{T}{T_{fr}}\right)^{10} \left(\frac{K_s}{K_f}\right)^{0.03} \varnothing^{0.66}$$

(1.3.7)

$$\mu_{nf} = \mu_f / \left(1 - 34.87 \left(\frac{d_s}{d_f)^{-0.3} \varnothing^{1.03}}\right)\right)$$

(1.3.8)

$$Re_B = \frac{d_s u_B \rho_f}{\mu_f}$$

(1.3.9)

$$u_B = \frac{2 k_b T}{\pi \mu_f d_s^2}$$

(1.3.10)

Here k_b, Pr, and T_{fr} depict the Boltzmann's constant, the Prandtl number of the base liquid, and the base liquid's freezing point, respectively. The nanofluid thermo-physical properties are demonstrated in Table 4. The vortices and stream functions are expressed as follows:

$$u = \frac{\partial \varphi}{\partial y}, v = -\frac{\partial \varphi}{\partial x}, w = \frac{\partial v}{\partial x} - \frac{\partial u}{\partial y}$$

(1.3.11)

describe the following dimensionless variables:

$$X = \frac{x}{L}, Y = \frac{y}{L}, \Omega = \frac{\omega L^2}{\alpha_f}, \quad = \frac{\varphi}{\alpha_f}, U = \frac{uL}{\alpha_f}, V = \frac{vL}{\alpha_f}, \theta = \frac{T - T_c}{T_h - T_c}$$

(1.3.12)

By applying the variables mentioned above, the governing equations reduce to non-dimensional form:

$$\frac{\partial}{\partial Y}\frac{\partial \Omega}{\partial x} - \frac{\partial}{\partial x}\frac{\partial \Omega}{\partial Y} = \frac{\mu_{nf}\rho_f}{\mu_f\rho_{nf}}\Pr\left(\frac{\partial^2 \Omega}{\partial x^2} + \frac{\partial^2 \Omega}{\partial y^2}\right) + \frac{\beta_{nf}}{\beta_f}RaPr\frac{\partial \theta}{\partial X}$$

$$+\frac{\rho_f}{\rho_{nf}}Ha^2\Pr\left(\frac{\partial^2 \varphi}{\partial Y^2}sin^2\gamma + 2\frac{\partial^2}{\partial x\partial Y}\sin\gamma\cos\gamma + \frac{\partial^2}{\partial X^2}cos^2\gamma\right) - \frac{\nu_{nf}}{\nu_f}\frac{Pr}{Da}\Omega$$

(1.3.13)

$$\frac{\partial}{\partial Y}\frac{\partial \theta}{\partial x} - \frac{\partial}{\partial x}\frac{\partial \theta}{\partial Y} = \frac{K_{nf}}{K_f}\frac{(\rho C_p)_f}{(\rho C_p)_{nf}}\cdot\left(\frac{\partial^2 \theta}{\partial x^2} + \frac{\partial^2 \theta}{\partial y^2}\right)$$

(1.3.14)

$$\frac{\partial^2}{\partial X^2} + \frac{\partial^2}{\partial Y^2} = 0$$

(1.3.15)

Subject to the boundary conditions:

θ = 1 on the inner circular wall
θ = 0 on the outer circular wall

$$\frac{\partial \theta}{\partial n} = 0, \quad = 0$$

(1.3.16)

The governing equations of this paper associated with the boundary conditions were solved numerically using the Control Volume Finite Element Method (CVFEM). To procure mesh independence, the average Nusselt number is calculated for different mesh sizes. Table 5 represents that the grid of 121 × 161 should be selected for the following study. Moreover, the present numerical solution is validated by comparing our FORTRAN code outcomes with other works. They are in a great compromise, as shown in Table 6.

The effects of controlling parameters such as Rayleigh number (Ra = 10^3, 10^4, and 10^5), Hartmann number (Ha = 0-50), Darcy number (Da = 10^{-1}, 10^{-2}, and 10^{-3}), and angle of magnetic field parameter (γ= 0°-60°) on the natural convection of magnetic nanofluid in the porous enclosure have been studied numerically in this work. The effect of Brownian motion is also taken into account. Figures 14 and 15 show the effects of the Rayleigh and Hartmann numbers on the streamlines and isotherms, respectively. Because the maximum absolute value of the stream function has a direct relationship with the Rayleigh number and an inverse relationship with the Hartmann number, the intensity of natural convection increases as the Rayleigh number increases and decreases as the Hartmann number increases. In fact, adding a magnetic field to the system increases Lorentz forces, which oppose buoyancy forces. As a result, magnetic fields can be used as a controlling parameter in various heat transfer applications.

Furthermore, ascending Ha causes the central streamline to stretch and the fluid to rise up along the outer circular wall. This is especially true for Ra = 10^5. It is worth noting that for low Rayleigh numbers, conduction is the dominant mode of heat transfer over convection. This is because, at low Ra, the viscous forces outnumber the buoyancy forces. However, at high Ra, buoyancy forces overcome vicious forces, and convection becomes the dominant mode. As a result, increasing Rayleigh number causes isotherms to distort, resulting in temperature stratification, especially at higher Ra and lower Ha. As shown, the thermal boundary layer forms along the inner and outer circular walls.

On the other hand, because conduction is the dominant mode of heat transfer at higher Hartmann numbers, ascending values of Ha result in isotherm parallelization. In other words, the thermal boundary layer will fade along the inner and outer circular walls. Figures 16 and 17 show the effects of the magnetic field's Darcy number and inclination angle on the streamlines and isotherms, respectively. It is worth noting that the Darcy number is directly related to the permeability of a porous medium, according to its definition. The permeability of a porous medium determines the flow intensity inside the enclosure. It is observed that increasing permeability, which results in an increase in the Darcy number, intensifies flow circulation in the enclosure.

In contrast, decreasing permeability, which results in a lower Darcy number, results in a weak flow circulation. The fluctuation of the local and average Nu number (Nu_{loc}. and Nu_{AVE}.) is shown in Figure 18 for various values of the Rayleigh number, Hartmann number, Darcy number, and magnetic field inclination angle. It is evident that the local Nusselt number's highest value always occurs close to the adiabatic wall at the bottom. The maximum value of the local Nusselt number rises with rising Rayleigh, Darcy, and magnetic field inclination numbers but falls with rising Hartmann numbers. By raising the Rayleigh number, the location of the Nu_{loc} maximum value can also be deduced.

Using CVFEM, a numerical study of magnetic nanofluid natural convection in the porous enclosure with Brownian motion is conducted. The comparison of CVFEM and other works shows that this method is valid. The findings show that the Rayleigh number, Darcy number, and magnetic field inclination angle directly correlate with the convective flow intensity. In contrast, the Hartmann number has a reverse relationship with it. Additionally, as the Rayleigh number rises, the site of the local Nusselt number's maximum value moves closer to the adiabatic bottom wall. Additionally, the average Nusselt number rises with rising Rayleigh, Darcy, and magnetic field inclination numbers while falling with rising Hartmann numbers.

Figure 13. (a) Physical model and coordinate system, (b) grid distribution

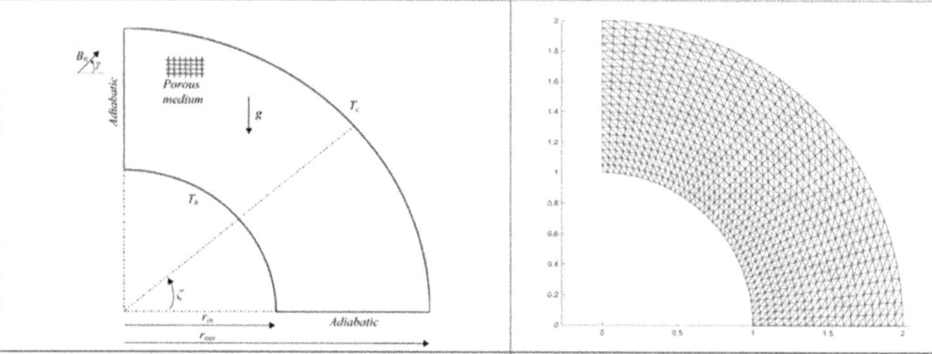

(a) (b)

Figure 14. Streamlines contours for different values of Rayleigh and Hartman numbers when $Da = 10^{-2}$, $\gamma = 30°$, and $\phi = 2\%$

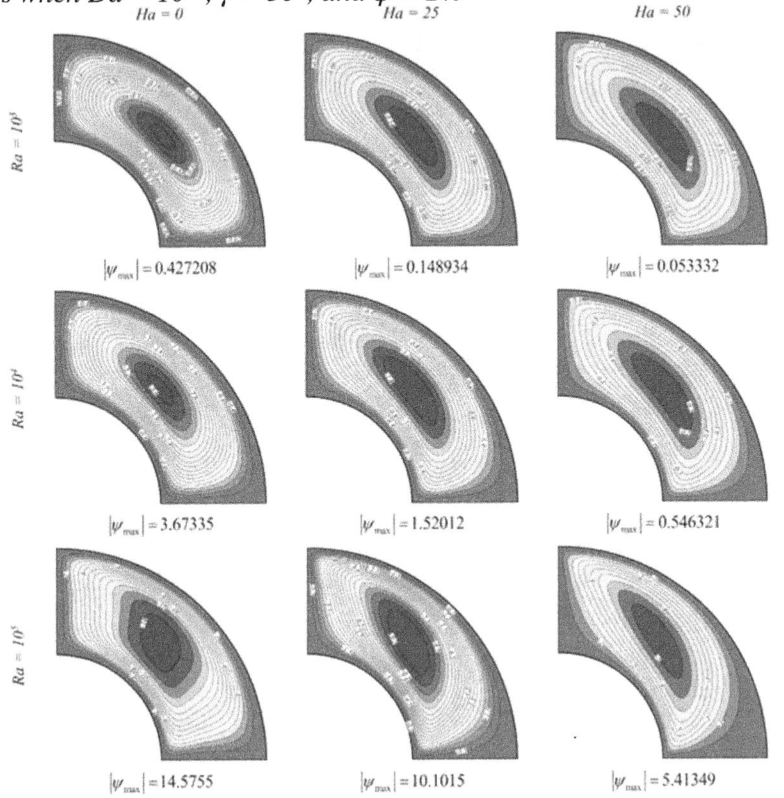

Figure 15. Isotherms contours for different values of Rayleigh and Hartmann numbers when $Da = 10^{-2}$, $\gamma = 30°$, and $\phi = 2\%$

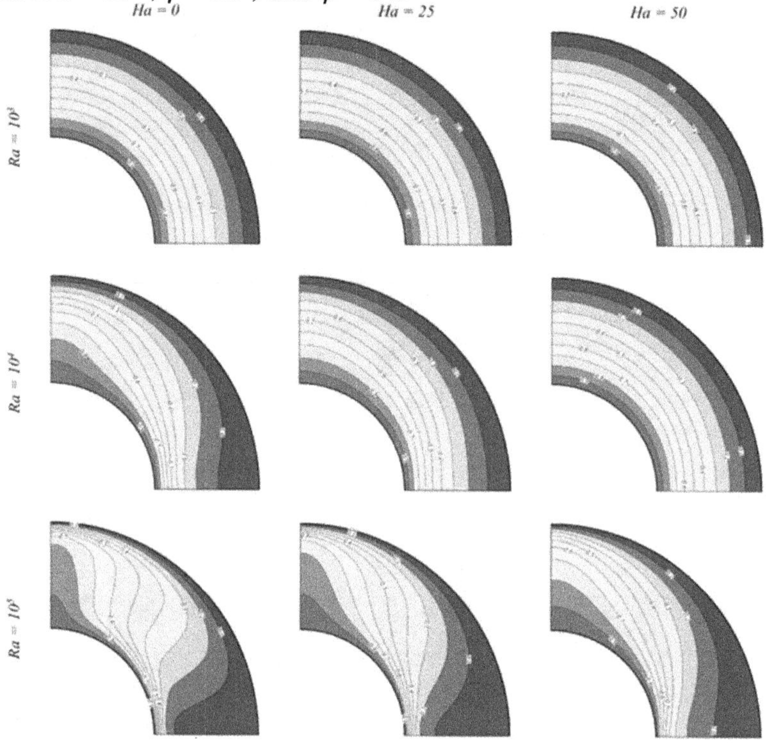

Figure 16. Streamlines contours for different values of Darcy number and angle of magnetic field parameter when $Ra = 10^4$, $Ha = 10$, and $\phi = 2\%$

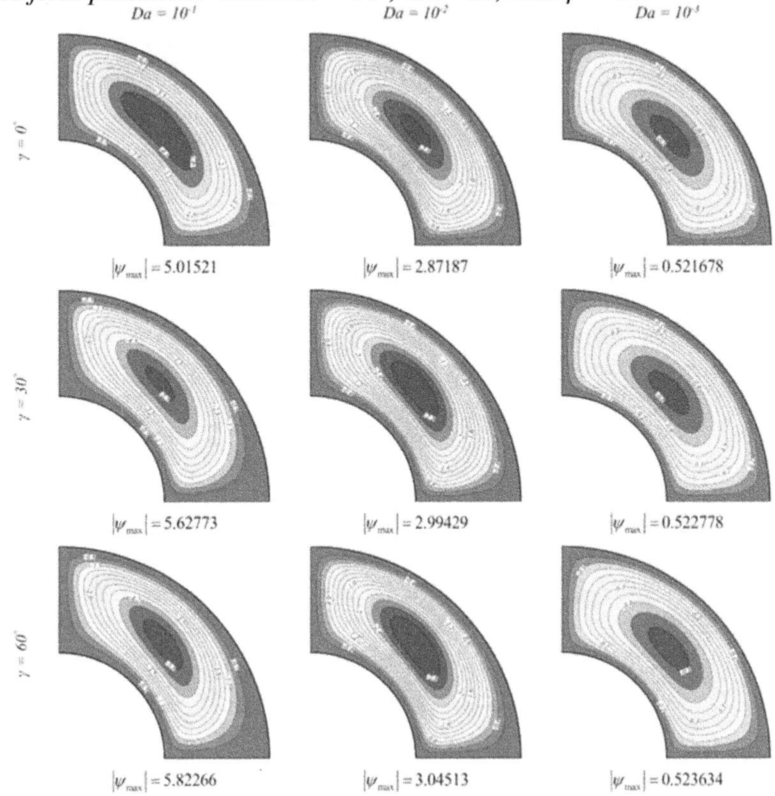

Figure 17. Isotherms contours for values of Darcy number and angle of magnetic field parameter when $Ra = 10^4$, $Ha = 10$, and $\phi = 2\%$

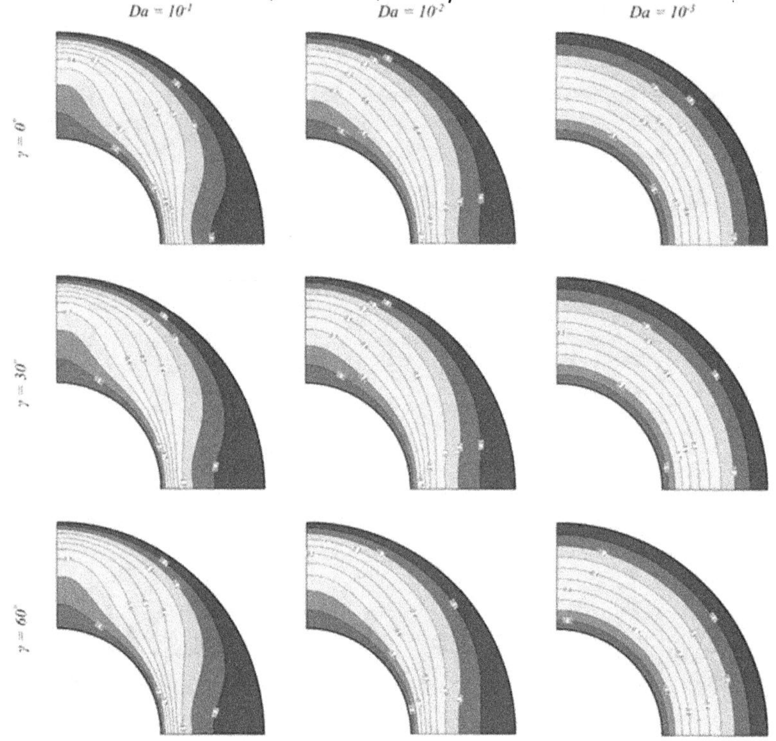

Figure 18. Local and average Nu numbers for values of Rayleigh number, Hartmann number, Darcy number, and angle of magnetic field parameter when $\phi = 2\%$

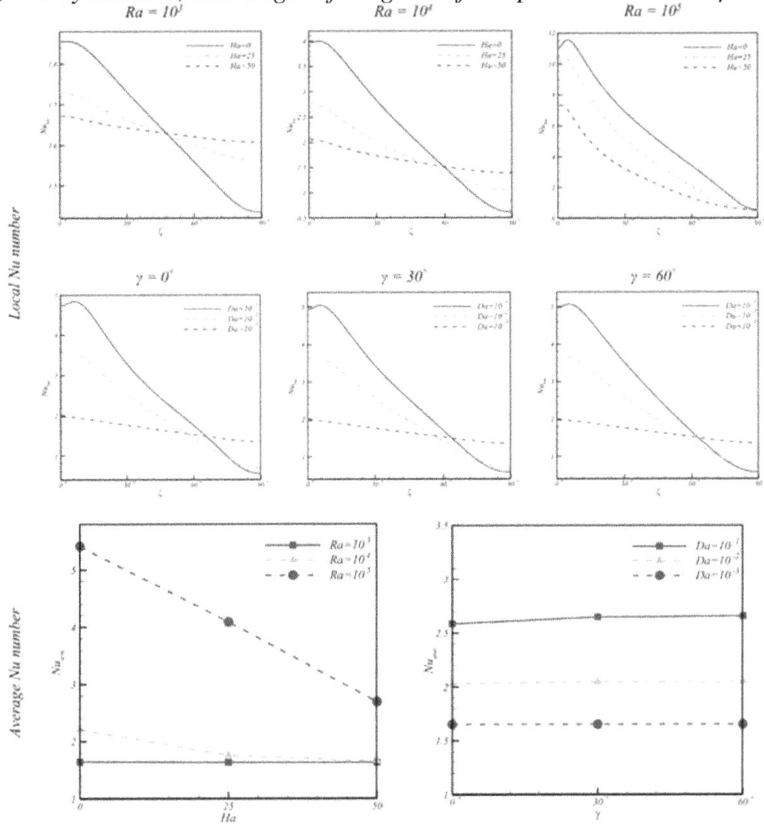

Table 4. Thermo-physical properties of water and nanoparticle at T = 310°K

	ρ (kg/ m³)	C_p (J / kg K)	k (W /m K)	$\beta \times 10^5 (K^{-1})$	$\mu \times 10^6$(kg $m^{-1}s^{-1}$)	d_s (nm)
Copper (Cu)	8933	385	401	1.67	–	25
Pure water	993	4178	0.628	36.2	695 (*Pr = 4.62*)	0.385

Table 5. Impact of grid size on the average Nusselt number (Nu_{avg}) when $Ra = 10^5$, $Ha = 50$, $\phi = 0.04$, $Da = 10^{-3}$ and $\gamma = \pi/3$.

Grid dimension (radial direction × angular direction) Nu_{avg}
91 × 131 2.049029
101 × 141 2.050515
111 × 151 2.051487
121 × 161 2.052133
131 × 171 2.052568
141 × 181 2.052862

Table 6. Comparison between present results and other works for the average Nusselt number (Nu_{avg})

Ra	Present work	Khanafer et al.	De Vahl Davis
10^3	1.1307	1.118	1.118
10^4	2.2674	2.245	2.243
10^5	4.5851	4.522	4.519
10^6	8.8341	8.826	8.799

1.4. Consider a Porous Artery With a Magnetic Field

The present study employs the Homotopy Perturbation Method to conduct an analytical study on blood flow containing nanoparticles through porous blood vessels in the presence of a magnetic field (HPM). Blood is classified as a third-grade non-Newtonian fluid that contains nanoparticles. Constant, Reynolds, and Vogel's models determine the viscosity of a nanofluid. Attempts have been made to demonstrate the reliability and performance of the current method in comparison to the numerical method, Runge-Kutta fourth-order. The results show that the HPM is capable of predicting the solutions to these problems. This study also declares the effect of some physical parameters on temperature, velocity, and nanoparticle concentration profiles, such as pressure gradient, Brownian motion parameter, thermophoresis parameter, magnetic field intensity, and Grashof number, (Basir *et al.*, 2021) (Mistry & Vadali, 2022) (Ajmal *et al.*, 2022) (Hamarat, Kuram, & Ozcelik,

2018) (Hossain, Acar, & Malalasekewwra, 2017) (Viswanathan, Sathiyamoorthy, & Ganji, 2018) (Timchenko *et al.,* 2020) (Borra & Das, 2021).

Consider a porous artery with a magnetic field and a steady, incompressible, non-Newtonian nanofluid. Figure 19 depicts a schematic of the problem's physics and coordinates. The blood flow is modeled as a one-dimensional problem in this study. One-dimensional blood flow modeling is widely acknowledged as a useful technique for analyzing arterial hemodynamics, dynamics, and cardiovascular diseases. Although blood flow is a 3D problem, modeling it at the moment is not feasible due to the lengthy computational time required to model the vessels' network as well as the fluid-solid interaction between the blood and vessel walls. For modeling such a structure, the nanofluids density, ρ should be defined as:

$$\rho = \varnothing . \rho_p + (1 - \varnothing)\rho_{f0} = \varnothing \rho_p + (1 - \varnothing)\{1 - \beta_T(\theta - \theta_{w})$$

(1.4.1)

Where is the base fluid's density, θ_w is the reference temperature, ρ_f is the base fluid's density at the reference temperature, is the density of the nanoparticles, and is the volumetric coefficient of expansion. Taking the density of base fluid as that of the nanofluid, density ρ in Eq. (1.4.1) becomes:

$$\rho = \varnothing \rho_p + (1 - \varnothing)\{\rho_f(1 - \beta_T(\theta - \theta_w))\}$$

(1.4.2)

Clearly, a viscous fluid is governed by continuity and Navier–Stokes equations, and when the fluid is considered to be incompressible, the conservation of momentum, total mass, thermal energy, and nanoparticles are as follows,

$$\rho_f\left(\frac{\partial V}{\partial t} + V.\nabla V\right) - divT + \frac{\mu\varphi}{k}\left(1 + \lambda_r\frac{\partial}{\partial t}\right)V$$

$$-\left[\varnothing \rho_p + (1 - \varnothing)\rho_f[1 - \beta_T(\theta - \theta_w)]\right]g + J.B = 0$$

(1.4.3)

$$\left(\rho c\right)_f\left(\frac{\partial \theta}{\partial t} + V.\nabla\theta\right) - k\nabla^2\theta - \left(\rho c\right)_p\left[D_b \ \nabla \ \varnothing . \nabla\theta + \frac{D_T}{\theta_w} \ \nabla \ \theta . \nabla\theta\right] = 0$$

(1.4.4)

$$\left(\frac{\partial \varnothing}{\partial t} + V.\nabla\varnothing\right) - D_b\nabla^2\varnothing - \frac{D_T}{\theta_w}\nabla^2\theta = 0$$

(1.4.5)

Since electrically conductive fluids are considered, the fourth term on the right side of Eq. (1.4.3) is considered. The following equation (Ohm's law) describes the origin of B:

$$J=\sigma(E+V.B)$$

(1.4.6)

In which E represents the electric field. On the right-hand side of equation (1.4.6), $B = B_0 + b$ is the total magnetic field, is the electrical conductivity, V is the velocity vector, and J is the electric current density. Because the induced magnetic field is ignored for small magnetic Reynolds numbers:

$$J.B= -\sigma B^2 V \qquad (1.4.7)$$

In this study, the following boundary conditions are used at $r = r_1$ and $r = r_2$ as below:

$$V(r_1) = V_0, V(r_2) = 0, \theta(r_1) = \theta_w, \theta(r_2) = 0, \varnothing(r_1) = \varnothing_w, \varnothing(r_2) = 0$$

(1.4.8)

Stress in a third-grade fluid is given by

$$T = -PI + \mu A_1 + \alpha_1 A_2 + \alpha_2 A_1^2 + \beta_1 A_3 + \beta_2(A_1 A_2 + A_2 A_1) + \beta_3(trA_1^2)A_1$$

(1.4.9)

Where α_1, α_2, β_1, β_2 and β_3 are the material modules and are considered to be functions of temperature in general. In Eq. (1.4.9), *-pI* shows the spherical stress due to the restraint of incompressibility, and the kinematical tensors A_1, A_2 and A_3 can be defined by the following equations:

$$A_1 = (\nabla V) + (\nabla V)^t$$

(1.4.10)

$$A_n = \frac{dA_{n-1}}{dt} + A_{n-1}(\nabla V) + \left(\nabla \ V \right)^t A_{n-1}, n = 2,3$$

(1.4.11)

Where V = [0, 0, v(r)] denotes the velocity field, grad is the gradient operator, and D/Dt is the material time derivative, which is defined by:

$$\frac{D(0)}{Dt} = \frac{\partial(0)}{\partial t} + [grad(0)]V$$

(1.4.12)

The specific Helmholtz free energy is minimal when the fluid is locally at rest.

$$\mu \geq 0, \alpha_1 \geq 0, |\alpha_1 + \alpha_2| \leq \sqrt{24\mu\beta_3}, \beta_1 = \beta_2 = 0, \beta_3 \geq 0$$

(1.4.13)

Ellahi et al. considered that the fluid is thermodynamically compatible, therefore Eq. (1.4.9) reduced to:

$$T = -PI + \mu A_1 + \alpha_1 A_2 + \alpha_2 A_1^2 + \beta_3 \left(trA_1^2\right) A_1$$

(1.4.14)

Eq. (14) can be rearranged as

$$T = -PI + \left[\mu + \beta_3\left(trA_1^2\right)\right] A_1 + \alpha_1 A_2 + \alpha_2 A_1^2$$

(1.4.15)

By introducing the following non-dimensional parameters:

$$\overline{V} = \frac{v}{v_0}, \overline{r} = \frac{r}{R}, \overline{\mu} = \frac{\mu}{\mu_0}, \overline{\theta} = \frac{\theta - \theta_0}{\theta_m - \theta_w}, \overline{\varnothing} = \frac{\varnothing - \varnothing_0}{\varnothing_m - \varnothing_w}$$

(1.4.16)

Where V_0, μ_0, θ_w, $\theta_{m,}$ and ϕ_m denote the reference velocity, viscosity, pipe temperature, fluid temperature, and mass concentration, respectively. Substituting Eq. (1.4.15) in the balance of linear momentum and using the non-dimensional quantities given in Eq. (1.4.16), the dimensionless forms of the governing equations ((1.4.3) to (1.4.5)), after dropping bars for simplicity, lead to the following no dimensional coupled equations,

$$N_b \frac{d\theta}{dr}\frac{d\varnothing}{dr} + \frac{1}{r}\frac{d\theta}{dr} + \alpha\frac{d^2\theta}{dr^2} + \alpha_1 N_t\left(\frac{d\theta}{dr}\right)^2 = 0$$

(1.4.17)

$$N_b\left(\frac{d^2\theta}{dr^2} + \frac{1}{r}\frac{d\theta}{dr}\right) + N_t\left(\frac{d^2\varnothing}{dr^2} + \frac{1}{r}\frac{d\varnothing}{dr}\right) = 0$$

(1.4.18)

$$\frac{d\mu}{dr}\frac{dv}{dr} + \frac{\mu}{r}\frac{dv}{dr} + \mu\frac{d^2v}{dr^2} + \frac{\wedge}{r}\left(\frac{dv}{dr}\right)^3 + 3\left(\frac{dv}{dr}\right)^2\frac{d^2v}{dr^2} = PV + M^2v + c - G_r\theta - B_r\varnothing$$

(1.4.19)

Where P is porosity and M is MHD parameters. With the boundary conditions as the form,

$$V(1) = 1, V(2) = 0, \theta(1) = 0, \theta(2) = 0, \varnothing(1) = 1, \varnothing(2) = 0$$

(1.4.20)

Gold nanofluids can be used to activate or inhibit blood vessel growth. Some drugs can be used to increase or decrease capillary growth in regular diseases, but most of these treatments are only advantageous for a short period of time.

Case I: Constant model $\mu = 1$ (1.4.21)
Case II: Reynolds' mode $\mu = e^{-B\theta}$ (1.4.22)
Case III: Vogel model

$$\mu = \mu_0 e^{\left(\frac{A}{B+\theta} - \theta_0\right)}$$

(1.4.24)

Where A and B are constants, θ is temperature, and subscript "0" denotes the ambient condition.

1.5. Homotopy Perturbation Method (HPM)

Consider the subsequent function:

$$A(u)\text{-}f(r)=0 \qquad (1.4.25)$$

With the boundary condition of:

$$B\left(u,\frac{\partial u}{\partial n}\right), \qquad (1.4.26)$$

where A (u) is determined as below:

$$A(u)=L(u)+N(u) \qquad (1.4.27)$$

The homotopy perturbation structure is represented as:

$$H(v,p) = L(V) - L(u0) + p\,.L(u0) + p[N(V) - f(r)] = 0 \qquad (1.4.28)$$

Where

$$V(r,p) : \Omega.[0,1] \rightarrow R \qquad (1.4.29)$$

Obviously, considering Eqs. (1.4.17) and (1.4.18):

H(V,0)=L(V)-L(u0)=0, H(V,1)=A(V)-f(r)=0(1.4.30)

The process of the changes in p from zero to unity is that of v(r, p) changing from u0 to u_r consider v as:

$$V = V_0 + p\,.V_1 + p^2 V_2 \qquad (1.4.31)$$

and the best access is

$$H(\theta,p) = (1-p)\left(\frac{1}{r}\frac{d\theta}{dr} + \alpha\frac{d^2\theta}{dr^2}\right) + p\left(N_b\frac{d\theta}{dr}\frac{d\varnothing}{dr} + \frac{1}{r}\frac{d\theta}{dr} + \alpha\frac{d^2\theta}{dr^2} + \alpha_1 N_t\left(\frac{d\theta}{dr}\right)^2\right)$$

(1.4.32)

$$H(\varnothing,p) = \left(1-p\right)\left(N_t\left(\frac{d^2\varnothing}{dr^2} + \frac{1}{r}\frac{d\varnothing}{dr}\right) + p\left(N_b\left(\frac{d^2\theta}{dr^2} + \frac{1}{r}\frac{d\theta}{dr}\right) + N_t\left(\frac{d^2\varnothing}{dr^2} + \frac{1}{r}\frac{d\theta}{dr}\right)\right)$$

(1.4.33)

Now an attempt is made to solve the equations. (1.4.33-34) as:

$$\theta(r) = \theta_0(r) + p\theta_1(r) + p^2\theta_2(r) + \dots$$

(1.4.34)

$$\varnothing(r) = \varnothing_0(r) + p\varnothing(r) + p^2\varnothing_2(r) + \dots$$

(1.4.35)

$$V(r) = V_0(r) + p\,V_1(r) + p^2 V_2(r) + \dots$$

(1.4.36)

According to the equation. (1.4.33-34) the initial approximation to satisfy the boundary condition is

$$\theta_0(r) = -1.44\ln(r) + 1$$

(1.4.37)

$$\varnothing_0(r) = -1.44\ln(r) + 1$$

(1.4.38)

$$V_0(r) = -0.5 + 0.09\,e^{(-1.414r+4.22)}$$

(1.4.39)

Substituting Eqs. (1.4.35-37) into Eqs. (1.4.38-40) yields

$$\frac{d^2}{dr^2}\theta_1(r) + \left(\frac{d}{dr}\theta_0(r)\right)^2 + \left(\frac{d}{dr}\varnothing_0(r)\right)\left(\frac{d}{dr}\theta_0(r)\right) + \frac{\frac{d}{dr}\theta_1(r)}{r} = 0$$

(1.4.40)

$$\frac{d^2}{dr^2}\theta_0(r) + \frac{d^2}{dr^2}\varnothing_1(r) + \frac{\frac{d}{dr}\varnothing_1(r)}{r} + \frac{\frac{d}{dr}\theta_0(r)}{r} = 0$$

(1.4.41)

$$\frac{d^2}{dr^2}\theta_0(r) + \frac{d^2}{dr^2}\varnothing_1(r) + \frac{\frac{d}{dr}\varnothing_1(r)}{r} + \frac{\frac{d}{dr}\theta_0(r)}{r} = 0$$

(1.4.42)

$$\frac{\frac{d}{dr}V_0(r)}{r} + 3\frac{d^2}{dr^2}V_0\left(r\right)\left(\frac{d}{dr}V_0(r)\right)^2$$
$$+\left(\frac{\frac{d}{dr}V_0(r))^3}{r} + \varnothing_0(r) + \theta_0(r) - 2V_1(r) + \frac{d^2}{dr^2}V_1(r) = 0\right.$$

(1.4.43)

Similarly, the remaining components are obtained using the Maple package. According to HPM:

$$\theta(r) = -1.44\ln(r) + 1 + 2.08\ln(r)(-\ln(r) + 0.69)$$

$$+3.00\,.\,10^{-10}\ln(r)(-2.4\,.\,10^{9} - 10^{10}\ln(r)^2 + 1.03\,10^{10}\ln(r))$$

(1.4.44)

$$\varnothing(r) = -1.44\ln(r) + 1 + 2.08\ln(r)(-\ln(r) + 0.69)$$

$$+3.00\,.\,10^{-10}\ln(r)(-2.4\,.\,10^{9} - 10^{10}\ln(r)^2 + 1.03\,10^{10}\ln(r))$$

(1.4.45)

$$V(r) = -0.5 + 0.09\,e^{(-1.41r+4.24)} + 0.03\,e^{(1.41r)} - 0.09\,e^{(1.41r-1.412)}$$

(1.4.46)

The behavior of blood in courses as a third review non-Newtonian liquid is examined systematically within the display. Homotopy annoyance strategy (HPM) is connected to unraveling the overseeing conditions, and the results are compared with numerical results. Figure 20 shows the precision and approval of HPM compared with the fourth arranged Runge-Kutta numerical strategy. As can be seen in these figures, HPM has an amazing understanding of numerical comes about and can be presented as a basic and effective explanatory strategy for these sorts of issues. Figure 21 shows the impact of weight gradient(c) on the velocity profile. The weight slope created between the blood vessel and the venous conclusion of the circulation is the driving drive causing bloodstream through the vessels, which is negative over the aortic valve all through the cardiac cycle. As the figure appears, an increment of short weight causes an increment of the velocity profile that contributes to an increment of the blood velocity in supply routes. As known, porosity and pore measure dispersions can be tuned to upgrade the change of different Nanomedicines. In this manner, understanding the impact of porosity parameters is of incredible significance. Figure 22 outlines the velocity profile conveyance over different values of the porosity parameter. The velocity conveyance isn't much influenced by a variety of diverse values of (P); however, the velocity profile somewhat increments with the decrease in porosity. When porosity increments, liquid can effortlessly move between the pores, resulting in velocity profile increments. The impact of the thermophoresis and Brownian movement parameters on the velocity profile is demonstrated in Figure 23 and Figure 24. Concurring with these figures, an increment of Nt and a decrease of Nb lead to continuous increments within the velocity profile. Distinctive values of these two parameters are compared and appear in these Figures. Figure 25 shows the thermophoresis parameter's impact on the non-dimensional temperature profile. It is self-evident that the temperature profile is an expanding work of Nt number. From the physical point of view, an increment within the thermophoresis number can create higher mass flux due to temperature angle, which thus raises the nanoparticle concentration. Figure 26 outlines the impact of the Brownian movement parameter on the temperature profile. Brownian movement is imperative in influencing the stream way of platelets or any other cells. This parameter can be negative or positive. Negative esteem shows that the concentration of nanoparticles on the divider surface is less than that of the boundary layer. This figure affirms that the Brownian movement parameter doesn't have predictable behavior. At Nb = 1 and Nb = 2, the temperature profile crosses each other when $0<\theta< 0.5$ and covers at $0.5<\theta< 1$. Separated from this perception at lower Nb, the temperature and boundary layer thickness appear to have a sensible drift at higher Brownian movement numbers. The impacts of the thermophoresis parameter and Brownian movement parameter on the nanoparticle concentration of the nanofluid are shown in Figures 27 and 28. Concurring to Figure 27, the concentration of nanoparticles has a coordinate effect on Nt, and it improves by expanding the thermophoresis number. In any case, the inverse slant is watched in Figure 28, suggesting that by expanding the Nb, the nanoparticle concentration of the nanofluid diminishes. Usually, likely due to the truth that the effect of heat conductivity increments by improvement in Brownian motion, which leads to the lessening within the concentration profile. It can be concluded that when the molecule measure limits to the nano-sized scale medication, the Brownian movement and its effect on the encompassing liquid play a noteworthy part within the heat exchange.

Figure 19. Schematic of blood flow through a porous artery in the presence of a magnetic field

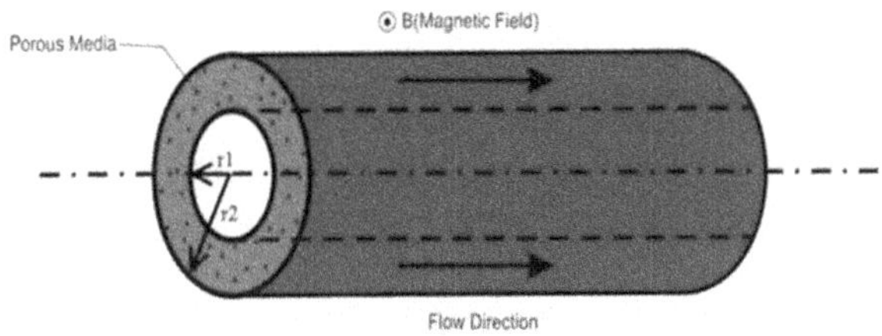

Figure 20. Comparison of HPM with NUM for θ, ϕ, V when Nb = Nt = α = Λ = P = M = C = Gr = Br =1, $\mu(r)$ = 1

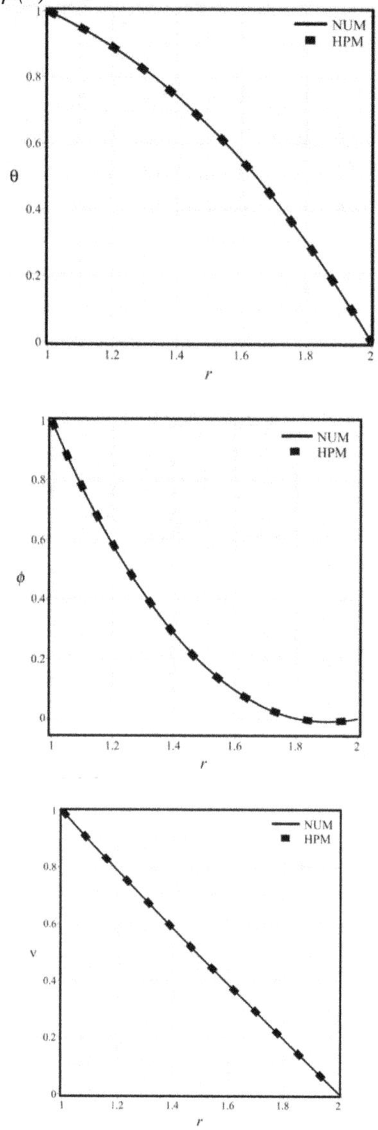

Figure 21. Effect of pressure gradient on the velocity profile

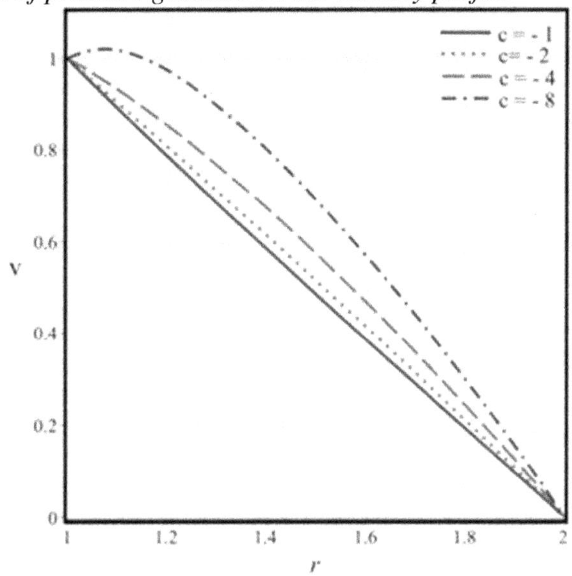

Figure 22. Effect of porosity parameter on the velocity profile

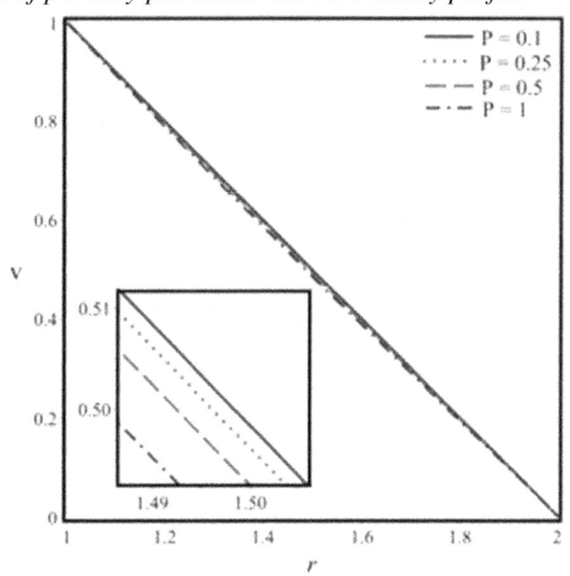

Figure 23. The effect of thermophoretic parameter on the velocity profile

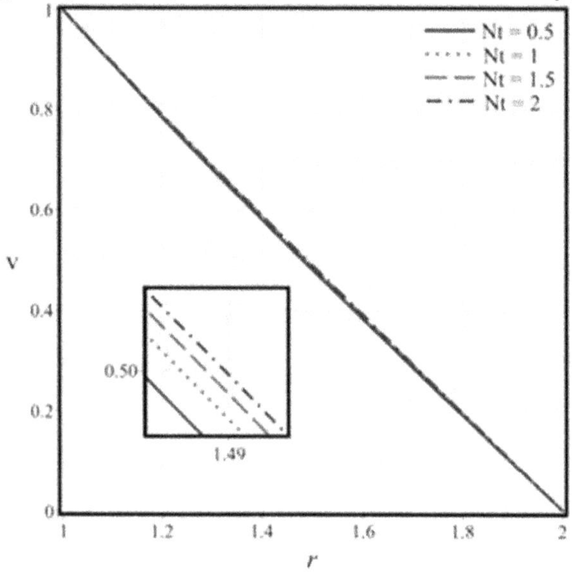

Figure 24. Effect of Brownian motion parameter on the velocity profile

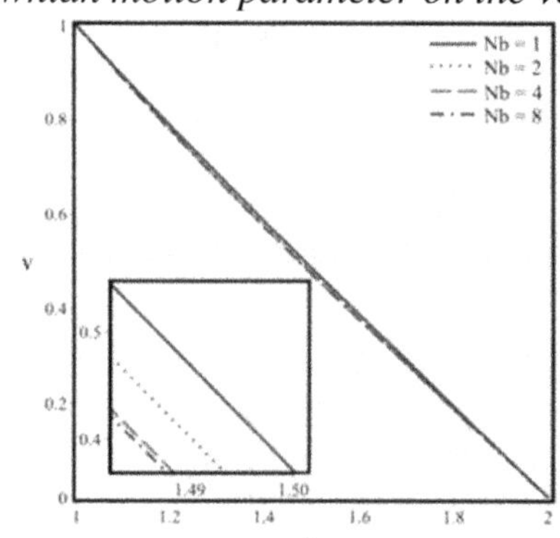

Figure 25. Effect of thermophoresis parameter on the temperature profile

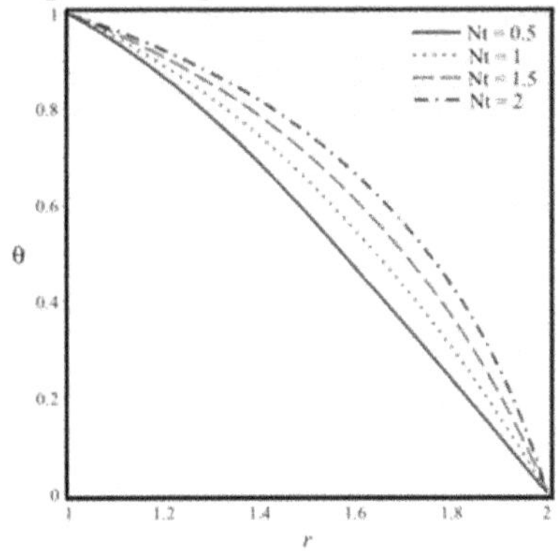

Figure 26. Effect of Brownian motion parameter on the temperature profile

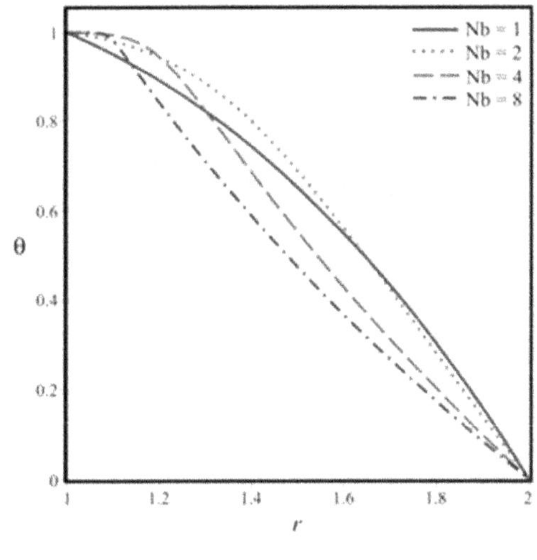

Figure 27. Effect of thermophoresis parameter on nanoparticles concentration profile

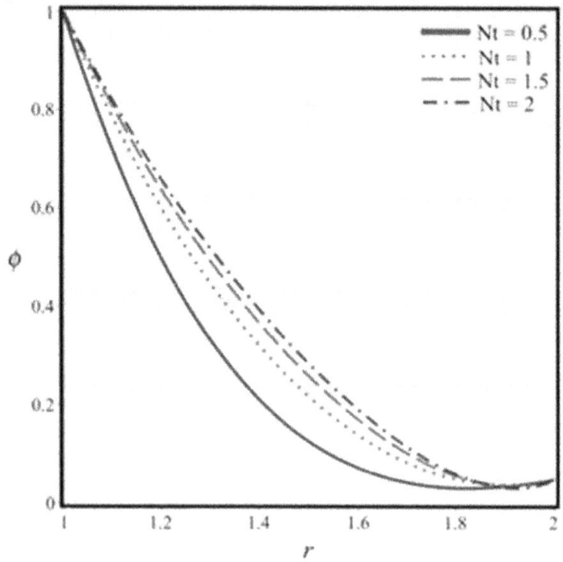

Figure 28. Effect of Brownian motion parameter on the nanoparticles concentration profile

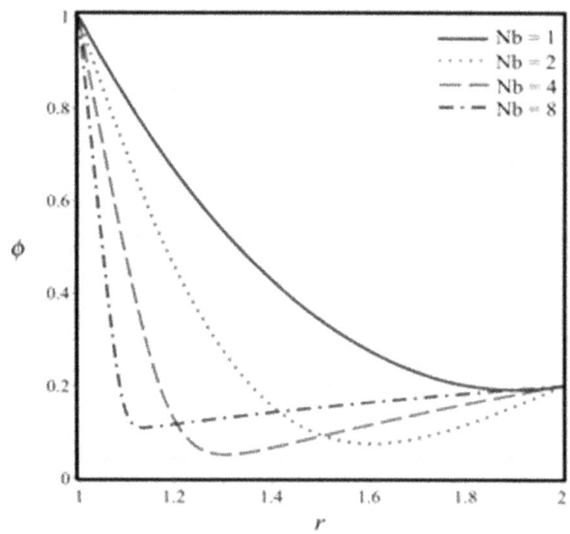

REFERENCES

Adamo, B. C., Junger, A. S., Bressan, L. P., da Silva, J. A. F., Poppi, R. J., & de Jesus, D. P. (2020). Fast and straightforward in-situ synthesis of gold nanoparticles on a thread-based microfluidic device for application in surface-enhanced Raman scattering detection. *Microchemical Journal*, 156, 104985. DOI: 10.1016/j.microc.2020.104985

Ajmal, T. S., Arya, S. B., Maurya, P., & Shariff, S. M. (2022). Effect of hydrodynamics and laser surface melting on erosion-corrosion of X70 steel pipe elbow in oilfield slurry. *International Journal of Pressure Vessels and Piping*, 199, 104687. DOI: 10.1016/j.ijpvp.2022.104687

Alizadeh, M., Nabizadeh, A., Fazlollahtabar, A., & Ganji, D. D. (2021). An optimization study of solidification procedure in a wavy-wall storage unit considering the impacts of NEPCM and curved fin. *International Communications in Heat and Mass Transfer*, 124, 105249. DOI: 10.1016/j.icheatmasstransfer.2021.105249

Bahmani, M., Babagoli, M., Jalili, P., Jalili, B., & Ganji, D. D. (2024). Numerical study of the porous cavity with different square size vanes with a high focus on Nusselt number. *International Journal of Thermofluids*, 100860.

Borra, S. R., & Das, S. (2021). Impact of permeability on heat and mass transfer of a magnetohydrodynamic nanofluid flow over an inclined plate with variable magnetic field. *Journal of Thermal Science*, 25(5), 1231–1242.

Dey, D., & Borah, R. (n.d.). Stability analysis on dual solutions of second-grade fluid flow with heat and mass transfers over a stretching sheet.

Ganji, D. D., Peiravi, M. M., & Abbasi, M. (2015). Evaluation of the heat transfer rate increases in retention pools nuclear waste. *International Journal of Nanodimension*, 6(4), 385–398.

Gao, X., Lv, J., Sun, B., & Liu, Y. (2022). Effects of bed permeability on the hydrodynamic characteristics in a channel with a vegetation patch: A modeling study. *Journal of Hydrology (Amsterdam)*, 612, 128046. DOI: 10.1016/j.jhydrol.2022.128046

Ghadikolaei, S. S., Hosseinzadeh, K., Hatami, M., & Ganji, D. D. (2018). MHD boundary layer analysis for micropolar dusty fluid containing Hybrid nanoparticles (Cu-Al_2O_3) over a porous medium. *Journal of Molecular Liquids*, 268, 813–823. DOI: 10.1016/j.molliq.2018.07.105

Gnaneswara Reddy, M., Punith Gowda, R. J., Naveen Kumar, R., Prasannakumara, B. C., & Ganesh Kumar, K. (2021). Analysis of modified Fourier law and melting heat transfer in a flow involving carbon nanotubes. *Proceedings of the Institution of Mechanical Engineers, Part E: Journal of Process Mechanical Engineering*, 09544089211001353. DOI: 10.1177/09544089211001353

Goud, B. S. (2020). Thermal radiation influences on MHD stagnation point stream over a stretching sheet with slip boundary conditions. *International Journal of Thermofluids Science and Technology*, 7(2). Advance online publication. DOI: 10.36963/IJTST.2020070201

Hamarat, I., Kuram, E., & Ozcelik, B. (2018). Investigation of the mechanical, rheological, and morphological properties of acrylonitrile butadiene styrene blends with different recycling number content. *Proceedings of the Institution of Mechanical Engineers. Part E, Journal of Process Mechanical Engineering*, 232(4), 449–458. DOI: 10.1177/0954408917717994

Hossain, M., Acar, M., & Malalasekera, W. (2017). Optimisation of a thermal heat sink for an automotive PCM storage system using porous media. *Applied Thermal Engineering*, 125, 923–931.

Jalili, B., Azar, A. A., Jalili, P., Liu, D., Abdelmohimen, M. A., & Ganji, D. D. (2024). Investigation of the unsteady MHD fluid flow and heat transfer through the porous medium asymmetric wavy channel. *Case Studies in Thermal Engineering*, 61, 104859. DOI: 10.1016/j.csite.2024.104859

Jalili, P., Asadi, Z., Shateri, A., Jalili, B., Ahmad, H., Albalwi, M. D., & Ganji, D. D. (2024). Thermal analysis of nanofluid magnetic flow on a rotating disk in the presence of radiation considering response surface method. *Modern Physics Letters B*, 38(24), 2450217. DOI: 10.1142/S0217984924502178

Md Basir, M. F., Mackolil, J., Mahanthesh, B., Nisar, K. S., Muhammad, T., Anuar, N. S., & Bachok, N. (2021). Stability and statistical analysis on melting heat transfer in a hybrid nanofluid with thermal radiation effect. *Proceedings of the Institution of Mechanical Engineers, Part E: Journal of Process Mechanical Engineering*. DOI: 10.1177/09544089211033161

Mistry, U., & Vadali, M. (2022). A steady-state semi-analytical approximation of melt pool evolution in pulsed laser surface melting. *Journal of Manufacturing Processes*, 74, 123–135. DOI: 10.1016/j.jmapro.2021.11.064

Muhammad, K., Hayat, T., & Alsaedi, A. (2021). Numerical study for entropy generation and melting heat in flow of modified hybrid nanomaterial (Ag+ MWCNTs+ SWCNTs+ Water). *Proceedings of the Institution of Mechanical Engineers, Part E: Journal of Process Mechanical Engineering.*

Nayak, M. K. (2019). HHR impact on 3D radiative stretched flow of Cu-H_2O nanofluid influenced by variable magnetic field and convective boundary condition. *International Journal of Thermofluids Science and Technology*, 6(2), 19060202. DOI: 10.36963/IJTST.19060202

Neuman, S. P., & Paul, A. W. (1970). Finite element method of analyzing steady seepage with a free surface. *Water Resources Research*, 6(3), 889–897. DOI: 10.1029/WR006i003p00889

Pasha, P., Mirzaei, S., & Zarinfar, M. (2021). Application of numerical methods in micropolar fluid flow and heat transfer in permeable plates. *Alexandria Engineering Journal.*

Peiravi, M. M., & Alinejad, J. (2019). Hybrid conduction, convection and radiation heat transfer simulation in a channel with rectangular cylinder. *Journal of Thermal Analysis and Calorimetry*, 140(6), 2733–2747. DOI: 10.1007/s10973-019-09010-0

Peiravi, M. M., & Alinejad, J. (2021). Nanoparticles distribution characteristics in multi-phase heat transfer between 3D cubical enclosures mounted obstacles. *Alexandria Engineering Journal*, 60(6), 5025–5038. DOI: 10.1016/j.aej.2021.04.013

Peiravi, M. M., Alinejad, J., Domairry Ganji, D., & Maddah, S. (2019). Numerical study of fins arrangement and nanofluids effects on three-dimensional natural convection in the cubical enclosure. *Transport Phenomena in Nano and Micro Scales*, 7(2), 97–112.

Peiravi, M. M., Alinejad, J., Domairry Ganji, D., & Maddah, S. (2019). 3D optimization of baffle arrangement in a multi-phase nanofluid natural convection based on numerical simulation. *International Journal of Numerical Methods for Heat & Fluid Flow*, 30(5), 2583–2605. DOI: 10.1108/HFF-01-2019-0012

Peiravi, M. M., Pasha, P. O. O. Y. A., & Ganji, D. D. (2021). Hydro magnetic effect on thread stretching surface with mounted obstacles in porous medium. *WSEAS Transactions on Heat and Mass Transfer*, 16, 95–105. DOI: 10.37394/232012.2021.16.12

Seyyedi, S. M., Dogonchi, A. S., Hashemi-Tilehnoee, M., Waqas, M., & Ganji, D. D. (2020). Investigation of entropy generation in a square inclined cavity using control volume finite element method with aided quadratic Lagrange interpolation functions. *International Communications in Heat and Mass Transfer*, 110, 104398. DOI: 10.1016/j.icheatmasstransfer.2019.104398

Sindhu, S., Gireesha, B. J., & Ganji, D. D. (2020). Simulation of Cu: Γ−ALO(OH)/ Water in a microchannel heat sink by dint of porous media approach. *Case Studies in Thermal Engineering*, 21, 100723. DOI: 10.1016/j.csite.2020.100723

Timchenko, V. (2020). Investigation of permeability and tortuosity of gas diffusion layers with respect to the fabrication method. *Journal of Power Sources*, 400, 267–275.

Vatani, M., Ghasemi, S. E., & Ganji, D. D. (2016). Investigation of micropolar fluid flow between a porous disk and a nonporous disk using efficient computational technique. *Proceedings of the Institution of Mechanical Engineers. Part E, Journal of Process Mechanical Engineering*, 230(6), 413–424. DOI: 10.1177/0954408914557375

Viswanathan, K. K., Sathiyamoorthy, N., & Ganji, D. D. (2018). Analysis of micropolar fluid flow on a stretching/shrinking surface with multiple slip effects. *Alexandria Engineering Journal*, 57(4), 3047–3055.

Waini, I., Ishak, A., & Pop, I. (2021). Melting heat transfer of a hybrid nanofluid flow towards a stagnation point region with second-order slip. *Proceedings of the Institution of Mechanical Engineers. Part E, Journal of Process Mechanical Engineering*, 235(2), 405–415. DOI: 10.1177/0954408920961213

Yen, C. Y., Kuo, P.-J., Lin, C.-Y., Nie-Shiuh Chang, N., Hsiao, H.-Y., Chin, Y.-T., Tsai, C.-C., & Lee, S.-Y. (2019). Accuracy of cone beam computed tomography in measuring thicknesses of hard-tissue-mimicking material adjacent to different implant thread surfaces. *Journal of Dental Sciences*, 14(2), 119–125. DOI: 10.1016/j.jds.2019.04.001 PMID: 31210886

Zar, P. M., Jalili, P., Jalili, B., & Ganji, D. D. (2024). Heat and mass transfer conduct in an unsteady two-dimensional stream between parallel sheets. *South African Journal of Chemical Engineering*, 50, 109–124. DOI: 10.1016/j.sajce.2024.07.011

Zhang, G., Ma, Z., Yin, Y., & Wei, X. (2022). Influence of initial permeability on seepage behavior of the lubricant stored in hydrogel surface. *Tribology International*, 173, 107666. DOI: 10.1016/j.triboint.2022.107666

Chapter 2
Entropy Analysis in MDD Fluid Flow

ABSTRACT

Entropy generation is the quantity of entropy that is produced in any irreversible processes, such as heat and mass transfer processes. It includes the movement of bodies, heat exchange, fluid flow, substances expanding or blending, and any irreversible thermodynamic cycle, including thermal machines such as power plants and air conditioners. Entropy production is one of the issues of interest in investigating magnetohydrodynamic flow. For example, entropy production is studied in the study of natural convection heat transfer in cavities, which is widely observed in many engineering fields, such as solar energy, electronic equipment, and heating and cooling structures. So far, considerable research has been focused on entropy analysis in various fields.

INTRODUCTION

In this chapter, entropy generation in MHD flow is investigated. First, the governing equations are rewritten in a non-dimensional format. The second step expresses the entropy generation equation in non-dimensional form. Then the non-dimensional governing expressions are solved by numerical techniques. Next, it examines the characteristics of flow, heat transfer, and entropy production for distinct values of different control parameters.

DOI: 10.4018/979-8-3693-5335-6.ch002

NONLINEAR EQUATIONS IN THE ENTROPY GENERATION AREA

2.1. Two-Dimensional Normal Convicted Laminar Stream Considering Entropy Optimization

The physical setup and work of the L-shaped walled area received in current CVFEM calculation are portrayed d within the Figure 1. The inward circle and external rectangular are held at settled temperatures, and ($T_c < T_h$) separately, while the other two dividers are protected thermally. In addition, it is expected that the uniform magnetism field with steady greatness is ($B = Bx_{ex} + By_{ey}$) $B_0 = B_{2x}$ utilized, where and indicates unit vectors in the Cartesian framework. The attractive field introduction $ex\ ey$ makes a point with an even pivot such that$\lambda = tan^{-1}\left(\frac{B_y}{B_x}\right)$.

Here two-dimensional normal convicted laminar stream considering entropy optimization is defined. Magnetized Al_2O_3-H_2O nanomaterial is considered. Thought of Boussinesq estimation compares to taking after administering expressions:

$$\frac{\partial u}{\partial x} + \frac{\partial v}{\partial y} = 0$$

(2.1.1)

$$u\frac{\partial u}{\partial x} + v\frac{\partial u}{\partial y} + \frac{1}{\rho_{nf}}.\frac{\partial p}{\partial x} - v_{nf}\left(\frac{\partial^2 u}{\partial x^2} + \frac{\partial^2 u}{\partial y^2}\right)$$

$$-\frac{\sigma_{nf}B_0^2}{\rho_{nf}}(v sin\lambda cos\lambda - u\, sin^2\lambda)$$

(2.1.2)

$$u\frac{\partial v}{\partial x} + v\frac{\partial v}{\partial y} + \frac{1}{\rho_{nf}}.\frac{\partial p}{\partial y} - v_{nf}\left(\frac{\partial^2 v}{\partial x^2} + \frac{\partial^2 v}{\partial y^2}\right)$$

$$-\frac{\sigma_{nf}B_0^2}{\rho_{nf}}(v sin\lambda cos\lambda - u\, cos^2\lambda) + g\beta_{nf}(T - T_c)$$

(2.1.3)

$$u\frac{\partial T}{\partial x} + v\frac{\partial T}{\partial y} - \alpha_{nf}\left(\frac{\partial^2 T}{\partial x^2} + \frac{\partial^2 T}{\partial y^2}\right) = 0$$

(2.1.4)

Where signify effective density, nanofluid heat capacitance, nanofluid thermal ρ_{nf} (ρCp) (β_{nf})expansion coefficient, (K_{nf}) nanofluid effective thermal conductivity, (k_{nf}) nanofluid thermal diffusivity, (α_{nf}). (Liu *et al.,* 2019) (Ghasemi *et al.,* 2022) (Azar *et al.,* 2024) (Sharma & Shaw, 2022) (Qiao *et al.*, 2022) (Alizadeh *et al.,* 2017) (Wang, Li, & Lu, 2022) (Zar *et al.,* 2024) (Fard *et al.,* 2024) (Khan *et al.*, 2021) (Pasha *et al.,* 2022) (Jalili *et al.*, 2024) (Waqas et al., 2021) (Mahboobtosi *et al.,* 2024) (Riehl & Mancin, 2022) (Santos *et al.*, 2022) (Çobanoğlu *et al.*, 2022) (Nabi *et al.*, 2022) (Jalili *et al.*, 2022) (Jalili *et al.,* 2021) (CFD Anaslysis) (Dogonchi & Ganji, 2017).

$$\rho_{nf} - \rho_f(1 - \varnothing) - \rho_s\varnothing = 0 \tag{2.1.5}$$

$$\left(\rho C_p\right)_{nf} - (\rho_f C_{pf}) . (1 - \varnothing) - \left(\rho C_p\right)_s\varnothing \tag{2.1.6}$$

$$\left(\rho\beta\right)_{nf} - \left(\rho\beta\right)_f(1 - \varnothing) - \left(\rho\beta\right)_s\varnothing \tag{2.1.7}$$

$$\frac{K_{nf}}{K_f} - \frac{K_s + (m-1)K_f - (m-1)\varnothing(K_f - K_s)}{K_s + (m-1)K_f + \varnothing(K_f - K_s)} = 0 \tag{2.1.8}$$

$$\alpha_{nf} = \frac{K_{nf}}{\left(\rho C_p\right)_{nf}} \tag{2.1.9}$$

$$\mu_{nf} = \frac{\mu_f}{(1 - \varnothing)^{2.5}} \tag{2.1.10}$$

$$\frac{\sigma_{nf}}{\sigma_f} - 1 - \frac{3\left(\frac{\sigma_s}{\sigma_f} - 1\right)\varnothing}{\left(\frac{\sigma_s}{\sigma_f} + 2\right) - \left(\frac{\sigma_s}{\sigma_f} - 1\right)\varnothing} = 0 \tag{2.1.11}$$

The stream function ( ), vortices(ω), and dimensionless variables in this study are defined as follows:

$$u = \frac{\partial}{\partial y}, v = -\frac{\partial}{\partial x} and$$

$$\omega = \frac{\partial v}{\partial x} - \frac{\partial u}{\partial y}, X = \frac{x}{L}, Y = \frac{y}{L}, AR = \frac{r_{in}}{L}, \theta = \frac{T - T_c}{T_h - T_c}, \Omega = \frac{wL^2}{\alpha_f}$$

, (2.1.12)

Thus, using Eqs. (2.1.12) and (2.1.13), the Eqs. (2.1.1-2.1.14) can be rewritten in dimensionless forms as:

$$\frac{\partial^2}{\partial X^2} + \frac{\partial^2}{\partial y^2} + \Omega = 0$$

(2.1.13)

$$\left[\frac{\partial}{\partial Y}\frac{\partial\theta}{\partial x} - \frac{\partial}{\partial x}\frac{\partial\Omega}{\partial y}\right] - \frac{\frac{\mu_{nf}}{\mu_f}}{\frac{\rho_{nf}}{\rho_f}} Pr_f\left(\frac{\partial^2\Omega}{\partial x^2} + \frac{\partial^2\Omega}{\partial y^2}\right) - \frac{\beta_{nf}}{\beta_f} Ra_f Pr_f \frac{\partial\theta}{\partial x}$$

$$-\frac{\frac{\sigma_{nf}}{\sigma_f}}{\frac{\rho_{nf}}{\rho f}} Ha^2 Pr_f\left(cos^2\beta\frac{\partial^2}{\partial x^2} + sin^2\beta\frac{\partial^2}{\partial y^2} + 2sin\beta cos\beta\frac{\partial^2}{\partial x\partial y}\right)$$

(2.1.14)

$$\left[\frac{\partial}{\partial Y}\frac{\partial\theta}{\partial x} - \frac{\partial}{\partial x}\frac{\partial\Omega}{\partial y}\right] - \frac{\alpha_{nf}}{\alpha_f}\left(\frac{\partial^2\theta}{\partial x^2} + \frac{\partial^2\theta}{\partial y^2}\right) = 0$$

(2.1.15)

In Eq. (2.1.15), Pr_f, Ra_f, and Ha_f are the Hartmann number, Rayleigh number, and Prandtl number, respectively, which are expressed as follows:

$$Ha_f = B_0 L\sqrt{\frac{\sigma_f}{\rho_f v_f}}, Ra_f = \frac{g\beta_f L^3(T_h - T_c)}{v_f\alpha_f}, Pr_f = \frac{v_f}{\alpha_f}$$

(2.1.16)

The boundary conditions revealed in Figure 1 are:

θ = 1.0 is on the inner boundary
θ = 0.0 is on the outer boundary

$\Psi = 0$ is on all solid boundaries

The local and average Nusselt numbers of the nanofluid along the hot wall can be represented as:

$$Nu_{local} = -\frac{k_{nf}}{k_f}\frac{\partial\theta}{\partial r}$$

(2.1.17)

$$Nu_{ave} = \frac{2}{\pi}\int Nu_{local}$$

(2.1.18)

Assessment of the entropy era, which concurs with the moment law of thermodynamics, can be valuable for superior planning of the vitality frameworks. The entropy era rate can be calculated by Eq (2.1.19).

$$s_{gen} = \frac{k_{nf}}{T_0^2}\left[\left\{\frac{\partial T}{\partial x}\right\}^2 + \left(\frac{\partial T}{\partial y}\right)^2\right]$$
$$+\frac{\mu_{nf}}{T_0}\left[2\left(\frac{\partial u}{\partial x}\right)^2 + 2\left(\frac{\partial v}{\partial x}\right)^2 + \left(\frac{\partial u}{\partial x} + \frac{\partial v}{\partial x}\right)^2\right] + \frac{\sigma_{nf}\{uB_0 sin\beta - vB_0 cos\beta}{T}$$

(2.1.19)

Eq. (2.1.21) is the nearby entropy era comprising three terms: nearby entropy era subjected to heat transport, thick dissemination, and attractive field, individually. The non-dimensional frame of the entropy era condition can be picked up as takes after:

$$N_{l,gen} = \frac{s_{gen}}{\left[\left(\frac{k_f}{v_0^2}\right)\left(\frac{\Delta T}{l}\right)^2\right]} = \frac{k_{nf}}{k_f}\left[\left(\frac{\partial\theta}{\partial x}\right)^2 + \left(\frac{\partial\theta}{\partial y}\right)^2\right]$$
$$+\frac{\mu_{nf}}{\mu_f}\varnothing_f\left[4.\left(\frac{\partial^2\psi}{\partial x\partial y}\right)^2 + \left(\frac{\partial^2\psi}{\partial y^2} - \frac{\partial^2\psi}{\partial x^2}\right)^2\right] + \frac{\sigma_{nf}}{\sigma_f}\left[\frac{\partial\psi}{\partial y} sin\lambda + \frac{\partial\psi}{\partial x} cos\lambda\right]^2$$

(2.1.20)

Where signifies the neighborhood entropy era number and indicates the irreversibility dissemination $N_{L,gen}$ Φ_f ratio communicated as follows:

$$\varnothing_f = \frac{\mu_f T_0}{k_f}\left(\frac{\alpha_f}{L\Delta T}\right)^2$$

(2.1.21)

Where is the mean temperature and is the temperature diversity within the $(T_0 = (T_h + T_c)/2) /\Delta T$ enclosure. $(\Delta T = T_h —— T_c)$.

Eq. (2.1.20) can be re-written as follows:

$$N_{L,gen} = N_{L,ht} + N_{L,FF} + N_{L,MF} \qquad (2.1.22)$$

where and separately demonstrate neighborhood entropy optimization number owing to heat transportation, NL, HT NL, FF NL, MF fluid contact, and attractive field:

$$N_{L,HT} = \frac{k_{nf}}{k_f}\left[\left(\frac{\partial\theta}{\partial x}\right)^2 + \left(\frac{\partial\theta}{\partial y}\right)^2\right]$$

(2.1.23)

$$N_{L,FF} = \frac{\mu_{nf}}{\mu_f}\varnothing_f\left\{4\left(\frac{\partial^2\psi}{\partial x\partial y}\right)^2 + \left(\frac{\partial^2\psi}{\partial Y^2} - \frac{\partial^2\psi}{\partial x^2}\right)^2\right\}$$

(2.1.24)

$$N_{L,FF} = \frac{\sigma_{nf}}{\sigma_f}Ha^2\varnothing_f\left[\frac{\partial\psi}{\partial Y}\sin\lambda + \frac{\partial\psi}{\partial x}\cos\lambda\right]^2$$

(1.2.25)

The whole entropy era is picked up by joining the nearby entropy era over the framework volume:

$$N_{T,HT} = \int N_{L,HT}dv \qquad (2.1.26)$$

$$N_{T,FF} = \int N_{L,FF}dv$$
(2.1.27)

$$N_{T,MF} = \int N_{L,MF}dv \qquad (2.1.28)$$

In 2006, the biological coefficient of execution (ECOP) was presented as a benchmark for assessing vitality frameworks such as control plants by Ust et al. In 2010, Seyyedi et al. amplified the definition of ECOP and connected it to assessing the heat execution of amplified surfaces (balances). They inferred an expression for ECOP in a stick blade.

A control volume limited component strategy (CVFEM) is applied in this paper. Saabas and Baliga presented the strategy. A FORTRAN code is set up to compute the overseeing expressions. Here, a calculation is displayed step by step to show the code based on the CVFEM.

To guarantee the grid-independent solution, the grid sizes of 51×151, 61×181, 71×211, 81×241 and 91×271 have been tested, ,, Ra = 105 Ha = 100,λ = 30° Pr = 6.2 ϕ = 0.04 AR = 0.75 and . $\Phi = 10^{-4}$. The upshots for entropy optimization and the normal Nusselt numbers appear (Ngen) and (Nu_{ave})in Table 1. In this manner, concurring with this table, the framework estimate has been chosen to be 81×241.

The current FORTRAN code is verified by comparing the picked-up results with other inquiries communicated within the writing. Figure 2 appears an uncommon compromise between current recreations and the results of Refs.

Here commonly convened heat transportation besides the entropy era in an L-shaped walled-in area stacked with an Al_2O_3 nanofluid is scrutinized numerically by CVFEM. The thermophysical characteristics of the Al_2O_3 nanoparticles and based fluid are shown in Table 2 Too, Table 3 depicts the values of the nanoparticles forming factor. Calculations are carried out for diverse estimations of Rayleigh number nanoparticles volume (Ra = 103,5 × 103,104,5 × 104, and 105), fraction; nanoparticle shape, Hartmann number (ϕ = 0 —— 0.06) (m = 3,4.8 and 5.7) (Ha = 0-100), magnetic field angle and aspect ratio (λ = 0°,15°,30°,45°,60°,75°,90°(. Figure 3 presents the entropy era number and normal Nusselt number against attractive field points for different estimations of Hartman number without nanoparticle (∅ = 0). The ϕ = Ra = 105 normal Nusselt number goes up through climbing the attractive field point, and there's the least esteem for entropy optimization number for each Hartmann number. Too, both of them diminish λ = 30° with expanding Hartmann number. It is discernible that expanding the normal Nusselt number is alluring though expanding the entropy optimization number is undesirable. Figure 4 discloses the entropy optimization number, normal Nusselt number, ECOP, and normal Bejan number versus Hartmann number for the diverse volume division of nanoparticles. Figure 5 presents the impact of the Rayleigh number on the normal Nusselt number and the entropy optimization number for two values of Hartmann number at and λ = 0°ϕ = 0.02. The figure appears that both increment as the Rayleigh number increments. The Nu_{ave} N_{gen} varieties of and for are more than those of when Rayleigh number Nu_{ave} N_{gen} Ha = Ha = 50 increases. Figure 6 depicts the entropy era and normal Nusselt numbers versus the Hartmann number for the distinctive values of shape calculated for m = 3, m = 4.8, and m = 5.7 λ = 30° and ϕ = 0.04. The figure discloses that both the entropy optimization and normal Nusselt Ra = 105 numbers develop when the shape figure goes up. Figure 7 shows the streamlines, the isotherms, and at NL, HT NL, FF NL, MF, $N_{L,gen,}$ $Be_{L,}$ Ra =10, and for two estimations of Hartmann number (and Pr = 6.2,ϕ = 0.04, AR

= 0.75,λ = 0° Ha = 5, Ha = 50). Figure 8 appears the streamlines, and for L-shape depth at NL, HT NL, FF NL, MF, $N_{L,gen,}$ $Be_{L,}$ Ra= 10, and for three estimations of perspective proportion 5 Pr = 6.2 ϕ = 0.04 Ha = 50 λ = 0° AR = 0.25, 0.50 and 0.75. The isotherms are parallel to the inward divider of depression when the angle proportion increments. The effect of viewpoint proportion on the normal Nusselt number and the entropy optimization number is delineated in Figure 9. In this figure and are considered. The figure presents λ = 30°, ϕ = 0.04 Ra = 105, the values of the entropy optimization and normal Nusselt numbers with regard to two reference values that are compared to Ha100. Figure 10 presents the normal Nusselt number in conjunction with the entropy optimization number versus nanoparticles volume division for and for at. The figure appears that normal Ha = 25 and Ha = 75 λ = 0° Ra = 105Nusselt number rises by means of bigger nanoparticles volume division while the entropy era number decreases as the nanoparticles volume division climbs. The varieties with regard to N_{gen} ϕ for is more than those of comparing to. For illustration, it diminishes from Ha = 25 Ha = 75 N_{gen} 53.09 to 45.05 (15.14% diminishing) while diminishes from 25.04 to 23.00 (8.15% Ha = 25 it is decreasing) for when increments from to 0.1.

The common convicted entropy era-based investigation was performed for an L-shape walled in an area stacked with Al_2O_3-H_2O nanofluid in the nearness of a situated attractive field. CVFEM was chosen as the fathoming strategy. The properties of dynamic factors were explored on the entropy optimization number. A straightforward financial investigation was presented to consider nanofluid taken a toll. The foremost results can be briefed as follows:

1. There's the least esteem for entropy optimization number for each Hartmann λ = 30° number (Figure 1).
2. The normal Nusselt number, the normal Bejan number, the ECOP increment, and the entropy optimization number decrease as the nanoparticle's volume division boosts (Figure 4).
3. Both increase as Rayleigh's number increases. Increases 56.8% and 11% Nu_{ave} N_{gen} Nu_{ave}

When increases from to for and, respectively (Figure 5).

Figure 1. Schematic and the grid of the problem

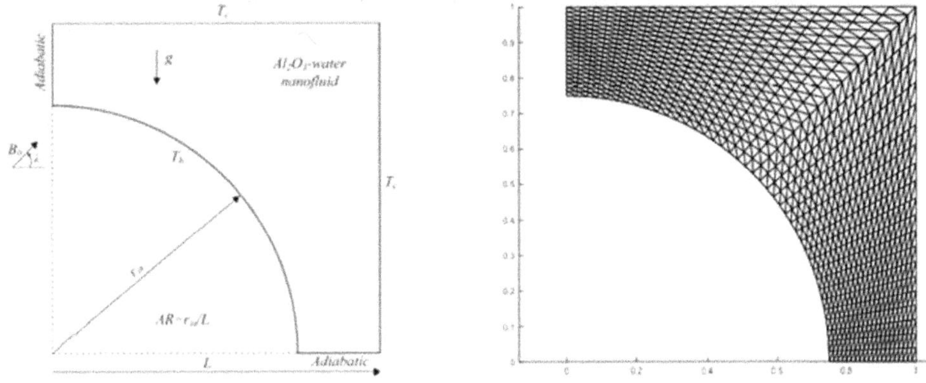

Figure 2. Comparison of the present results with those reported by references; (a)the average Nusselt number for different Hartmann numbers at and when the entropy Ra = 104, Ra = 105, Pr = 0.733 generation number and the average Bejan number for at and when Ha = 0, Ra = 103, Ra = 105, Pr = 0.71

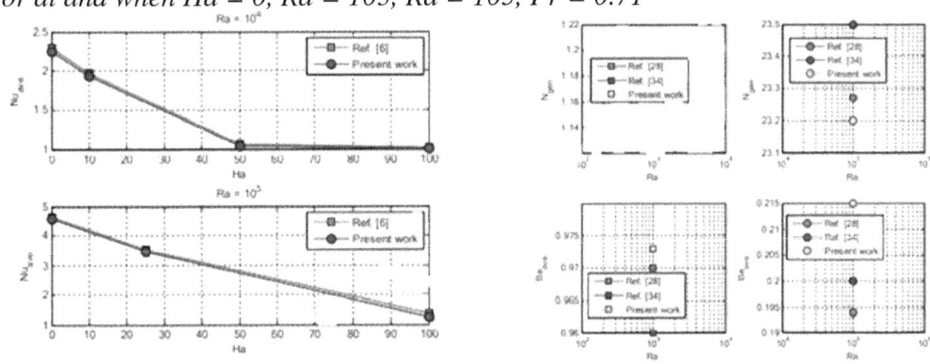

(a) (b)

Figure 3. Average Nusselt number and entropy generation number versus angle of the magnetic field for different Hartmann numbers for pure water at ($\phi = 0$) Ra = 105

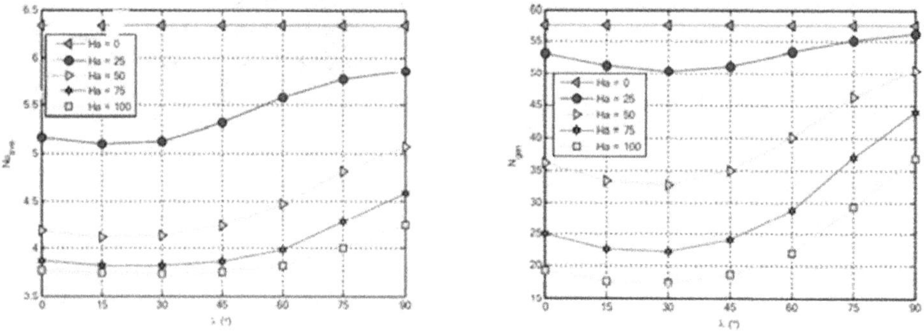

Figure 4. Average Nusselt number and entropy generation number versus Hartmann number for the different volume fraction of nanoparticles at $\lambda = 30$ and ° Ra = 105

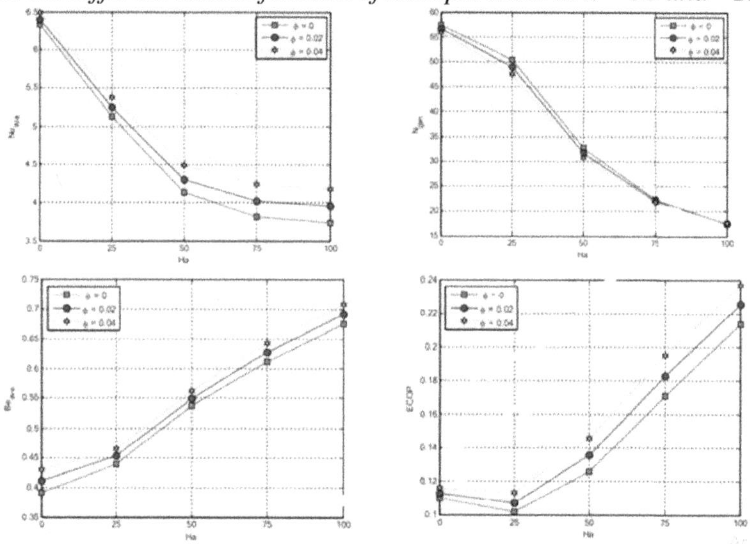

Figure 5. Average Nusselt number and entropy generation number versus Rayleigh number for and Ha = 0, Ha = 50 at and λ = 0° ϕ = 0.02

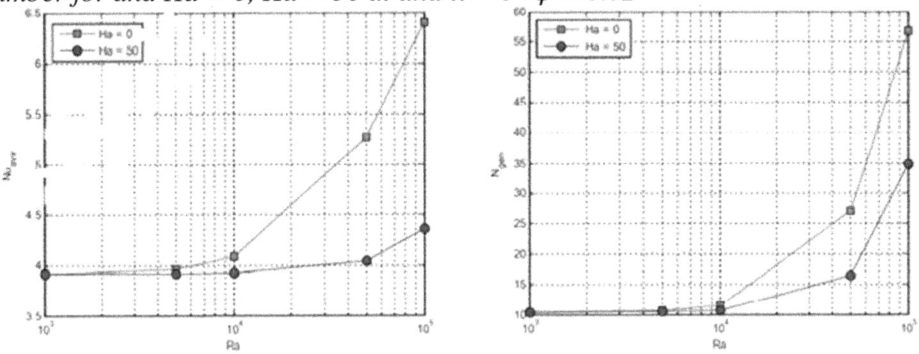

Figure 6. Average Nusselt number and entropy generation number versus Hartmann number for at, λ = 30° ϕ = 0.04 and Ra = 105

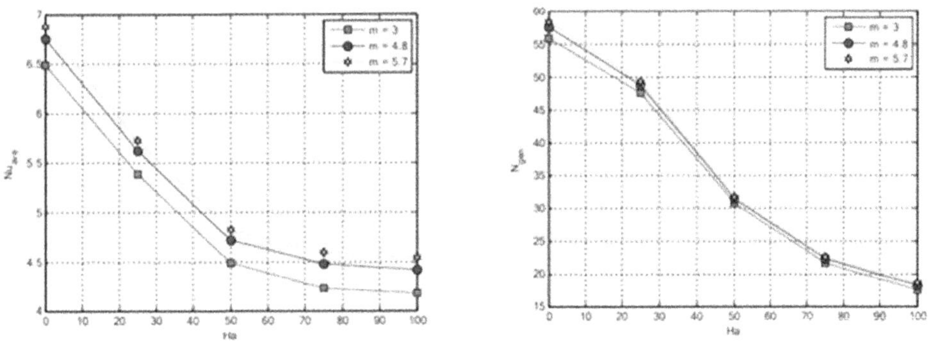

Figure 7. Streamlines the isotherms contours of and at, (ψ) (θ) NL, HT, NL, FF, NL, MF, NL, gen BeL Ra = 105 Pr = 6.2 ϕ = 0.04, for an AR = 0.75 λ = 0° Ha = 5 Ha = 50

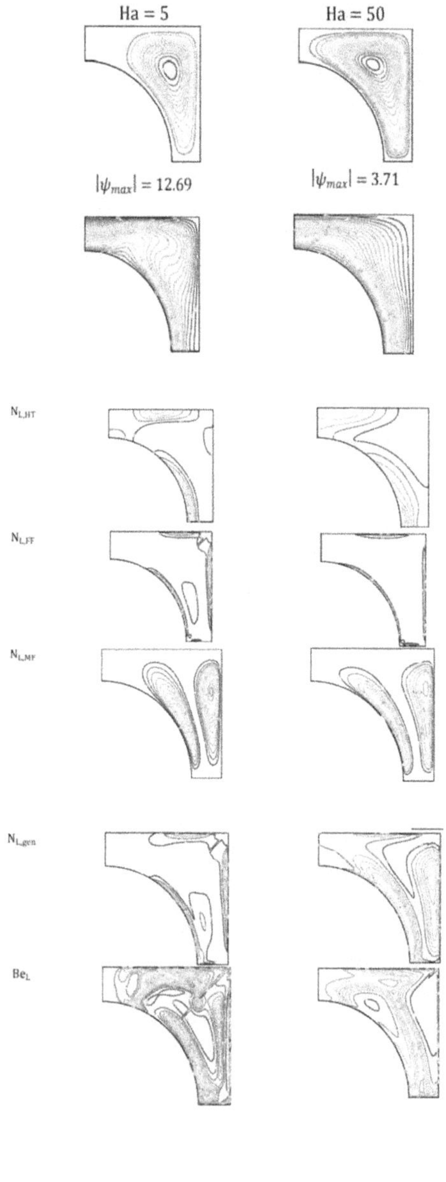

Figure 8. Streamlines, the isotherms, contours of and at, (ψ) (θ) NL, HT, NL, FF, NL, MF, NL, gen BeL Ra = 105 Pr = 6.2 ϕ = 0.04, for Ha = 50, λ = 0°, AR = 0.25, AR = 0.50 and AR = 0.75

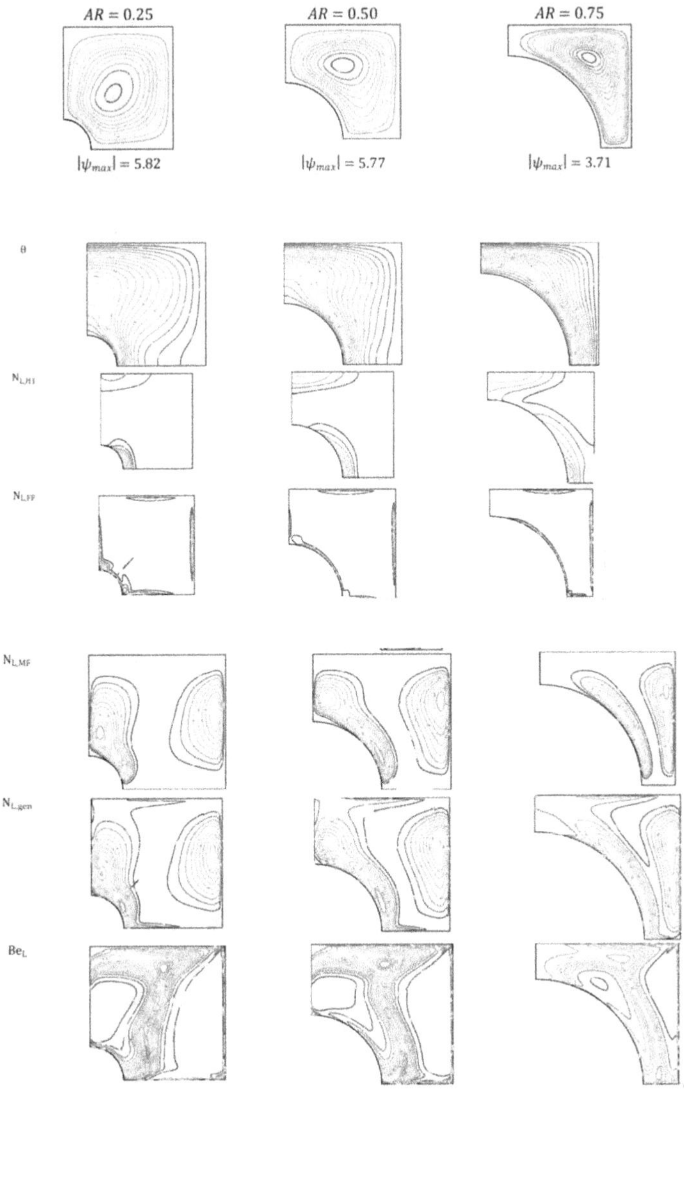

Figure 9. The average Nusselt number and entropy generation number for different values of aspect ratio at various values of Hartman number for $\lambda = 30°$

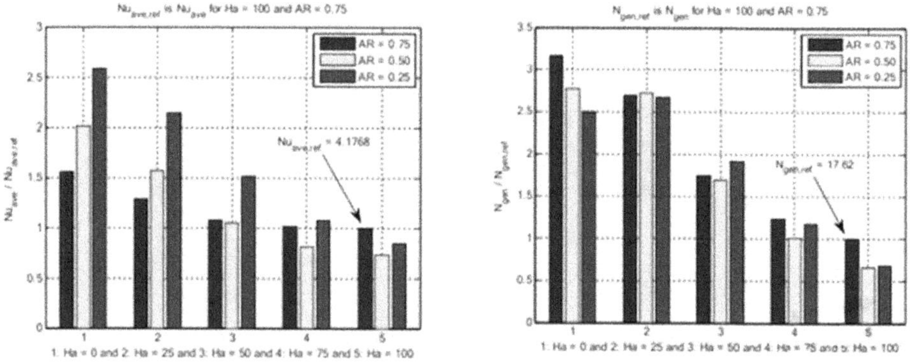

Figure 10. The normal Nusselt number and entropy era number versus volume division of nanoparticles for two values of Hartman number for $\lambda = 0$ at ° $Ra = 10^5$

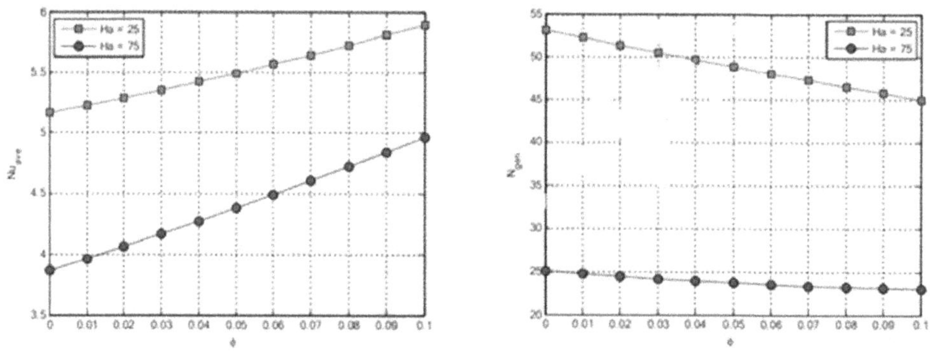

Table 1. Comparison of the average Nusselt number and entropy generation number for different (Nu_{ave}) and (N_{gen}) grid resolution at, and Ra = 105, Ha = 100, = 30° $\phi = 0.04$ $\Phi = 10^{4}$

Mesh size	51 × 151	61 × 181	71 × 211	81 × 241	91 × 271
Nu_{ave}	4.1811	4.1788	4.1768	4.1751	4.1737
N_{gen}	17.6698	17.6365	17.6190	17.6093	17.6039

Table 2. Thermo-physical properties of water and alumina

	(kg/m3)	(J/kgK)	(W/mK)	(1/K)	(/cm)
Alumina (Al_2O_3)	3970	765	40	0.85×10^{-5}	1×10^{-10}
Pure water	997.1	4179	0.613	21×10^{-5}	0.05

Table 3. Shape factor values for diverse nanoparticle shapes

Name	m
Spherical	3
Cylinder	4.8
Platelet	5.7

Table 4. Nanofluid cost for different values of nanoparticles

ϕ	Base (EUR)	Nano(EUR)	T (EUR)
0	0.0478	0	0.0478
0.01	0.0478	72.96	73.01
0.02	0.0478	145.92	145.97
0.03	0.0478	218.88	218.93
0.04	0.0478	291.48	291.89
0.05	0.0478	364.80	364.85

REFERENCES

Alizadeh, M., Seyyedi, S. M., Rahni, M. T., & Ganji, D. D. (2017). Three-dimensional numerical simulation of rising bubbles in the presence of cylindrical obstacles, using lattice Boltzmann method. *Journal of Molecular Liquids*, 236, 151–161. DOI: 10.1016/j.molliq.2017.04.009

Azar, A. A., Jalili, P., Moziraji, Z. P., Jalili, B., & Ganji, D. D. (2024). Analytical solution for MHD nanofluid flow over a porous wedge with melting heat transfer. *Heliyon*, 10(15). PMID: 39166020

Çobanoğlu, N., Banisharif, A., Estellé, P., & Karadeniz, Z. H. (2022). The developing flow characteristics of water-ethylene glycol mixture based Fe3O4 nanofluids in eccentric annular ducts in low temperature applications. *International Journal of Thermofluids*, 14, 100149. DOI: 10.1016/j.ijft.2022.100149

Dogonchi, A. S., & Ganji, D. D. (2017). Impact of Cattaneo-Christov heat flux on MHD nanofluid flow and heat transfer between parallel plates considering thermal radiation effect. *Journal of the Taiwan Institute of Chemical Engineers*, 80, 52–63. DOI: 10.1016/j.jtice.2017.08.005

Fard, M. S., Torabiyan, A., Jalili, P., Jalili, B., & Ganji, D. D. (2024). Investigating the magnetohydrodynamics non-Newtonian fluid movement on a tensile plate affected by variable thickness with dufour and soret effects: Akbari Ganji and finite element methods. *International Journal of Electrochemical Science*, 19(8), 100701. DOI: 10.1016/j.ijoes.2024.100701

Ghasemi, S. E., Mohsenian, S., Gouran, S., & Zolfagharian, A. (2022). A novel spectral relaxation approach for nanofluid flow past a stretching surface in presence of magnetic field and nonlinear radiation. *Results in Physics*, 32, 105141. DOI: 10.1016/j.rinp.2021.105141

Jalili, B., Aghaee, N., Jalili, P., & Ganji, D. D. (2022). Novel usage of the curved rectangular fin on the heat transfer of a double-pipe heat exchanger with a nanofluid. *Case Studies in Thermal Engineering*, 35, 102086. DOI: 10.1016/j.csite.2022.102086

Jalili, B., Jalili, P., Sadighi, S., & Ganji, D. D. (2021). Effect of magnetic and boundary parameters on flow characteristics analysis of micropolar ferrofluid through the shrinking sheet with effective thermal conductivity. *Zhongguo Wuli Xuekan*, 71, 136–150. DOI: 10.1016/j.cjph.2020.02.034

Jalili, B., Sadighi, S., Jalili, P., & Ganji, D. D. (2019). Characteristics of ferrofluid flow over a stretching sheet with suction and injection. *Case Studies in Thermal Engineering*, 14, 100470. DOI: 10.1016/j.csite.2019.100470

Jalili, P., Azar, A. A., Jalili, B., & Ganji, D. D. (2024). A Novel analytical investigation of a swirling fluid flow and a rotating disk in the presence of uniform suction. *Arabian Journal for Science and Engineering*, 49(8), 10453–10469. DOI: 10.1007/s13369-023-08391-7

Khan, S. A., Waqas, H., Muhammad, T., Farooq, U., & Alghamdi, M. (2021). Recent progress in melting phenomenon for magnetized hybrid nanofluid flow over a stretching surface with temperature dependent viscosity: A comparative study. *Journal of Materials Research and Technology*, 15, 3965–3973. DOI: 10.1016/j.jmrt.2021.09.099

Liu, Z., Zhang, T., Zhao, Y., & Bi, S. (2019). Time-varying stiffness model of spur gear considering the effect of surface morphology characteristics. *Proceedings of the Institution of Mechanical Engineers. Part E, Journal of Process Mechanical Engineering*, 233(2), 242–253. DOI: 10.1177/0954408918775955

Mahboobtosi, M., Ganji, A. M., Jalili, P., Jalili, B., Ahmad, I., Hendy, A. S., Ali, M. R., & Ganji, D. D. (2024). Investigate the influence of various parameters on MHD flow characteristics in a porous medium. *Case Studies in Thermal Engineering*, 59, 104428. DOI: 10.1016/j.csite.2024.104428

Nabi, H., Pourfallah, M., Gholinia, M., & Jahanian, O. (2022). Increasing heat transfer in flat plate solar collectors using various forms of turbulence-inducing elements and CNTs-CuO hybrid nanofluids. *Case Studies in Thermal Engineering*, 33, 101909. DOI: 10.1016/j.csite.2022.101909

Pasha, P., Alibak, A. H., Nabi, H., & Tat Shahdost, F. (2022). Examination of heat transfer on extending sheet by variation iteration method strategy and investigation of arrangements for optimizing liquid properties. *Engineering Reports*, 4(10), 12505. DOI: 10.1002/eng2.12505

Qiao, L., Chen, H. E., Wang, M., Zhu, Y., Li, X., Tong, Z., Xie, S., Chen, H., & Chen, Z. (2022). Numerical simulation methods for motion-induced eddy current testing signals based on Ar formulation and edge finite elements. *NDT & E International*, 129, 102651. DOI: 10.1016/j.ndteint.2022.102651

Riehl, R. R., & Mancin, S. (2022). Estimation of thermophysical properties for accurate numerical simulation of nanofluid heat transfer applied to a loop heat pipe. *International Journal of Thermofluids, 100158.*

Santos, R., Ribeiro, A. P. C., Moreira, A. L. N., & Moita, A. S. (2022). Effect of alumina nanofluids on bubble dynamics and heat transfer under quiescent conditions. *International Journal of Thermofluids, 100168.*

Sharma, R. P., & Shaw, S. (2022). MHD Non-Newtonian fluid flow past a stretching sheet under the influence of non-linear radiation and viscous dissipation. *Journal of Applied and Computational Mechanics*, 8(3), 949–961.

Wang, K., Li, G., & Lu, Z. (2022). Numerical simulation of dynamic fractures in 2D FGMs using the numerical manifold method. *Engineering Analysis with Boundary Elements*, 140, 32–47. DOI: 10.1016/j.enganabound.2022.04.002

Waqas, H., Farooq, U., Bukhari, F. F., Alghamdi, M., & Muhammad, T. (2021). Chemically reactive transport of magnetized hybrid nanofluids through Darcian porous medium. *Case Studies in Thermal Engineering*, 28, 101431. DOI: 10.1016/j.csite.2021.101431

Zar, H., Jalili, B., Jalili, P., & Ganji, D. D. (2024). Thermal study of magnetohydrodynamic nanofluid flow and brownian motion between parallel sheets. *International Journal of Thermofluids*, 23, 100806. DOI: 10.1016/j.ijft.2024.100806

Section 2

Advanced Topics in MHD Nanofluids and Analytical Approaches

This section focuses on MHD nanofluids, their behavior, and various analytical techniques used to solve complex MHD problems.

Chapter 3
MHD Nanofluid Flow

ABSTRACT

Increasing heat transfer in the field of thermal management is one of the essential features of engineering studies. In many cases, the use of nanoparticles leads to increased heat transfer. Hence, this analysis uses nanoparticles in magnetohydrodynamic fluid flow to increase heat transfer. The main objective of this chapter is to increase the heat transfer in magnetohydrodynamic nanofluid flow. Governing equations have been solved using the finite element method, and results have been shown with contours and graphs.

NONLINEAR EQUATIONS IN THE ENTROPY GENERATION AREA

3.1. Two-Dimensional Stream of a Micropolar/Classical Ferrofluid

On an extending sheet with twelve numbers, Distinctive confuses within the y > course, a steady-state two-dimensional stream of a micropolar/classical Ferrofluid is expected in Figure 1, Table 1. The triangular and rectangular astounds are utilized on the surfaces. The stream of nanofluid is entered at a 20-degree temperature and precise velocity from four bearings and passes through the perplexes and surface. The surface temperature is 30 degrees, and the surface is extended by ($U_w(x)=ax$, $a>0$). The amount of velocity within the x-direction is 2m/s. This article explores the impact of attractive constraints on the surface (Mirzaei *et al.*, 2024) (Bhatti & Rashidi, 2017) (Ellahi, Hassan, & Zeeshan, 2016) (Ellahi et al., 20170 (Ganji *et al.*, 2024) (Dib, Haiahem, & Bou-said, 2015) (Bahmani *et al.*, 2024) (Rashidi *et al.*, 2024) (Alsavery *et al.*, 2017) (Alsabery *et al.*, 2019) (Jalili *et al.*, 2024) (Zeeshan *et al.*, 2017) (Hajizadeh *et al.*, 2024) (Dogonchi *et al.*, 2018) (Dogonchi & Ganji,

DOI: 10.4018/979-8-3693-5335-6.ch003

2018) (Chari *et al.*, 2024) (Dogonchi *et al.*, 2019a) (Waqas *et al.*, 2017) (Dogonchi *et al.*, 2019b).

The finite element method is utilized to fathom this article. Within the underneath, the overseeing condition for Ferrofluid is:

$$\frac{d}{dt}(\rho_{nf}) = \nabla.(\rho_{nf}v)$$

(3.1.1)

$$\rho_{nf}\left(\frac{dv}{dt}\right) = -\nabla_{p} + \left(2\mu_{nf} + \aleph\right)\ \nabla\ (\nabla.v)$$

$$-\left(\mu_{nf} + \aleph\right)\ \nabla\ .(\nabla.v) + \aleph(\nabla.N) + j.B + \rho_{nf}g$$

(3.1.2)

$$\rho_{nf}j\left(\frac{dN}{dt}\right) = \left(\varphi + \lambda + Y_{nf}\right)\ \nabla\ .\left(\rho_{nf}.N\right) - Y_{nf}\ \nabla\ .(\nabla.N) + \aleph(\nabla.v) - 2\aleph N + \rho_{nf}I$$

(3.1.3)

Brinkman, Bourantas, and Loukopoulos present effective density, effective dynamic viscosity, and spin gradient viscosity, respectively:

$$\rho_{nf} = (1 - \varphi)\rho_{f} + \varnothing\rho_{s}$$

(3.1.4)

$$\mu_{nf} = \frac{\mu_{f}}{(1 - \varphi)^{2.5}}$$

(3.1.5)

$$Y_{nf} = \left(\mu_{nf} + {}^{\aleph}/_{2}\right)j$$

(3.1.6)

In expansion, the stream is incompressible within the nonappearance of outside powers. Hence, conditions will be troublesome to fathom.

$$\rho_{nf}\left(\frac{dv}{dt} + (v.\nabla)v\right) = \left(\mu_{nf} + \aleph\right)\nabla^{2}v + \aleph(\nabla.N) + j.B$$

(3.1.7)

$$\rho_{nf} j\left(\frac{dN}{dt} + (N.\nabla)N\right) = Y_{nf}.\nabla^2 N + \aleph(\nabla.v) - 2\aleph N$$

(3.1.8)

The body force proposed by Hussanan et al. is:

$$j.B = -\sigma_{nf} B_0^2 v$$

(3.1.9)

The Ferrofluid electric conductivity represents:

$$\sigma_{nf} = \left[1 + \frac{3(\sigma - 1)\varnothing}{(\sigma + 2) - (\sigma - 1)\varnothing}\right]\sigma_f$$

(3.1.10)

And the governing equations go to:

$$\rho_{nf}\left(u\frac{du}{dx} + v\frac{du}{dy}\right) = \left(\mu_{nf} + \aleph\right)\frac{d^2u}{dy^2} + \aleph\frac{dN}{dy} - \sigma_{nf} B_0^2 u$$

(3.1.11)

$$\rho_{nf}\left(u\frac{du}{dx} + v\frac{du}{dy}\right) = \left(\mu_{nf} + \aleph\right)\frac{d^2u}{dy^2} + \aleph\frac{dN}{dy} - \sigma_{nf} B_0^2$$

$$\rho_{nf} j\left(u\frac{dN}{dx} + v\frac{dN}{dy}\right) = Y_{nf}\frac{d^2N}{dy^2} - \aleph\left(2N + \frac{du}{dy}\right)$$

(3.1.12)

The specific heat capacity and the effective thermal conductivity researched by Din and Khan et al.

$$C_{p,nf} = \frac{C_p)s + (1 - \varnothing)(\rho C_p)f}{\rho_{nf}}$$

(3.1.13)

$$\frac{k_{nf}}{k_f} = \frac{k_s + 2k_f - 2\varnothing\left(k_f - k_s\right)}{k_s + 2k_f + \varnothing\left(k_f - k_s\right)}$$

(3.1.14)

The energy equation is represented as follows:

$$u\frac{dT}{dx} + v\frac{dT}{dy} = \frac{k_{nf}}{C_p)nf}\frac{d^2T}{dy^2} - \frac{dq_r}{dy}$$

(3.1.15)

It is stated according to Rosseland's approximation as below:

$$u\frac{dT}{dx} + v\frac{dT}{dy} = \frac{1}{C_p)nf}\left(k_{nf} + \frac{16\sigma^* T_\infty^3}{3k^*}\right)\frac{d^2T}{dy^2}$$

(3.1.16)

The boundary conditions are:

$$u = au_w(x), v = v_w, at\, y = 0, u \leftrightarrow 0, y \rightarrow \infty$$

(3.1.17)

$$N = -\delta\frac{du}{dy} at\, y = 0, N \leftrightarrow 0, y \rightarrow \infty$$

(3.1.18)

$$T = T_w, at\, y = 0, T \leftrightarrow T_w, y \rightarrow \infty$$

(3.1.19)

The positive and negative values of the surface mass transfer velocity, v_w represent the suction and injection. The angular velocity is denoted by N. μ is dynamic viscosity, and δ is a constant in the 0-1 range. When set to 0, the microelement is specified. If δ=1/2, the concentration of microelements was shown to be weak. Furthermore, when the turbulent boundary layer flow is required, δ=1, j represents microelement inertia per unit mass, where is spin gradient viscosity. To solve these nonlinear differential equations comfortably:

$$\eta = y\sqrt{\frac{a}{v_f}}, u = axF'(\eta), v = -\sqrt{av_f}F(\eta), N = ax\sqrt{\frac{a}{v_f}}G\left(\eta, \theta(\eta) = \frac{T - T_\infty}{T_w - T_\infty},\right.$$

(3.1.20)

y substituting the variables (3.1.20) into Eq. (3.1.11), (3.1.12), (3.1.16):

$$\left(\frac{1}{(1-\varnothing)^{2.5}}+k\right)F'''(\eta)+\left(1-\varnothing+\varnothing\frac{\rho_s}{\rho_f}\right)F(\eta)F''(\eta)$$

$$-\left(1-\varnothing+\varnothing\frac{\rho_s}{\rho_f}\right)F'^2(\eta)-M\left(1+\frac{3(\sigma-1)\varnothing}{\sigma+2-(\sigma-1)\varnothing}\right)F'(\eta)+kG'(\eta)=0$$

(3.1.21)

$$\left(\frac{1}{(1-\varnothing)^{2.5}}+\frac{k}{2}\right)G''(\eta)$$

$$+\left(1-\varnothing+\varnothing\frac{\rho_s}{\rho_f}\right)F(\eta)G'(\eta)-\left(1-\varnothing+\varnothing\frac{\rho_s}{\rho_f}\right)F'(\eta)G(\eta)-k(2G(\eta)+F''(\eta))=0$$

(3.1.22)

$$\frac{1}{Pr}\left(\frac{k_s+2k_f-2\varnothing(k_f-k_s)}{k_s+2k_f+\varnothing(k_f-k_s)}+R\right)\theta''\left(\eta\right)+\left(1-\varnothing+\frac{C_p)s}{C_p)f}\right)F(\eta)\theta'(\eta)=0$$

(3.1.23)

And here are the new boundary conditions:

$$F(\eta)=S,\ F'(\eta)=\alpha,\ G(\eta)=-\delta F''(\eta),\ \theta(\eta)=1,\ at\ \eta=0$$

(3.1.24)

Limited Component Strategy is a critical numerical strategy. One of the viable applications for this strategy is the FlexPDE computer program, which illuminates the nonlinear halfway differential conditions and conventional differential conditions. The yield of the results FlexPDE is additionally a 'problem-solving environment' that executes the complete extent of capacities required to resolve fractional differential condition frameworks.

The sum of computational blunders in our work is exceptionally moo compared to others. The greatest number of mistakes happened in η=1.5 and the least number of mistakes in η=1, Table (3.1.10).

The impact of velocity changes within the x-direction around the perplexes more esteem than the encompassing space. Concurring to Figure 2, the sum of stream velocity is diminished by expanding the distance from the surface within the y- y-heading. Moreover, the rate of nanofluids is tall, and the stream is adjusted on both sides of the surface. Figure 3 spoke to various precise velocities of the nanofluid stream through the surface. For the best astounds of the surface, the precise velocity values around that surface are exceptional moo, and the most noteworthy velocity values are watched at the conclusion of the surface. Concurring to Figure 4, the precise velocity of the nanofluid is exceptionally tall around the perplexes within the lower portion of the surface, and the liquid stream particles circle around the surface at tall velocities. The liquid flow velocity is adjusted within the whole space around the astounds. Figure 5 shows the nanofluid temperature changes around the surface and the astounds. Agreeing to Figures (a and b), the volume of the nanofluid stream at the start of the surface and around the primary astound is exceptionally tall and has come to its most extreme. The liquid stream is diminished marginally, and the temperature is adjusted around the other edges and surface space as the liquid stream surpasses the astound. At the conclusion of the page, the temperature gets greatest. Agreeing to figs (c and d), the nanofluid temperature is exceptionally tall around the perplexes, and by expanding to remove the confuses, the sum of temperature is decreased. In Figure 6, the vector framework of temperature has been watched. The movement vectors of the nanofluid temperature around the surface have moo velocity and moo temperature. In Figure 7, with rectangular and triangular confusion, the nanofluid temperature is diminished from the extreme of 35 degrees to 26 degrees. In Figure a-(7) and b-(7), the sum of liquid temperature is moo at the starting. By passing the nanofluid stream from astounds, the stream temperature diminished from 35 degrees to 25 degrees. In Figures c-(6) and d-(6), the nanofluid temperature is consistent all over the confuses. The sum of temperature is 35 degrees. In Figure 8 with rectangular and triangular perplexes, the stream rate of nanofluid is moo at to begin with, and by passing through the other perplexes, the velocity of nanofluid is higher and from an esteem of comes to an esteem of 2. Figures a-8) and b-(8) expand the nanofluid velocity within the y-direction. The sum of nanofluid velocity is changed from u=0.1 to u=2. According to Table 2 the most extreme nanofluid velocity is observed in rectangular perplexes. The liquid velocity for nanofluid on the rectangular astounds is 19.5% greater than other confuses. Concurring to table (3.1.12), the greatest of nanofluid temperature is watched in rectangular and triangular perplexes. The liquid temperature for nanofluid on the rectangular and triangular confuses is 50% greater than other astounds.

This paper explores various temperatures and velocities within the x-direction and the precise velocity of the nanofluid stream through triangular and rectangular baffles in the presence of a uniform, attractive field. The development of this paper is to explore the parameters of the nanofluid stream passing from the diverse astounds on the extending surface. The stream of attraction influences the surface. The limited Component Strategy is chosen for tackling administering conditions.

Figure 1. Schematic of the threaded extending surface

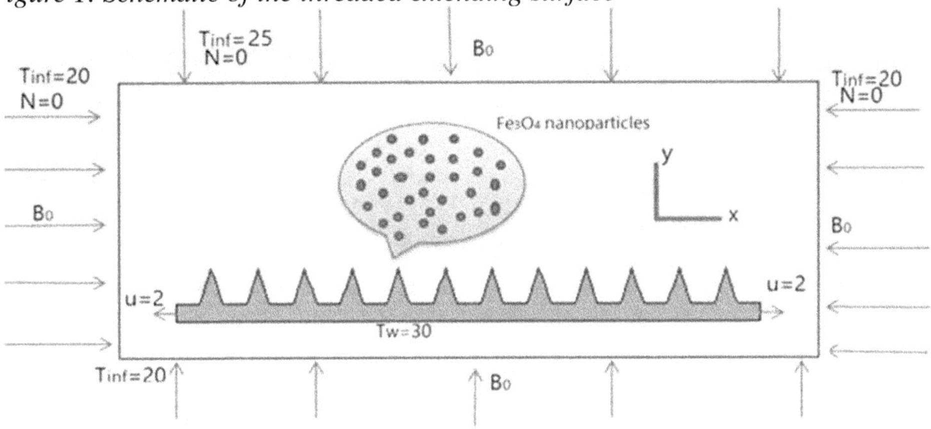

Figure 2. Comparison of work geometry for threaded extending surface

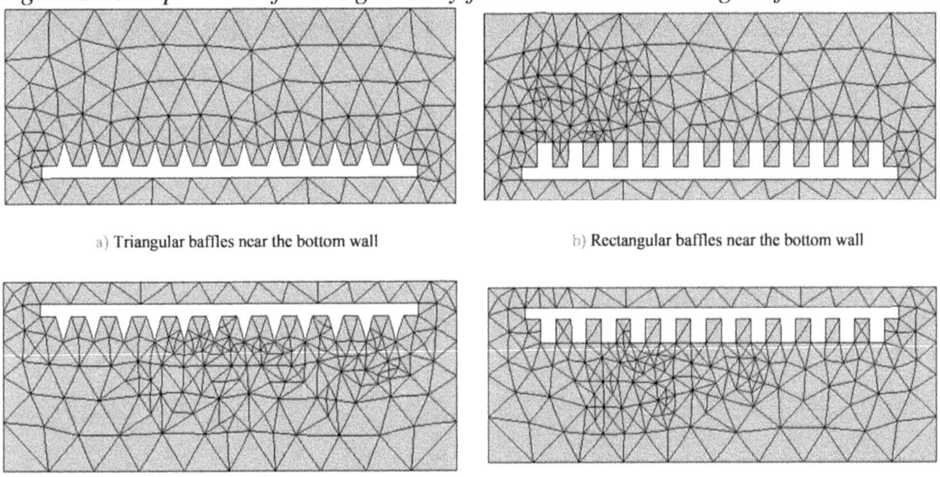

a) Triangular baffles near the bottom wall

b) Rectangular baffles near the bottom wall

c) Inverse triangular baffles near the top wall

d) Inverse rectangular baffles near the top wall

Figure 3. Comparisons of velocity changes within the X course in several, astounds for threaded extending surface

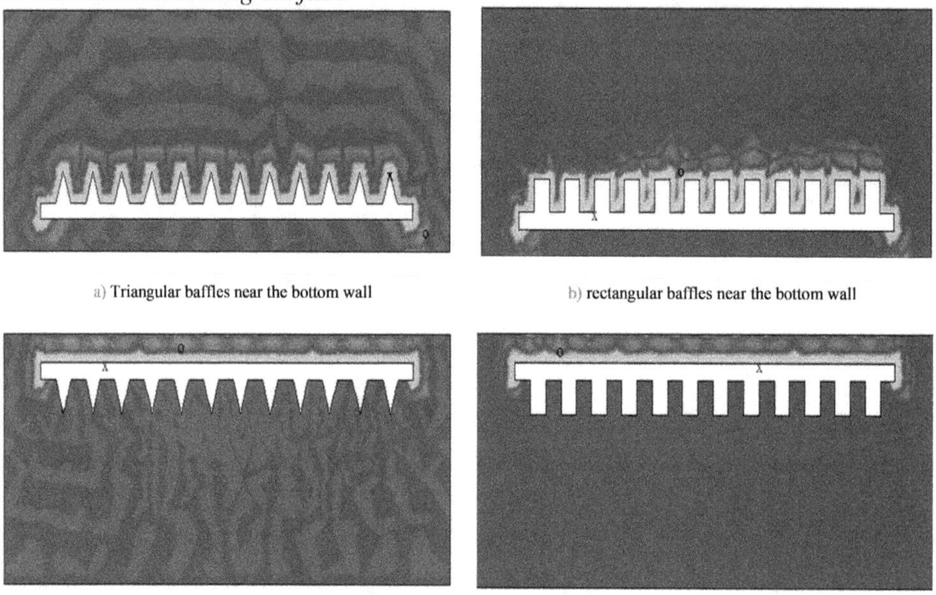

a) Triangular baffles near the bottom wall

b) rectangular baffles near the bottom wall

c) Inverse triangular baffles near the top wall

d) Inverse rectangular baffles near the top wall

Figure 4. Comparison of precise velocity changes totally different, confuses for threated extending surface

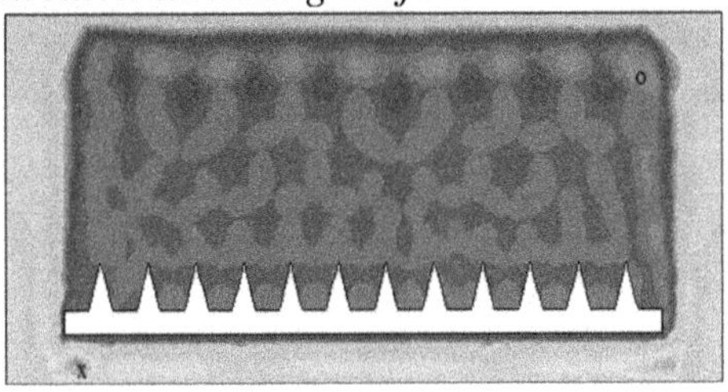

a) Triangular baffles near the bottom wall

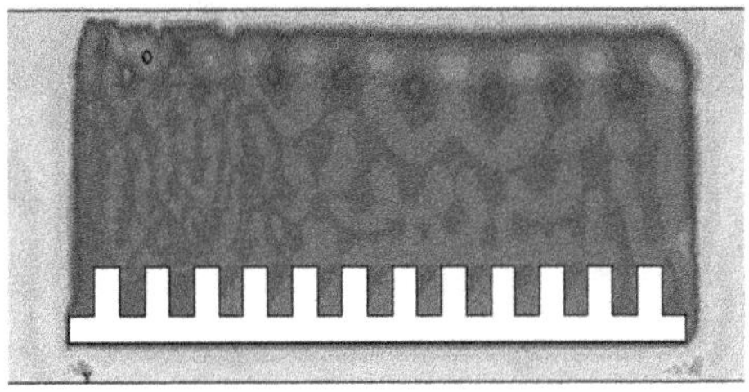

b) Rectangular baffles near the bottom wall

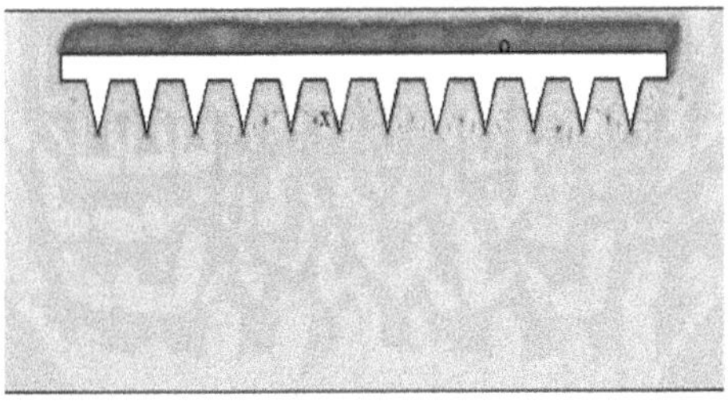

c) Inverse triangular baffles near the top wall

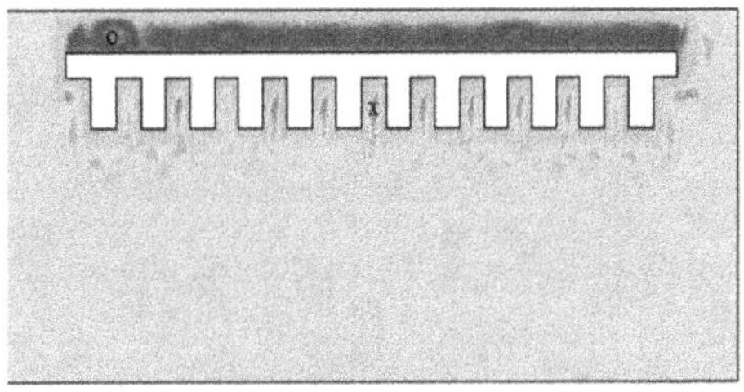

d) Inverse rectangular baffles near the top wall

Figure 5. Comparison of temperature profiles for threated extending surface with diverse perplexes

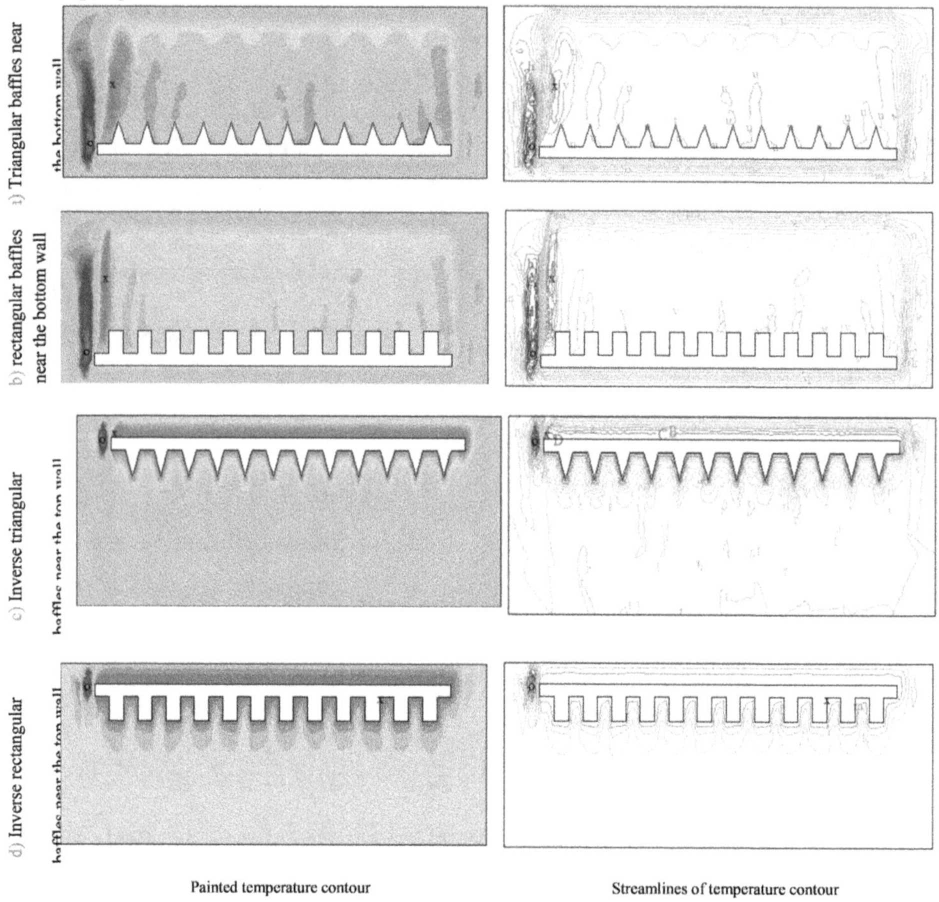

Figure 6. Comparison of vector framework temperature changes in several perplexes for threated extending surface

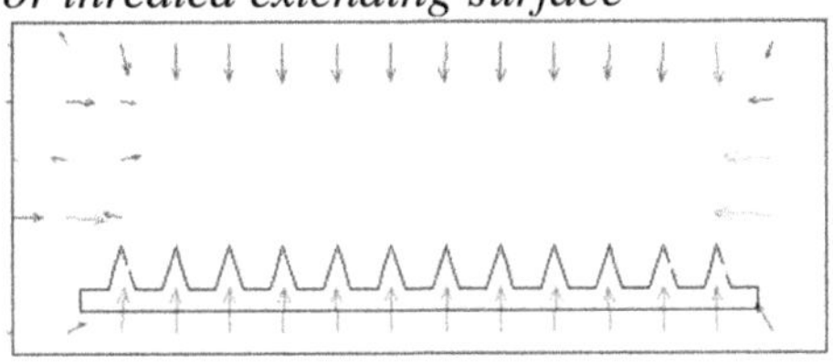

a) Triangular baffles near the bottom wall

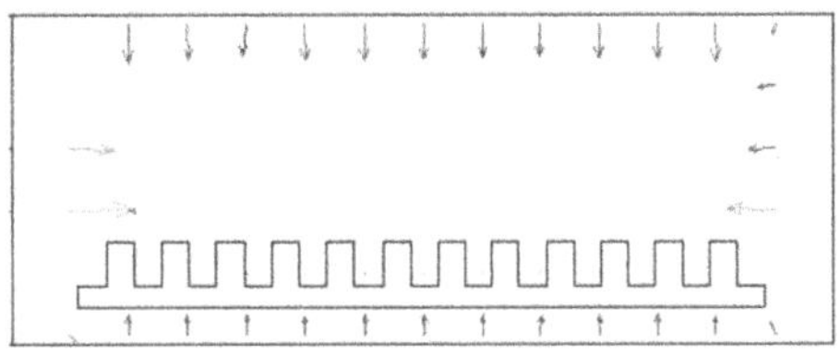

b) Rectangular baffles near the bottom wall

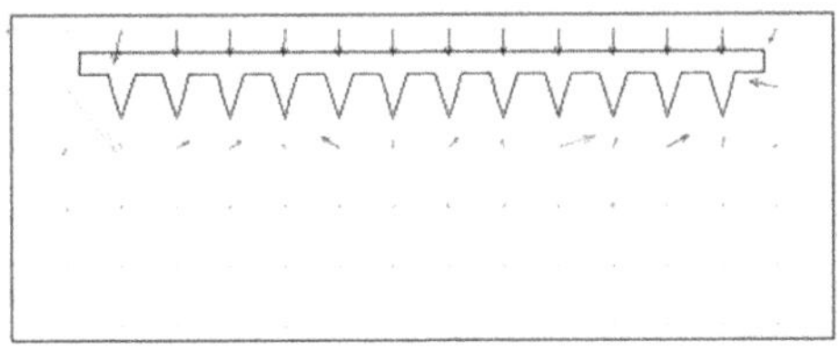

c) Inverse triangular baffles near the top wall

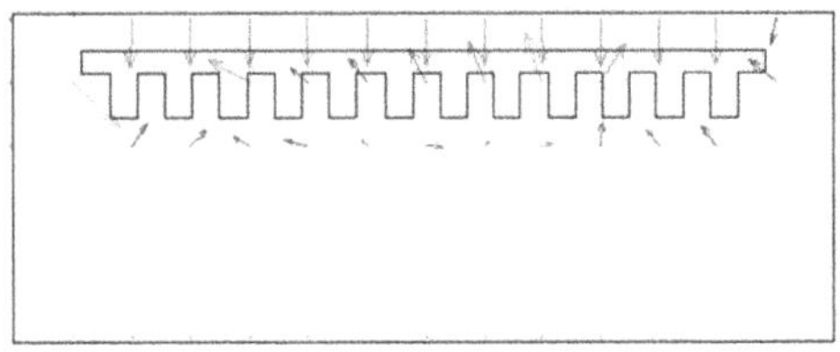

d) Inverse rectangular baffles near the top wall

Figure 7. Comparison of temperature changes in several confuses for threated extending surface

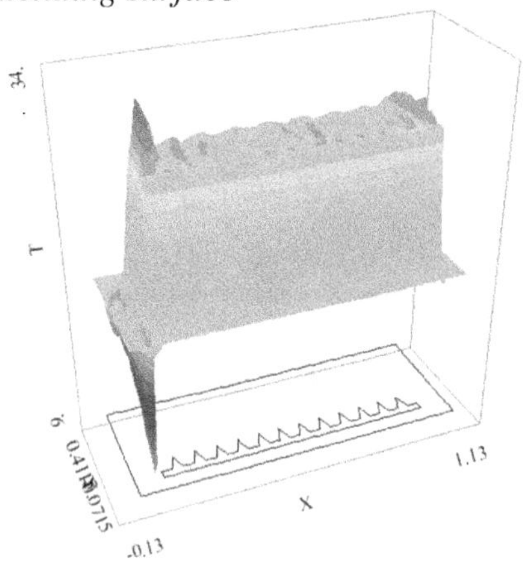

a) Triangular baffles near the bottom wall

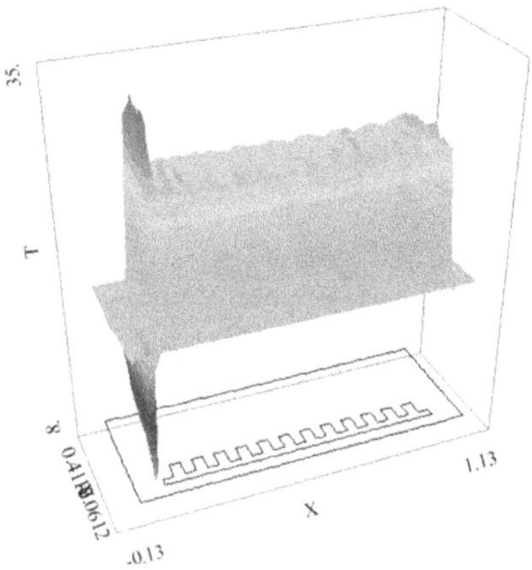

b) Rectangular baffles near the bottom wall

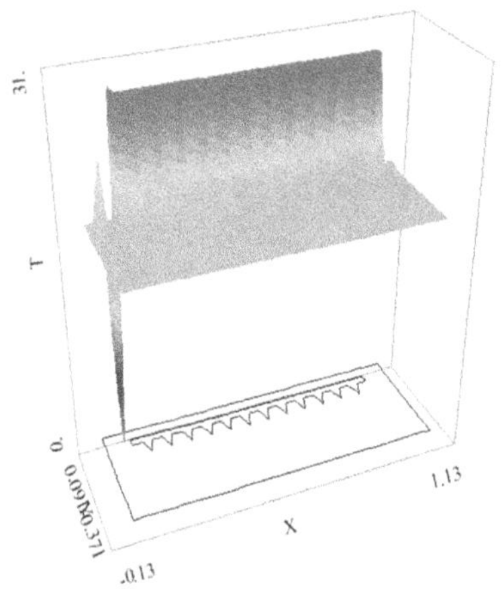

c) Inverse triangular baffles near the top wall

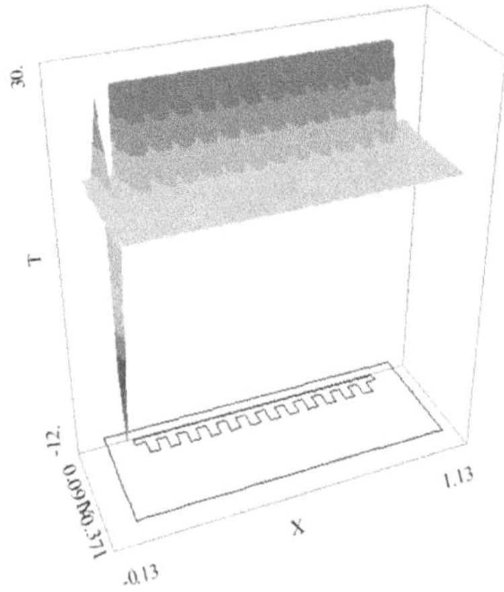

d) Inverse rectangular baffles near the top wall

Figure 8. Comparison of velocity (u) changes completely different perplexes for threated extending surface

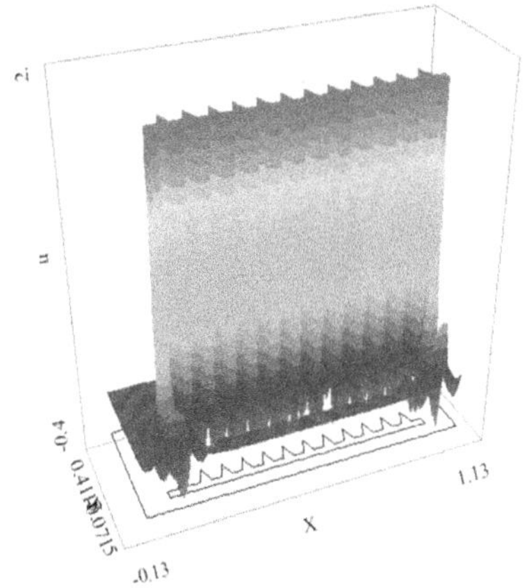

a) Triangular baffles near the bottom wall

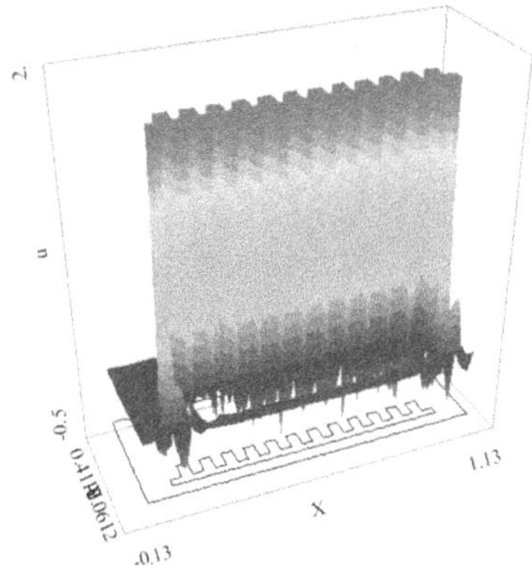

b) Rectangular baffles near the bottom wall

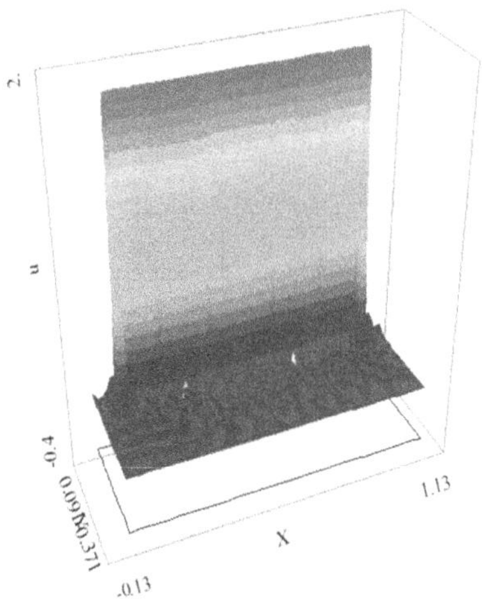

c) Inverse triangular baffles near the top wall

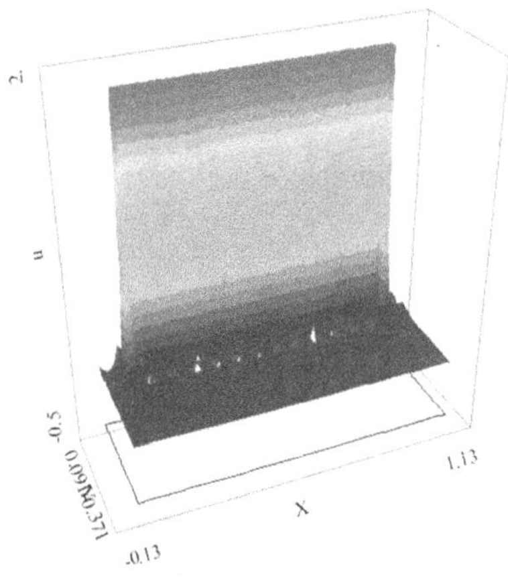

d) Inverse rectangular baffles near the top wall

Table 1. Properties of water and Fe_3O_4 nanofluid at 20 degree

	ρ (kg/m³)	K(w/m.k)	C_p (j/kg.k)	σ
Fe_3O_4	5180	9.7	670	25000
H_2O	997	0.613	4179	0.05

Table 2. Comparison of temperature for present work and Jalili work at K=10, δ=0.5, ϕ =0.03

	=0.0	=1	=1.5	=2	=2.5	=3
Present work	1	1.5	0.72	0.05	0.01	0
Jalili	1	1.7	0.75	0.03	0.01	0

Table 3. Numerical comparison nanofluid temperature between distinctive perplexes

x	$u_{Triangular\ baffle}$	$u_{Rectangular\ baffle}$	$u_{inverwe\ rectangular}$	$u_{inverse\ triangular}$
0.0	0	0	0	0
0.2	0.28	1.17	0.34	1.10
0.3	1.55	1.95	1.43	1.54
0.5	0.02	0.42	0.12	0.35
0.8	0.17	1.36	0.19	1.26
1	0	0	0	0

Table 4. Numerical comparison nanofluid temperature between distinctive astounds

x	$T_{Triangular\ baffle}$	$T_{Rectangular\ baffle}$	$T_{inverwe\ rectangular}$	$T_{inverse\ triangular}$
0.0	22.07	22.08	20.01	20.01
0.2	27.32	27.26	26.66	22.30
0.3	27.66	27.65	30.01	29.10
0.5	28.53	28.64	21.05	21.18
0.8	29.72	29.72	27.30	21.46
1	30.21	30.16	20.70	20.99

3.2. Magnetohydrodynamic (MHD) Nanofluid Flow in a Circular Penetrable Space

The geometry considered in this ponder is a Cassini oval with a circular permeable walled-in area, as appeared in Figure 9. The oval was subjected to steady heat flux. The permeable media are filled with MHD nanofluid. The shape of the Cassini oval is made by embracing the taking after equation (Table 1):

$$\left[(x-a)^2+(y-b)^2\right]\bullet\left[(x+a)^2+(y-b)^2\right] = c^4.$$

(3.2.1)

For the nanofluid flow within the enclosure, the conservation of mass, momentum, and energy can be expressed as (Roshani *et al.,* 2024) (Alizadeh, Dogonchi, & Ganji, 2018) (Seyyedi *et al.,* 2015) (Alsabery *et al.,* 2018) (Sheremet, Revnic, & Pop, 2017) (Hatami, Song, & Jing, 2016) (Bessa, Ali, & Said, 2017) (Sheremet & Pop, 2014) (Revnic, Sheremet, & Pop, 2016) (Mahmud & Pop, 2016):

$$\frac{\partial u}{\partial x} + \frac{\partial v}{\partial y} = 0,$$

(3.2.2)

$$\frac{\mu_{nf}}{K} u = -\frac{\partial P}{\partial x} + \sigma_{nf} B_0^2 \left[v \bullet sin(\gamma) \bullet cos(\gamma) - u \bullet sin(\gamma)^2 \right],$$

(3.2.3)

$$\frac{\mu_{nf}}{K} v = -\frac{\partial P}{\partial y} + \sigma_{nf} B_0^2 \left[u \bullet sin(\gamma) \bullet cos(\gamma) - v \bullet cos(\gamma)^2 \right] + (T - T_c) g \rho_{nf} \beta_{nf},$$

(3.2.4)

$$\left(u\frac{\partial T}{\partial x} + v\frac{\partial T}{\partial y} \right) = -\frac{1}{\left(\rho C_p\right)_{nf}} \frac{\partial q_r}{\partial y} + \frac{k_{nf}}{\left(\rho C_p\right)_{nf}} \left(\frac{\partial^2 T}{\partial x^2} + \frac{\partial^2 T}{\partial y^2} \right),$$

$$T^4 \cong 4T_c^3 T - 3T_c^3,\ q_r = -\frac{4\sigma_e}{3\beta_R} \frac{\partial T^4}{\partial y}.$$

(3.2.5)

The constitutive relations used in this model are:

$$\Delta = \frac{\sigma_s}{\sigma_f}, \frac{\sigma_{nf}}{\sigma_s} = 1 + \frac{3\varphi(\Delta - 1)}{(\Delta + 2) - \varphi(\Delta - 1)},$$

(3.2.6)

$$\left(\rho C_p\right)_{nf} = \varphi \left(\rho C_p\right)_s + (1 - \varphi) \left(\rho C_p\right)_f,$$

(3.2.7)

$$(\rho\beta)_{nf} = \varphi (\rho\beta)_s + (1 - \varphi)(\rho\beta)_f$$

(3.2.8)

$$\rho_{nf} = \varphi\rho_s + (1 - \varphi)\rho_f,$$

(3.2.9)

By considering the shape factor effect, the nanofluid thermal conductivity k_{nf} and viscosity μ_{nf} can be presented as:

$$\frac{k_{nf}}{k_f} = \frac{-m\varphi(k_f - k_p) + \varphi(k_p - k_f) + mk_f + k_p + k_f}{mk_f + \varphi(k_f - k_p) + k_f + k_p}$$

(3.2.10)

$$\frac{\mu_{nf}}{\mu_f} = \frac{1}{(1-\varphi)^{2.5}}.$$

(3.2.11)

To simplify the complex mathematical model developed in this study, non-dimensional analysis is conducted. Dimensionless quantities are introduced:

$$\Psi = \frac{\psi}{\alpha_{nf}}, \Theta = \frac{T - T_c}{\Delta T}, \Delta T = \frac{Lq''}{k_f}, X = \frac{x}{L}, Y = \frac{y}{L},$$

(3.2.12)

Here are the final equations:

$$\frac{\partial\Psi}{\partial Y}\frac{\partial\Theta}{\partial X} - \frac{\partial\Psi}{\partial X}\frac{\partial\Theta}{\partial Y} = \frac{\partial^2\Theta}{\partial X^2} + \left[1 + Rd\left(\frac{k_{nf}}{k_f}\right)^{-1}\right]\frac{\partial^2\Theta}{\partial Y^2},$$

(3.2.13)

$$\frac{\partial^2\Psi}{\partial X^2} + \frac{\partial^2\Psi}{\partial Y^2} = -Ha\frac{A_6}{A_5}\left[\frac{\partial^2\Psi}{\partial X^2}cos(\gamma)^2 + \frac{\partial^2\Psi}{\partial Y^2}sin(\gamma)^2 + 2\frac{\partial^2\Psi}{\partial X\partial Y}sin(\gamma)cos(\gamma)\right] - \frac{A_2A_3}{A_4A_5}Ra\frac{\partial\Theta}{\partial X}.$$

(3.2.14)

where parameters and A_i, i=1 to 6 are defined as:

$$A_1 = \frac{\rho_{nf}}{\rho_f}, A_2 = \frac{(\rho C_p)_{nf}}{(\rho C_p)_f}, A_3 = \frac{(\rho\beta)_{nf}}{(\rho\beta)_f}, A_4 = \frac{k_{nf}}{k_f}, A_5 = \frac{\mu_{nf}}{\mu_f}, A_6 = \frac{\sigma_{nf}}{\sigma_f},$$

(3.2.15)

$$Ha = \frac{\sigma_f K B_0^2}{\mu_f}, Ra = \frac{gK(\rho\beta)_f L\Delta T}{\mu_f \alpha_f}, Rd = \frac{16\sigma_e T_c^3}{3\beta_R k_f}.$$

(3.2.16)

Furthermore, the boundary conditions can be stated as follows:

Ψ=0 @ All surfaces, (3.2.17)

Θ=0 @ Outer surface,
(3.2.18)

$\frac{\partial\Theta}{\partial n} = 1$ @

Inner surface, (3.2.19)

Nu_{loc} and Nu_{ave} are determined from:

$$Nu_{loc} = \left[1 + Rd\left(\frac{k_{nf}}{k_f}\right)^{-1}\right]\left(\frac{k_{nf}}{k_f}\right)\frac{1}{\Theta},$$

(3.2.20)

$$Nu_{ave} = \frac{1}{S}\int_0^S Nu_{loc}\, ds.$$

(3.2.21)

Different nanoparticle shape variables were explored to decide the most noteworthy normal Nusselt number (Table 2). Eqs. (20, 21))) were utilized to get the Nu_{ave} on the hot Cassini oval. As appeared in Table 3 the platelet shape figure has the foremost Nu_{ave} among the shape components. Hence, it is chosen for the calculations.

The ultimate demonstration, represented by Eqs. (3.2.13-14) alongside, the comparing boundary conditions (3.2.17-19), are unraveled by embracing the limited Component Strategy (FEM). A test has been conducted to guarantee mesh independence by efficiently expanding the number of networks until no critical distinction within the comes about is watched. Limited component strategies comprise the taking-after methods: (a) Spaces are discretized into components. (b) Component conditions are inferred. (c) Component conditions are amassed. (d) Boundary conditions are connected. (e) Gathered conditions are unraveled. The lattice freedom is displayed in Table 4.

For the effect of various Hartmann numbers (Ha=0–100), Rayleigh numbers ($Ra=10^3$–10^4), and solid nanoparticle volume fractions (φ=0.01–0.2) with inclined angles ($\gamma=0$–$\pi/2$) are evaluated in this study. To assist discussion, numerical results will be presented in streamlines and isotherms, commonly adopted for similar studies. Prior to assessing the impact of examined key parameters on the stream behavior and heat exchange characteristics of the MHD nanofluid, it is basic to confirm the exactness of the created show by comparing it with the information from the already distributed writing. Worldwide and neighborhood comparisons between the demonstrated expectation and the comparing esteem from Grosan et al. are displayed in Figure 23 and Table 4 Figure 24 (a, b) shows streamlines and isotherms that contrast due to Ha. The convection component of the nanofluid stream diminishes with the rise in Lorentz drive, causing the conduction to extend. The center of the primary eddies floated upwards within the nonattendance of the Lorentz constrain, and a lean heat boundary layer was watched over the hot oval. As a result of forcing a solid, attractive field, conduction in heat exchange becomes more overwhelming. In expansion, this causes the isotherms to parallelize, whereas the vortexes got to be weaker. Given that the Lorentz powers are contrary to the buoyancy strengths, an attractive field could be a controlling parameter for heat exchange applications. In comparison to Θ_{max}, Ψ_{max} has ended up more changed. Expanding Ha from 100 diminished Ψ_{max} by 88.87%. Ra impacts the streamlines and isotherms, as shown in Figure 12 (a, b). Vortexes tend to be more grounded with expanding buoyancy strengths, as appeared in Figure 12(a). Expanding the Rayleigh number disturbs the structures of the isotherms, consequently driving the heat tuft to rise. In expansion, the two fundamental centers coordinated into one more grounded center vortex and tend towards the beat. In expansion, when buoyancy strengths are expanded, convection fortifies. At tall Ra, the buoyancy strengths rule the viscous strengths, which suggests convection will be the overwhelming heat exchange mode. Ra was expanded from 10^3 to 10^4, which comes about in a 1092.71% increment in Ψ_{max}. As discussed in various thinks about, the volume division of nanoparticles suspended within the nanofluid plays a critical part in deciding stream behavior and heat exchange characteristics. Hence, the impact of the MHD nanofluid stream in this ponder is altogether examined by changing its esteem from 1% to 20%. The streamlines and isotherms of Fig. (3.2.5) (a, b) are affected by φ. As the rate of MHD nanoparticles increments, the center of the vortex partitions gets weaker. Expanding the volume division of nanoparticles causes an increment within the thickness of the nanofluid. Subsequently, the control of the vortex diminishes due to the increase in grinding between nanoparticles. Expanding grinding between nanoparticles also causes the temperature to rise, and more grounded conduction causes the isotherm lines to end up parallel. It takes after that an increment in φ has a reverse and coordinate connection with Ψ and Θ, individually.

Magnetohydrodynamic (MHD) nanofluid flow and its comparing heat exchange in a circular penetrable space, including a Cassini oval depth subjected to buoyancy and Lorentz strengths, have been examined by utilizing non-dimensional investigation and limited component strategy. The demonstration was created based on the Darcy demonstration. A great understanding was accomplished between the demonstrated forecast and its partners from an already distributed consideration, affirming the exactness and legitimacy of the created show. The show was at that point utilized to assess different key parameters influencing the flow behavior and heat exchange characteristic.

Figure 9. (a) Geometry and the boundary conditions. (b) Considered mesh of the domain

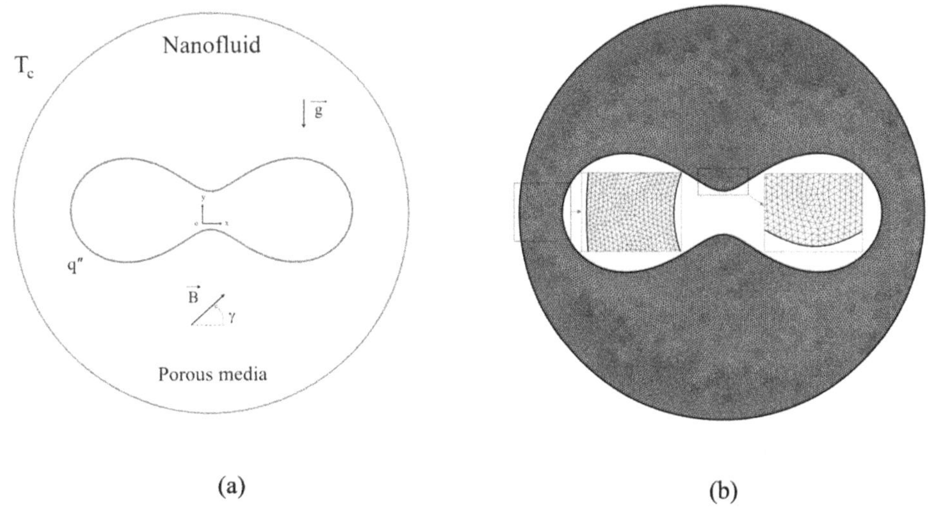

Figure 10. Streamline and Isothermal for Ra $= 10^5$, Ha $= 10$, $\gamma = \frac{\pi}{6}$.

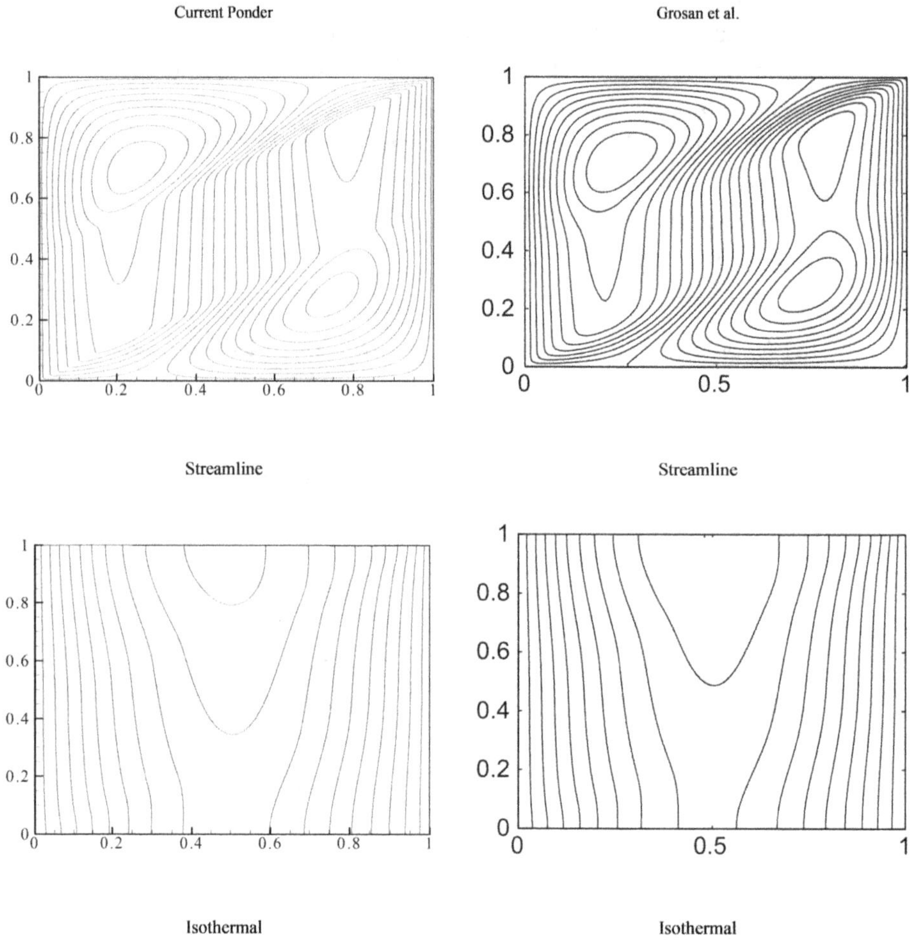

Figure 11. Influence of Ha on Ψ and T when $\varphi = 0.1$, $Ra = 2 \times 10^3$, $\gamma = \frac{\pi}{2}$ *&* $Rd = 0.7$ *at (a) Ha=0 and (b) Ha=10*

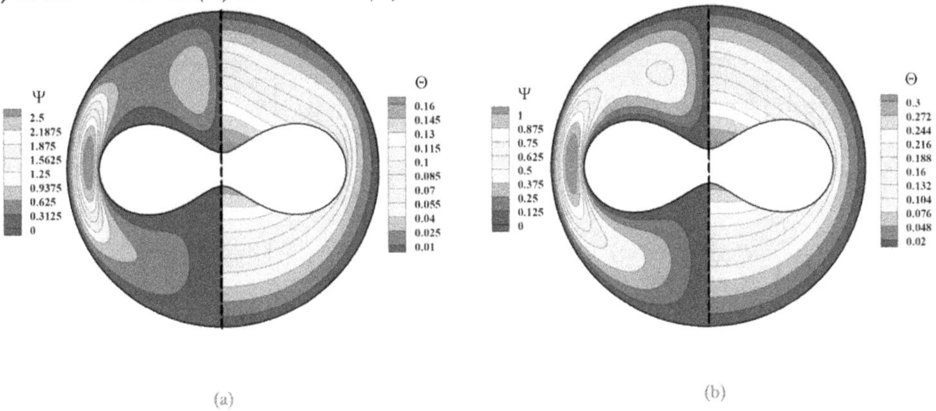

3.3. Nanofluid Flow and Heat Transfer in a Vertical Channel

The current study explores nanofluid flow and heat transfer in a vertical channel, focusing on the fully developed laminar regime. An external pressure gradient and buoyancy force drive the flow. In Figure 13, the physical model depicts the central location of coordinates within the channel. The gravitational acceleration vector, g, is at a right angle to the y-axis. The velocities in the x and y directions are labeled as vx and vy, respectively. The temperature distribution for the volume fraction of nanoparticles on and inside the walls is also provided.

$$Tw(x) = T_0 + A_1 x \, , \, C_w(x) = C_0 + A_2 x$$

(3.3.1)

The equation provided contains a few fixed values. The reference temperature at the channel's beginning is called 'T_0'. Then, A_1 is a constant, C_0 represents the nanoparticle volume fraction at the far field and A_2 is a positive constant, (Sheremet & Pop, 2014) (Hussein, Qasem, & Sheremet, 2016) (Abu-Nada & Oztop, 2009) (Hatami, Ganji, & jing, 2016) (Waqas et al., 2020) (Ellahi, Hassan, & Zeeshan, 2017) (Reddy et al., 2021) (Hayat et al., 2020) (Shehzad, Hayat, & Rashidi, 2016) (Mohammadi et al., 2018)(Liu et al., 2018) (Jallili et al., 2024) (Sheremet, Ghasemi, & Ellahi, 2020) (Revnic, Ghasemi, & Pop, 2019) (Husssein, Qasem, & Sheremet, 2017) (Sheremet & Pop, 2017).

$$\frac{\partial u}{\partial x} + \frac{\partial v}{\partial y} = 0$$

(3.3.2)

$$\frac{\partial p}{\partial x} = \mu\frac{\partial^2 u}{\partial y^2} + \left[1 - C_0 \cdot \rho_f \cdot \beta \cdot (T - Tw)(\rho_{s\_}\rho_f) \cdot (C - C_\omega)\right]g$$

(3.3.3)

$$\alpha\frac{\partial^2 T}{\partial y^2} + \tau\left\{D_B\left(\frac{\partial C}{\partial x}\frac{\partial T}{\partial x} + \frac{\partial C}{\partial y}\frac{\partial T}{\partial y}\right) + \left(\frac{D_T}{T_0}\right)\left[\left(\frac{\partial T}{\partial x}\right)^2 + \left(\frac{\partial T}{\partial y}\right)^2\right]\right\} = u\frac{\partial T}{\partial x}$$

(3.3.4)

$$D_B\frac{\partial^2 C}{\partial y^2} + \left(\frac{D_T}{T_0}\right)\left(\frac{\partial T}{\partial y}\right)^2 = u\frac{\partial C}{\partial x}$$

(3.3.5)

With boundary conditions:

$$u(-L) = 0, u(+L) = 0, T_W(x) = T_0 + A_1 x\, , C_W(x) = C_0 + A_2 x$$

(3.3.6)

The equation given includes several variables that are worth mentioning. These variables consist of the dynamic viscosity, 'μ', the thermal diffusivity of the nanofluid, referred to as 'α', and the volumetric expansion coefficient, denoted as 'β'. Additionally, 'ρ' represents the density of the base fluid and 'ρp' represents the nanoparticle's mass density. Now, let's explore some physical explanations and real-world applications for these conservation equations.

1. The principle of Conservation of Total Mass is expressed by this equation, which captures the equilibrium of mass within the system. It takes into consideration the inflow and outflow of mass in the channel, as well as any generation or consumption of mass due to chemical reactions or other phenomena. When dealing with nanofluids, this mass balance encompasses both the mass of the fluid and the suspended nanoparticles.
2. The concept of Conservation of Momentum is expressed by this equation, which accounts for the balance of momentum within the system. It considers the forces acting on both the fluid and nanoparticles in the channel. The momentum equation is crucial for comprehending the flow characteristics of nanofluids in

the channel and offers valuable insights into phenomena like turbulence and drag.

3. The concept of thermal energy conservation is captured in this equation, which considers the nanofluid's movement and its effects on temperature distribution within the channel. This equation is crucial in understanding the heat transfer properties of the nanofluid and can be applied to enhance heat transfer efficiency in a range of engineering applications, such as heat exchangers and cooling systems.
4. The conservation of the nanoparticle volume fraction equation is concerned with maintaining a balance of the amount of nanoparticles within the system. It considers how the nanoparticles behave within the nanofluid suspension and how they are transported within the channel. This equation is crucial in understanding the process of nanoparticle sedimentation and aggregation within the fluid.

These equations have numerous real-world applications, such as optimizing heat transfer in heat exchangers, cooling systems, and other thermal management applications. They are also relevant to manufacturing nanocomposites and other materials that incorporate nanoparticles. Furthermore, these equations help us understand how nanofluids behave in microfluidic devices, which have many applications in fields like drug delivery and lab-on-chip systems. In studies on channel flow, treating the mass flow rate as a predetermined value is common practice. This is denoted by 'Q', which represents the mean fluid velocity in the specific section of the channel.

$$U_m = \frac{1}{2L}\int_{-l}^{l} u(y)\, dy = \frac{1}{L}\int_{0}^{d} u(y)\, dy \tag{3.3.7}$$

Dimensionless variables are as follows:

$$x = \frac{x}{L}, y = \frac{y}{L}, U(y) = \frac{\mathrm{Re}\cdot Pr}{U_m}u, \theta(y) = \frac{T - T_w}{A_1 L}, \tag{3.3.8}$$

By substituting dimensionless variables from equation (3.3.8) into Equations (3.3.2) through (3.3.5), the resulting equations can be obtained:

$$U'' + (G_r \cdot p_r)\theta - N_r\varphi + \sigma = 0 \tag{3.3.9}$$

$$\theta'' + N_b\theta'\varphi' + N_t(\theta')^2 + N_b + N_t - U = 0$$

(3.3.10)

$$\varphi'' + \frac{N_t}{N_b}\theta'' - L_e U = 0$$

(3.3.11)

Considering the boundary conditions

$$U(1) = U(-1) = 0, \theta(1) = \theta(-1) = 0, \varphi(1) = \varphi(-1) = 0$$

(3.3.12)

$$\int_0^1 U\, dy = \mathrm{Re}Pr$$

(3.3.13)

The variables used to describe the characteristics of the system are as follows: the pressure parameter, denoted by s; the Grashof number, represented by G_r; the buoyancy ratio, described by N_r; and the Brownian motion parameter, denoted by N_b. Furthermore, N_t represents the thermophoresis parameter and the Lewis number is represented by L_e, and the Prandtl number is represented by P_r.

$$Gr = \frac{(1 - C_0)\beta g A_1 L^4}{\nu^2}, N_r = \frac{(\rho_C - \rho_f)A_2 g L^4}{\alpha\mu}, \sigma = \frac{\rho_c}{\rho_f}Pr\mathrm{Re}^2\frac{dp}{dx},$$

$$N_b = \frac{(\rho c)_s}{(\rho c)_f}D_B\frac{A_2 L}{\alpha}, N_t = \frac{(\rho c)_s D_r A_1 L}{(\rho c)_f T_0 \alpha}, L_e = \frac{\alpha}{D_B}, Pr = \frac{\nu}{\alpha}$$

(3.3.14)

The proposed method is described in detail, including mathematical equations and derivations of the HPM (Homotopy Perturbation Method) and AGM (Adomian Decomposition Method), which are presented in the appendices. Appendix A provides an overview of HPM and AGM and the necessary definitions to understand the mathematical derivations. In Appendix B, we demonstrate the application of both methods to the specific problem, presenting their respective equations.

In summary, according to the core concepts and practical implementation described in the appendix, it is evident that the Homotopy Perturbation Method (HPM) is a powerful technique for addressing the issue of Mixed Convection Flow of Nanofluid in a Vertical Channel. The step-by-step framework for applying this method can be summarized as follows:

$$H(U,P) = (1-P)(U'') + P\left(U'' + (Gr^{*}Pr)\theta - N_r\varphi + \sigma\right)$$

(3.3.15)

$$H(\theta,p) = (1-P)(\theta'') + P\left(\theta'' + N_b\theta'\varphi' + N_t(\theta')^2 + N_b + N_t - U\right)$$

(3.3.16)

$$H(\varphi,p) = (1-P)(\varphi'') + p\left(\varphi'' + \frac{N_t}{N_b}\theta'' - L_e U\right)$$

(3.3.17)

One can now try to obtain a solution of Eqs. (3.3.15) – (3.3.17) in the form of

$$(y) = U_0(y) + pU_1(y) + P^2U_2(y) + \ldots$$

(3.3.18)

$$\theta(y) = \theta_0(y) + p\theta_1(y) + p^2\theta_2(y) + \ldots \text{ (3.3.19)}$$

$$\varphi(y) = \varphi_0(y) + p\varphi_1(y) + p^2\varphi_2(y) + \ldots \text{ (3.3.20)}$$

Functions yet to be determined are represented by $v_i(y)$ for =1,2,3,... Based on Eqs. (3.3.15) - (3.3.17), the initial approximation that satisfies the boundary condition is given by:

$U_0(y)=(1-y^2)Pr\,Re$ (3.3.21)

$\theta_0(y)=0,$

(3.3.22)

$\varphi_0(y)=0.$ (3.3.2)

Substituting Eqs. (3.3.18) - (3.3.20) into Eqs. (3.3.15) - (3.3.17) yields

$$Gr \times \theta_0\left(y\right) \times \Pr - \varphi_0\left(y\right) \times N_r + \frac{\partial^2 U_1(y)}{\partial y^2} = 0$$

$$N_b \times \frac{\partial \varphi_0}{\partial y} \times \frac{\partial \theta_0}{\partial y} + \left(\frac{\partial \theta_0}{\partial y}\right)^2 N_t + \frac{\partial^2 \theta_1}{\partial y^2} - U_0\left(y\right) + N_b + N_t = 0$$

(3.3.24)

$$\frac{\partial^2 \varphi_1}{\partial y^2} + \frac{\left(\frac{\partial^2 \theta_0}{\partial y^2}\right) \times N_t}{N_b} - 10\, U_0\left(y\right) = 0$$

The solutions of Eq. (3.3.24) may be written as follows:

$$U_1(y)=0, \qquad (3.3.25)$$

$$\theta_1(y) = -\text{RePr} \times (0.083\, y^4 - 0.5\, y^2)$$

$$+ (0.5(-N_b - N_t))\, y^2 - 0.416\text{RePr} + 0.5\, N_b + 0.5\, N_t$$

, (3.3.26)

$$\varphi_1(y) = -0.83\text{RePr} \times y^4 + 5.\text{RePr} \times y^2 - 4.17\text{RePr}$$

. (3.3.27)

In the same manner, the rest of the components were obtained by using Python calculations. According to the HPM, (G_r=σ=0, N_r=N_b=N_t=0.1, L_e=10, and R_e=P_r=1)

$$U(y) = 0.0049 + 3.34 \times 10^{-11} y^{12} + 0.001\, y^4 - 0.0000147\, y^6$$

$$-3.02 \times 10^{-9} y^{10} - 0.0059\, y^2 + 5.498 \times 10^{-7} y^8$$

(3.3.28)

$$\theta(y) = 8.1 \times 10^{-11} y^{12} + 0.098 - 0.0004\, y^4 + 0.000022\, y^6$$

$$+6.78 \times 10^{(-9)} y^{10} - 0.098\, y^2 + 3.948 \times 10^{-7} y^8 + 2.29 \times 10^{-13} y^{14}$$

(3.3.29)

$$\varphi(y) = -1.92\times10^{(-10)}y^{12} - 0.118 - 0.0046y^4 + 0.0003177438823y^6$$

$$+5.648\times10^{-8}y^{10} + 0.12y^2 - 0.000003y^8 + 2.3\times10^{-12}y^{14}$$

(3.3.30)

And so on. Similarly, the rest of the components of the iteration formula can be obtained.

By applying the core principles of AGM and utilizing them practically as explained in the appendix, determine the boundary conditions needed to compute the constant coefficients for equations (3.3.31) through (3.3.33).

$$U(y) = \sum_{i=0}^{5} a_i y^i = a_0 + a_1 y + a_2 y^2 + a_3 y^3 + a_4 y^4 + a_5 y^5$$

(3.3.31)

$$\theta(y) = \sum_{i=0}^{5} b_i y^i = b_0 + b_1 y + b_2 y^2 + b_3 y^3 + b_4 y^4 + b_5 y^5$$

(3.3.32)

$$\varphi(y) = \sum_{i=0}^{5} c_i y^i = c_0 + c_1 y + c_2 y^2 + c_3 y^3 + c_4 y^4 + c_5 y^5$$

(3.3.33)

In the present study, the boundary conditions for the original differential equations have been applied. Specifically, equations (3.3.9) – (3.3.11) are considered, along with their derivatives, after substituting equations (3.3.31) – (3.3.33) into the main differential equations.

$$U(y) = Gr\left(b_0 + b_1 y + b_2 y^2 + b_3 y^3 + b_4 y^4 + b_5 y^5\right)$$

$$-N_r\left(c_0 + c_1 y + c_2 y^2 + c_3 y^3 + c_4 y^4 + c_5 y^5\right) + \sigma$$

(3.3.34)

$$+2\left(a_2 + 3a_3 y + 6a_4 y^2 + 10a_5 y^3\right) = 0$$

$$\theta(y) = N_b\left(b_1 + 2b_2 y + 3b_3 y^2 + 4b_4 y^3 + 5b_5 y^4 + c_1 + 2c_2 y + 3c_3 y^2 + 4c_4 y^3 + 5c_5 y^4\right)$$

$$+N_2 + N_t(b_1 + 2b_2 y + 3b_3 y^2 + 4b_4 y^3 + 5b_5 y^4)^2 + N_t - a_0 - a_1 y - a_2 y^2 - a_3 y^3$$

$$-a_4 y^4 - a_5 y^5 + 2(b_2 + 3b_3 y + 6b_4 y^2 + 10b_5 y^3) = 0$$

(3.3.35)

$$\varphi(y) = -L_e(a_0 + a_1 y + a_2 y^2 + a_3 y^3 + a_4 y^4 + a_5 y^5)$$

$$+2(c_2 + 3c_3 y + 6c_4 y^2 + 10c_5 y^3)\frac{2N_t(b_2 + 3b_3 y + 6b_4 y^2 + 10b_5 y^3)}{N_b} = 0$$

(3.3.36)

Based on our observations, the HPM and AGM techniques have been successfully applied to effectively tackle the problem at hand. Figure 14 is divided into two sections, illustrating the contrasting outcomes obtained using the numerical approach (specifically, Runge-Kutta 4th-order), as well as the HPM and AGM methods.

$$\%\ \text{Error} = \left|\frac{u(\eta)_{NUM} - u(\eta)_{HPM} - u(\eta)_{AGM}}{u(\eta)_{NUM}}\right| \times 100 \quad (3.3.37)$$

In this particular scenario, where NUM indicates the outcome of a numerical method, and HPM and AGM symbolize the values acquired from HPM and AGM, respectively, it is important to note that the convergence problem is illustrated in both Figure 15 and 16. Based on the findings, it can be deduced that the calculated error would diminish by incorporating more terms into the suggested HPM and AGM solutions. For a more comprehensive examination of this issue, Figure 15 provides a comparison of the outcomes generated by AGM, HPM, and the Numerical method for various values. The results demonstrate a high level of proximity to each other, indicating the accuracy of the proposed methods. Additionally, Table 5 presents a detailed comparison of the numerical results obtained from the Akbari-Ganji Method (AGM), the Homotopy Perturbation Method (HPM), and a numerical method.

The table indicates that the HPM method produces highly accurate results, resembling the numerical solution. Conversely, the AGM method demonstrates reasonable agreement with the numerical solution but falls slightly short compared to the HPM method. Table 1 offers further insights into the numerical magnitudes of each method's velocity profiles (U) and temperature (θ).

The findings unequivocally demonstrate that the AGM and HPM techniques yield precise numerical outcomes. This affirmation of their results further underscores their potential for effectively analyzing and resolving intricate engineering challenges. Figure 14 illustrates the impact of various values on the dimensionless velocities for both AGM and HPM methodologies. Examining Figure 16, when the buoyancy force supports the flow (2.5, 5), the velocities near the centerline of the channel exhibit a consistent decline as it increases. Conversely, velocities near the walls exhibit a continuous increase in this scenario.

This phenomenon can be attributed to the buoyancy force, which amplifies fluid movement in the boundary layer near the walls, where viscosity is dominant. Figures 16 (b) and 16 (c) illustrate the temperature distribution and nanoparticle volume fraction profiles for scenarios involving buoyancy assistance and opposition, respectively.

According to Figure 16 (b), the absolute amplitude for the buoyancy-assisted case increases as it decreases, while it increases with increasing due to fluid convection. Figure 16 (c) shows that it decreases with increasing for both cases and significantly affects the volume fraction of nanoparticles and the temperature distribution profiles. Additionally, for a better understanding of the results, Figure 17 presents 2D plots illustrating the dimensionless parameters: (a) velocity profiles, (b) temperature profiles, and (c) nanoparticle volume fraction profiles for different values of when and. Furthermore, Figure 18 depicts the impact on the flow for HPM and AGM methods, with the observation that it decreases significantly as changes from 1 to 10. $\theta(Y)$ behaves differently near the walls and near the center of the channel. The changing trend for $\varphi(Y)$ is similar to $\theta(Y)$ near the channel center. That is, $\varphi(Y)$ increases as Pr increases, but more significantly than $\theta(Y)$, as shown in Figure 18. Figure 19 shows 2D plots of the dimensionless parameters for different values of Pr, with Nb=Nt=Nr=0.1, Gr=5, Re=1, Le=10, and $\sigma=0$. Figure 20 shows the results of HPM and AGM for the dimensionless parameters at different values of Re, with Nb=Nt=Nr=0.1, Gr=5, Pr=1, Le=10, and $\sigma=0$. Figure 21 showcases 2D plots of the dimensionless parameters at different values of a certain variable, with certain values for other variables. The figures also display the outcomes obtained using two alternative methods, AGM and HPM, and compare them. Both methods are in line with the numerical method, which demonstrates the accuracy and efficacy of the proposed approaches. Figures 22, 23, and 24 present the optimized results for variables a, b, and c for different values of x, y, and z. The 3D plots visually illustrate the variations in these profiles for the investigated parameters. The optimized results offer valuable insights into the characteristics of velocity, temperature profiles, and nanoparticle volume fraction profiles. These insights serve as a valuable foundation for future research in this field and can contribute to a better understanding of the complex behavior of fluid flow in these systems.

This research paper thoroughly investigates a nanofluid's laminar mixed convection flow in a vertical channel, which is driven by an external pressure gradient and buoyancy force. To solve the governing partial differential equations, similarity variables were employed to transform them into ordinary differential equations. The homotopy perturbation method (HPM) and Abari-Ganji's method (AGM) were then utilized to solve these equations.

Figure 12.

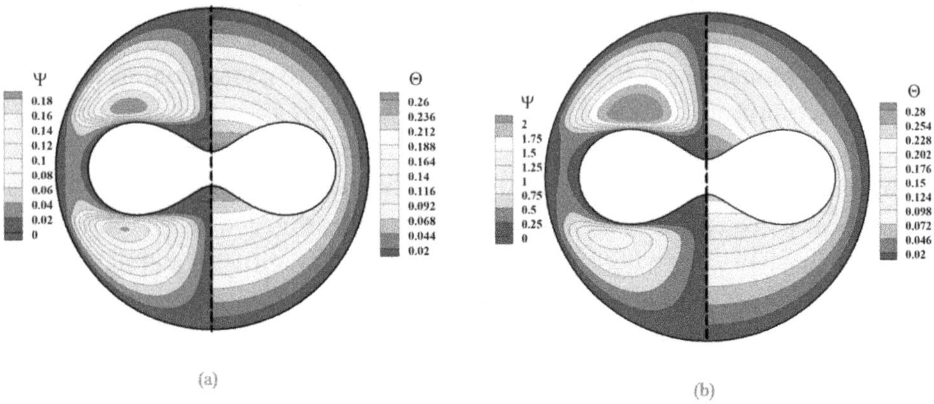

Figure 13. Geometry of the problem

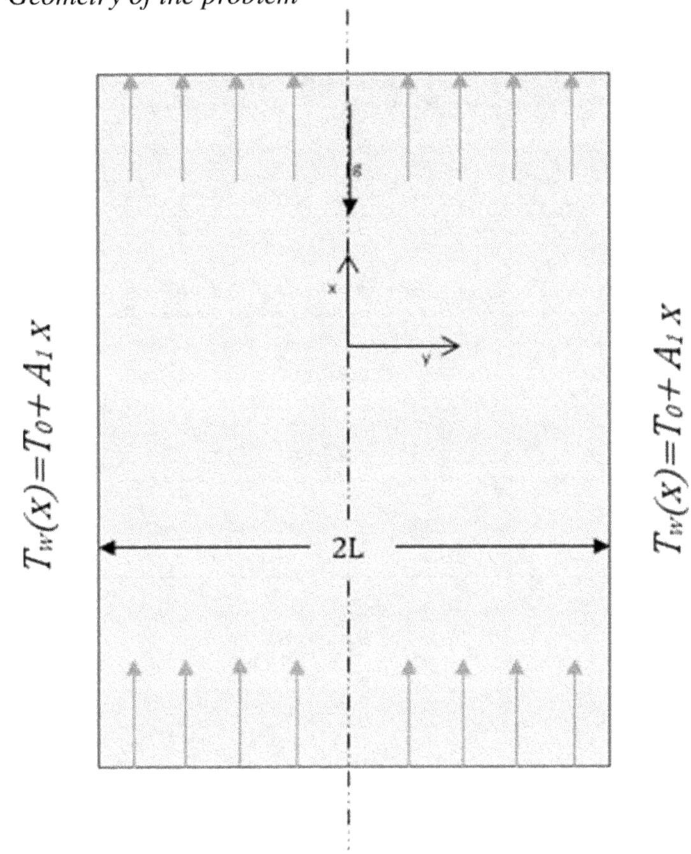

Figure 14. Convergence figures (a) for f(η) in various terms of semi-analytical techniques and (b) for Θ(η) in various terms of HPM and AGM

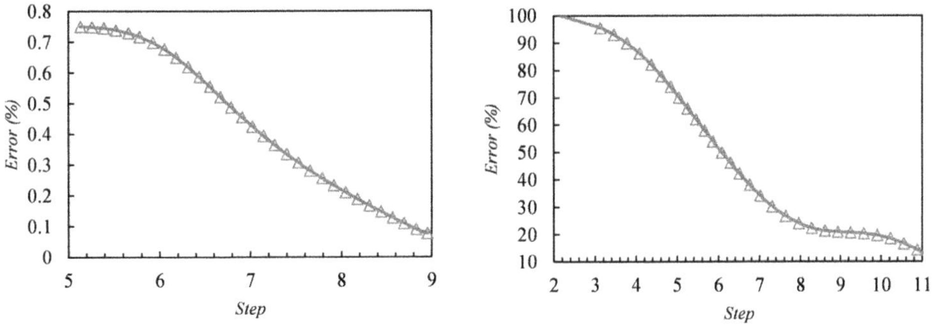

Figure 15. Comparison between HPM, AGM, and Numerical method for U, θ and φ

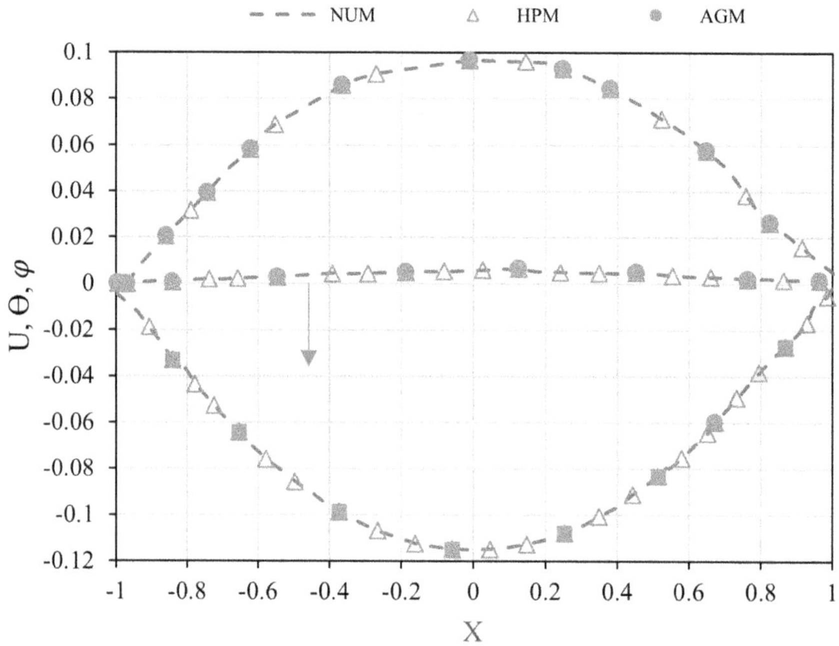

Figure 16. The HPM and AGM results for the dimensionless parameters (a) velocity profiles, (b) temperature profiles, and (c) nanoparticle volume fraction profiles for various values of G_r when N_b=N_t=N_r=0.1, P_r=R_e=1, L_e=10 and σ=0

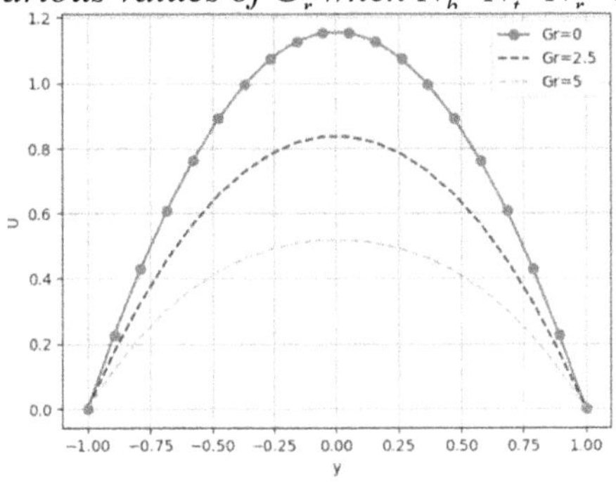

(a)

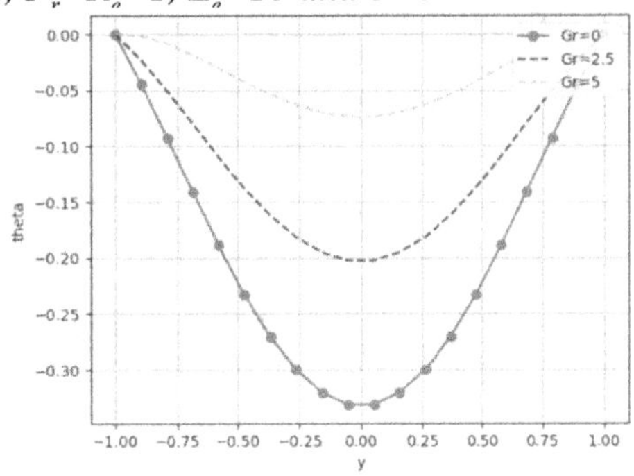

(b)

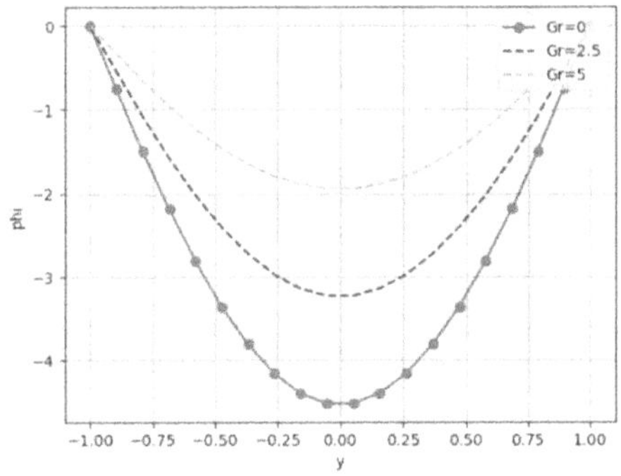

(c)

Figure 17. 2D plots for the dimensionless parameters(a) U, (b) Θ and (c) φ for various values of G_r when N_b=N_t=N_r=0.1, P_r=R_e=1, L_e=10 and σ=0

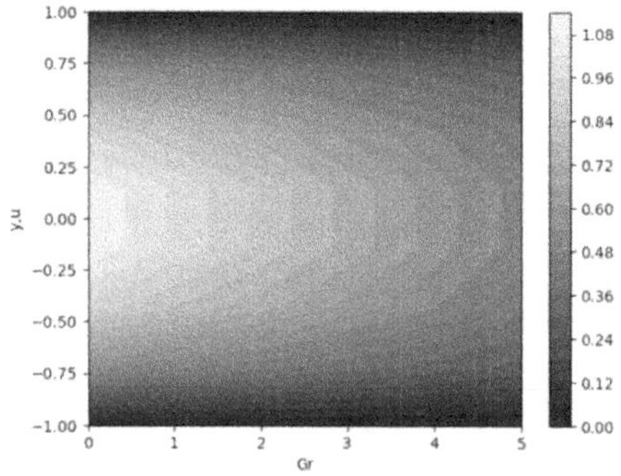

(a)

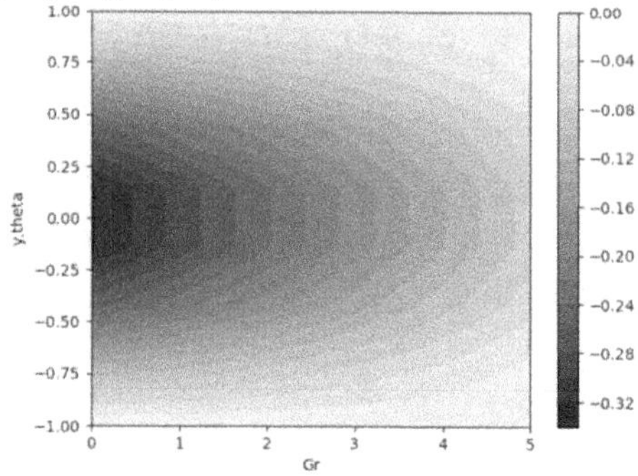

(b)

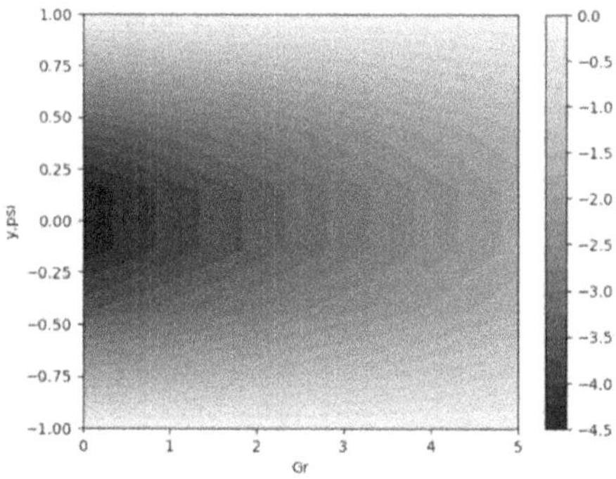

(c)

Figure 18. The HPM and AGM results for the dimensionless parameters (a) U, (b) Θ and (c) φ for various values of P_r when N_b=N_t=N_r=0.1, G_r=5, R_e=1, L_e=10 and σ=0

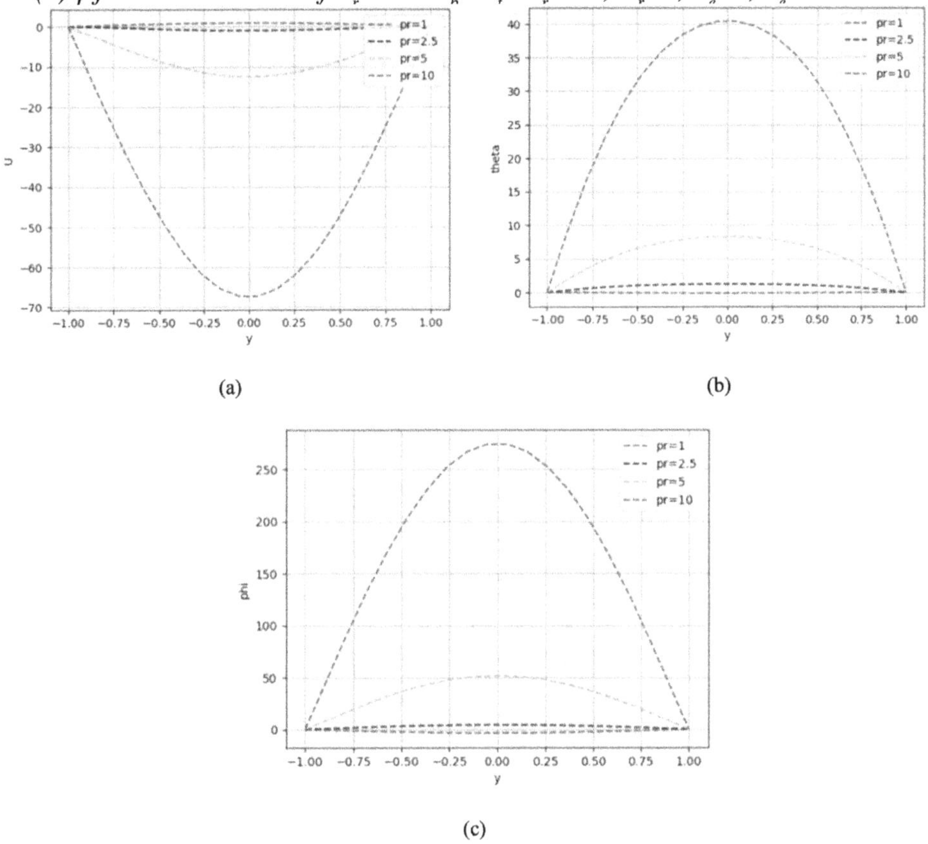

Figure 19. 2D plots for the dimensionless parameters (a) U, (b) Θ and (c) φ for various values of P_r when N_b=N_t=N_r=0.1, G_r=5, R_e=1, L_e=10 and σ=0

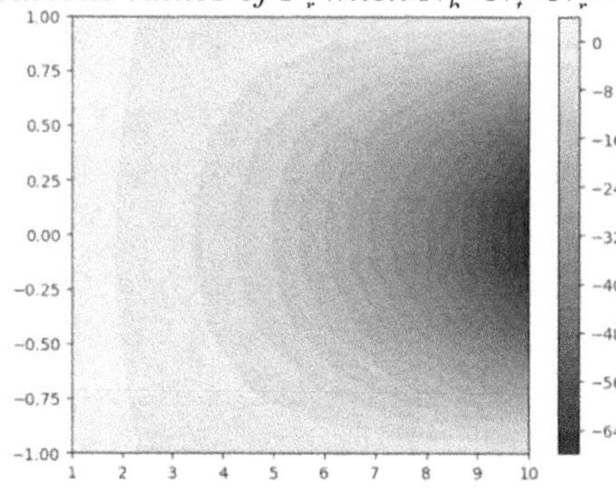

(a)

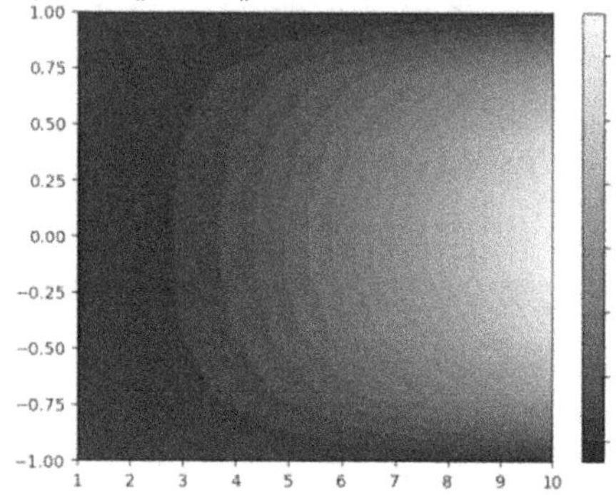

(b)

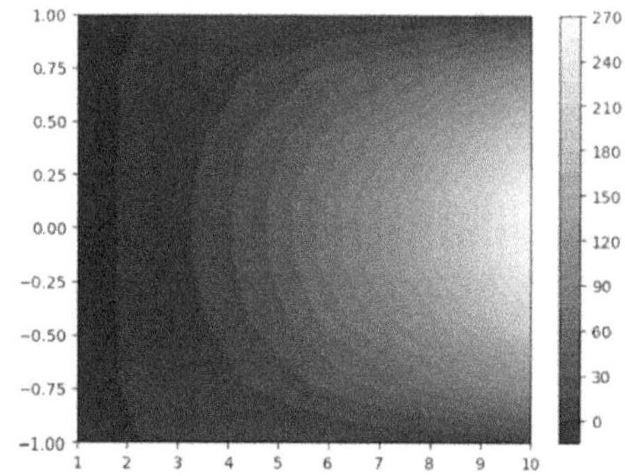

(c)

Figure 20. The HPM and AGM results for the dimensionless parameters (a) U, (b) Θ and (c) φ for various values of R_e when N_b=N_t=N_r=0.1, G_r=5, P_r=1, L_e=10 and σ=0

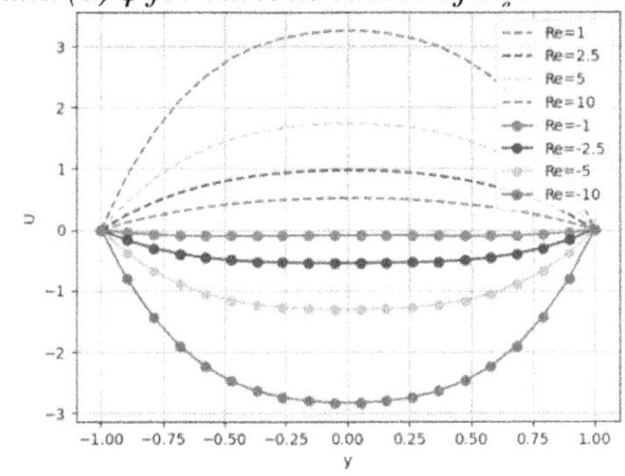

(a)

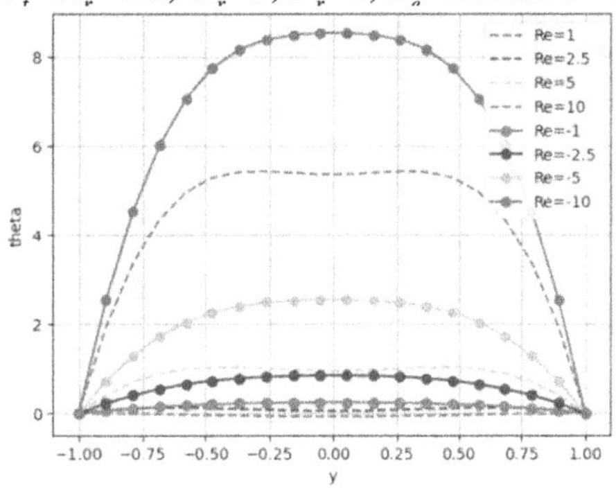

(b)

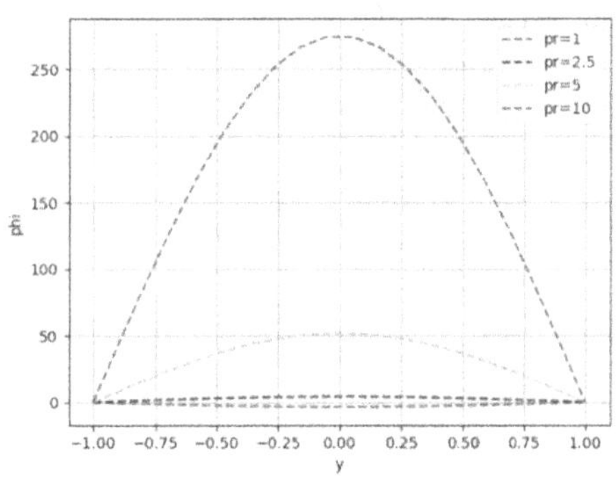

(c)

Figure 21. 2D plots for the dimensionless parameters (a) U, (b) Θ and (c) φ for various values of R_a when N_b=N_t=N_r=0.1, G_r=5, P_r=1, L_e=10 and σ=0

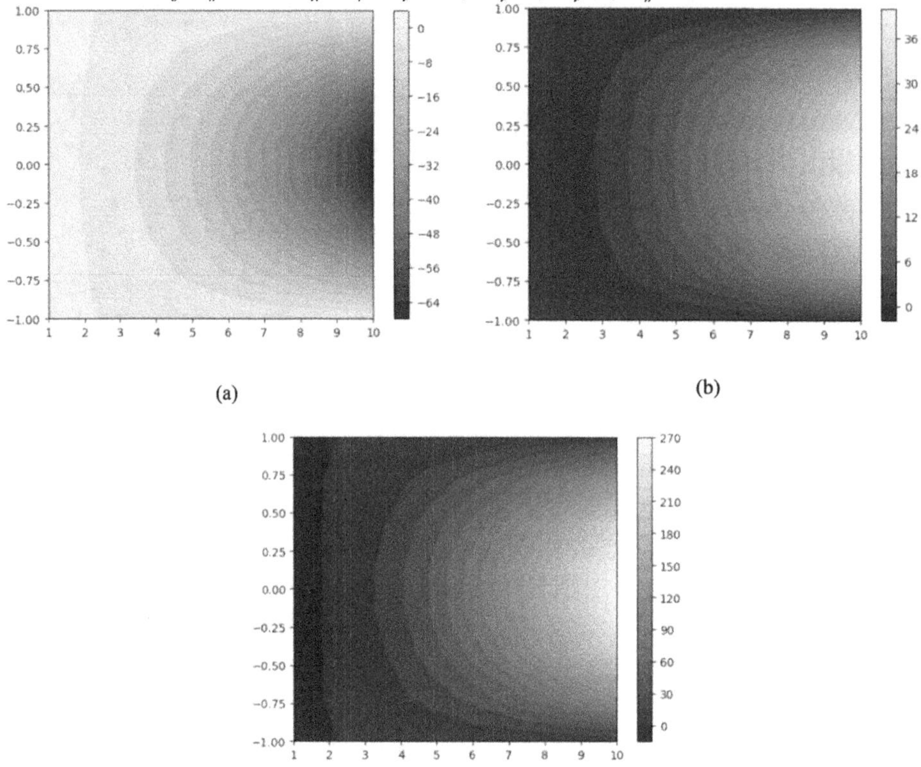

(a) (b) (c)

Figure 22. 3D surface of (a) U, (b) Θ and (c) φ for various values of G_r

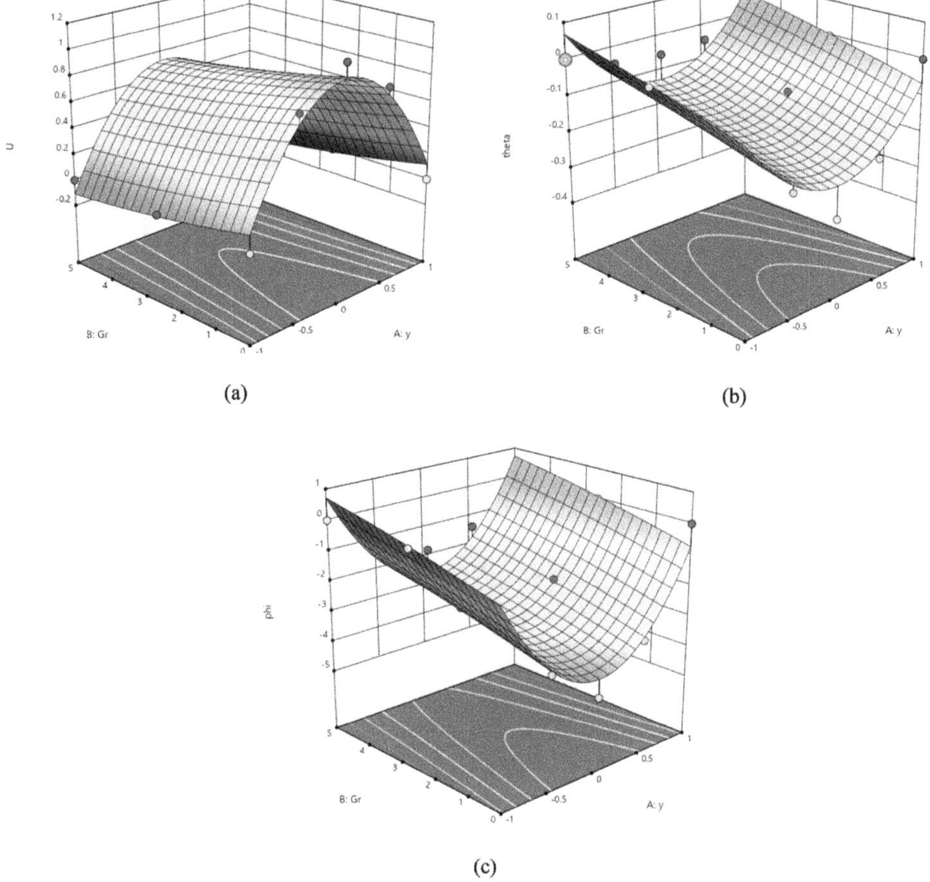

Figure 23. 3D surface of (a) U, (b) Θ and (c) φ for various values of P_r

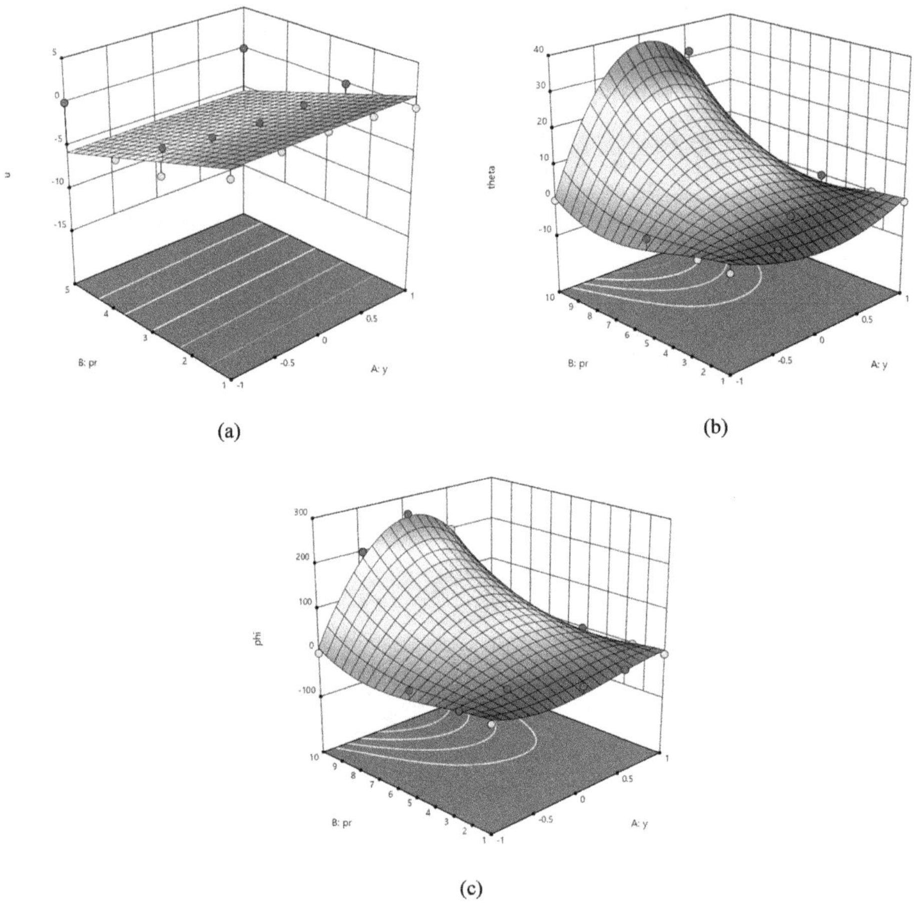

Figure 24. 3D surface of (a) U, (b) Θ and (c) φ for various values of R_n

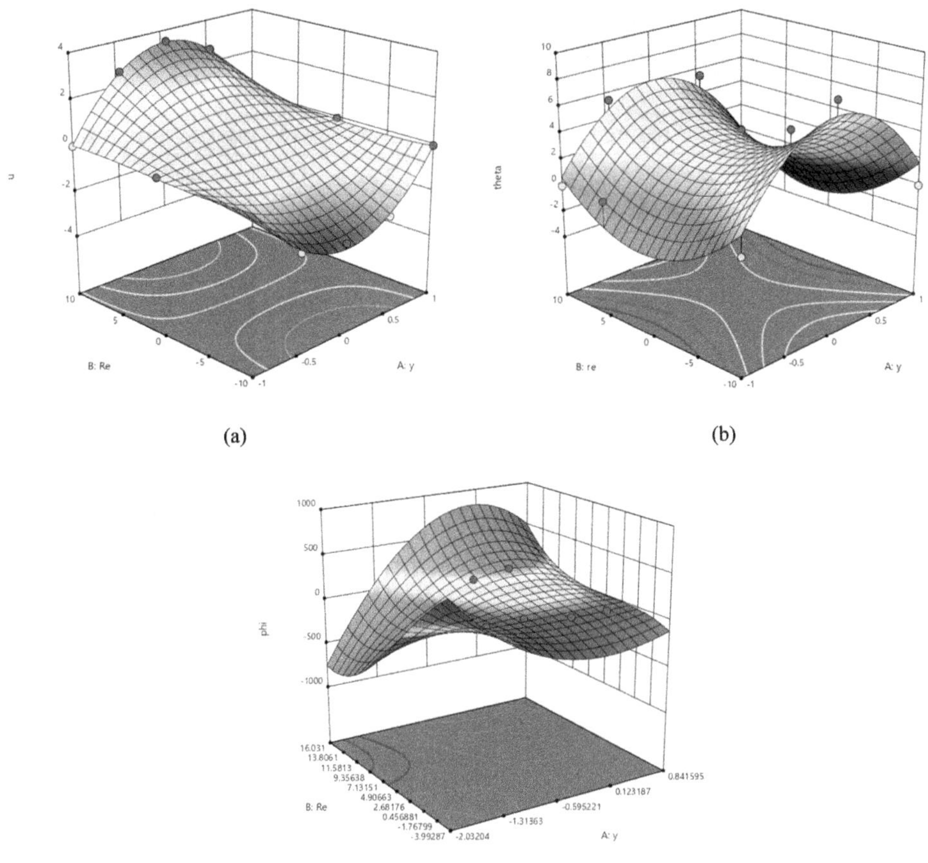

(a) (b)

(c)

Table 5. Comparison between proposed methods and numerical method

y	U_{HPM}	U_{NUM}	U_{AGM}	NUM	HPM	AGM
-1.0	0.0000000000	0.0000000000	0.0000000000	1.0000000000	1.0000000000	1.0000000000
-0.8	0.0015259450	0.0015322789	0.0015253205	0.0353488532	0.0353430994	0.0353422582
-0.6	0.0029135494	0.0029013226	0.0029419816	0.0627767997	0.0627741121	0.0627729563
-0.4	0.0039407397	0.0039752697	0.0039337401	0.0823313032	0.0823308634	0.0823289112
-0.2	0.0046227190	0.0046583907	0.0046018428	0.0940487521	0.0940466490	0.0940447483
0.0	0.0048624361	0.0048925940	0.0048489649	0.0979519717	0.0979518612	0.0979513971

continued on following page

Table 5. Continued

y	U_{HPM}	U_{NUM}	U_{AGM}	$_{NUM}$	$_{HPM}$	$_{AGM}$
0.2	0.0046301889	0.0046583907	0.0046189534	0.0940487521	0.0940486922	0.0940481934
0.4	0.0039391620	0.0039752697	0.0039274841	0.0823313032	0.0823312861	0.0823311183
0.6	0.0028680251	0.0029013226	0.0028435319	0.0627767997	0.0627765390	0.0627760192
0.8	0.0015258029	0.0015322789	0.0015258428	0.0353488532	0.0353486906	0.0353482483
1.0	0.0000000000	0.0000000000	0.0000000000	0.0000000000	0.0000000000	0.0000000000

REFERENCES

Abu-Nada, E., & Oztop, H. F. (2009). Effect of inclination angle on natural convection in enclosures filled with nanofluids. *International Journal of Heat and Fluid Flow*, 30(4), 669–678. DOI: 10.1016/j.ijheatfluidflow.2009.02.001

Alizadeh, M., Dogonchi, A. S., & Ganji, D. D. (2018). Micropolar nanofluid flow and heat transfer between penetrable walls in the presence of thermal radiation and magnetic field. *Case Studies in Thermal Engineering*, 12, 319–332. DOI: 10.1016/j.csite.2018.05.002

Alsabery, A. I., Hashim, I., Chamkha, A. J., Saleh, H., & Chanane, B. (2017). Effect of spatial side-wall temperature variation on transient natural convection of a nanofluid in a trapezoidal cavity. *International Journal of Numerical Methods for Heat & Fluid Flow*, 27(6), 1365–1384. DOI: 10.1108/HFF-11-2015-0488

Alsabery, A. I., Sheremet, M. A., Chamkha, A. J., & Hashim, I. (2018). MHD convective heat transfer in a discretely heated square cavity with conductive inner block using a two-phase nanofluid model. *Scientific Reports*, 8(1), 7410. DOI: 10.1038/s41598-018-25749-2 PMID: 29743641

Alsabery, A. I., Sheremet, M. A., Chamkha, A. J., & Hashim, I. (2019). Impact of nonhomogeneous nanofluid model on transient mixed convection in a double lid-driven wavy cavity involving a solid circular cylinder. *International Journal of Mechanical Sciences*, 150, 637–655. DOI: 10.1016/j.ijmecsci.2018.10.069

Bahmani, M., Babagoli, M., Jalili, P., Jalili, B., & Ganji, D. D. (2024). The numerical study on the MHD natural convection trend of square/circle corrugated porous media. *Journal of Engineering Research*. Advance online publication. DOI: 10.1016/j.jer.2024.05.012

Bessa, W. M., Ali, M. E., & Said, S. A. (2017). Nanofluid flow and heat transfer inside an annular finned cylinder subject to constant wall temperature. *Powder Technology*, 316, 504–514.

Bhatti, M. M., & Rashidi, M. M. (2017). Numerical simulation of entropy generation on MHD nanofluid towards a stagnation point flow over a stretching surface. *International Journal of Applied and Computational Mathematics*, 3(3), 2275–2289. DOI: 10.1007/s40819-016-0193-4

Chari, F. N., Jalili, P., Jalili, B., & Ganji, D. D. (2024). MHD Casson non-Newtonian fluid flow in a channel with expanding/contracting porous walls in the presence of thermal radiation. *Proceedings of the Institution of Mechanical Engineers, Part E: Journal of Process Mechanical Engineering*.

Dib, A., Haiahem, A., & Bou-said, B. (2015). Approximate analytical solution of squeezing unsteady nanofluid flow. *Powder Technology*, 269, 193–199. DOI: 10.1016/j.powtec.2014.08.074

Dogonchi, A. S., Armaghani, T., Chamkha, A. J., & Ganji, D. D. (2019). Natural convection analysis in a cavity with an inclined elliptical heater subject to shape factor of nanoparticles and magnetic field. *Arabian Journal for Science and Engineering*, 44(9), 7919–7931. Advance online publication. DOI: 10.1007/s13369-019-03956-x

Dogonchi, A. S., Chamkha, A. J., Hashemi-Tilehnoee, M., Seyyedi, S. M., Rizwan-Ul-Haq, , & Ganji, D. D. (2019). Effects of homogeneous–heterogeneous reactions and thermal radiation on magneto-hydrodynamic Cu–water nanofluid flow over an expanding flat plate with non-uniform heat source. *Journal of Central South University*, 26(5), 1161–1171. DOI: 10.1007/s11771-019-4078-7

Dogonchi, A. S., Chamkha, A. J., Seyyedi, S. M., & Ganji, D. D. (2018). Radiative nanofluid flow and heat transfer between parallel disks with penetrable and stretchable walls considering the Cattaneo–Christov heat flux model. *Heat Transfer, Asian Research*, 47(5), 735–753. DOI: 10.1002/htj.21339

Dogonchi, A. S., & Ganji, D. D. (2018). Effects of Cattaneo–Christov heat flux on buoyancy MHD nanofluid flow and heat transfer over a stretching sheet in the presence of Joule heating and thermal radiation impacts. *Indian Journal of Physics and Proceedings of the Indian Association for the Cultivation of Science*, 92(6), 757–766. DOI: 10.1007/s12648-017-1156-2

Ellahi, R., Hassan, M., & Zeeshan, A. (2016). Aggregation effects on water-based Al2O3 nanofluid over permeable wedge in mixed convection. *Asia-Pacific Journal of Chemical Engineering*, 11(2), 179–186. DOI: 10.1002/apj.1954

Ellahi, R., Hassan, M., & Zeeshan, A. (2017). Effects of magnetic field on water-based nanofluid containing Al2O3 ferroparticles: Mixed convection, a rotating permeable wedge. *Applied Thermal Engineering*, 118, 552–563.

Ellahi, R., Tariq, M. H., Hassan, M., & Vafai, K. (2017). On boundary layer nano-ferroliquid flow under the influence of a low oscillating stretchable rotating disk. *Journal of Molecular Liquids*, 229, 339–345. DOI: 10.1016/j.molliq.2016.12.073

Ganji, D. D., Mahboobtosi, M., Jalili, B., & Jalili, P. (2024). Heat transfer analysis of magnetized fluid flow through a vertical channel with thin porous surfaces: Python approach. *Case Studies in Thermal Engineering*, 60, 104643. DOI: 10.1016/j.csite.2024.104643

Hajizadeh, S., Jalili, P., Jalili, B., & Ganji, D. D. (2024). The new analytical and numerical analysis of 2D stretching plates in the presence of a magnetic field and dependent viscosity. *Advances in Mechanical Engineering*, 16(4), 16878132231220361. DOI: 10.1177/16878132231220361

Hatami, M., Ganji, D. D., & Jing, D. (2016). Optimization of a circular-wavy cavity filled by nanofluid under the natural convection heat transfer condition. *International Journal of Heat and Mass Transfer*, 98, 758–767. DOI: 10.1016/j.ijheatmasstransfer.2016.03.063

Hatami, M., Song, D., & Jing, D. (2016). Optimization of a circular-wavy cavity filled by nanofluid under the natural convection heat transfer condition. *International Journal of Heat and Mass Transfer*, 98, 758–767. DOI: 10.1016/j.ijheatmasstransfer.2016.03.063

Hayat, T., Ullah, A., Waqas, M., & Khan, M. I. (2020). Nonlinear radiative three-dimensional MHD nanofluid slip flow towards a rotating surface. *International Journal of Numerical Methods for Heat & Fluid Flow*, 30, 2791–2808.

Hussein, A. K., Qasem, N. A., & Sheremet, M. A. (2016). Free convection of a nanofluid in a porous square cavity filled with a nanofluid and subjected to a magnetic field. *Journal of Heat Transfer*, 138, 101–114.

Hussein, A. K., Qasem, N. A., & Sheremet, M. A. (2017). Natural convection of a nanofluid in a porous enclosure with discrete heating. *Advances in Heat Transfer*, 49, 1–48.

Jalili, P., Mirzaei, A., Jalili, B., Shateri, A., Ganji, D. D., Ozsahin, D. U., & Ahmad, H. (2024). Thermal analysis of transverse fluid flow in a gradient porous media with the exponentially boundary conditions. *Modern Physics Letters B*, 38(10), 2350229. DOI: 10.1142/S0217984923502299

Jalili, P., Sharif Mousavi, S. M., Jalili, B., Pasha, P., & Ganji, D. D. (2024). Thermal evaluation of MHD Jeffrey fluid flow in the presence of a heat source and chemical reaction. *International Journal of Modern Physics B*, 38(08), 2450113. DOI: 10.1142/S0217979224501133

Liu, Z., Gao, Z., Li, Y., & Sheikholeslami, M. (2018). Buoyancy-driven MHD nanofluid convection inside a wavy enclosure under magnetic field. *Journal of Thermal Analysis and Calorimetry*, 131, 3085–3095.

Mahmud, S., & Pop, I. (2016). Thermal and mass transfer characteristics of a porous rectangular cavity saturated with nanofluid under various thermal boundary conditions. *International Journal of Numerical Methods for Heat & Fluid Flow*, 26, 1730–1757.

Mirzaei, A., Jalili, B., Jalili, P., & Ganji, D. D. (2024). Free convection in a square wavy porous cavity with partly magnetic field: A numerical investigation. *Scientific Reports*, 14(1), 14152. DOI: 10.1038/s41598-024-64850-7 PMID: 38898150

Mohammadi, A., Li, Z., Sheikholeslami, M., & Chamkha, A. J. (2018). Numerical simulation of MHD effects on nanofluid flows. *Journal of Molecular Liquids*, 252, 218–232.

Pop, I., & Ingham, D. B. (2001). *Convective Heat Transfer: Mathematical and Computational Modelling of Viscous Fluids and Porous Media*. Pergamon.

Rashidi, M. M., Ganesh, N. V., Abdul Hakeem, A. K., & Ganga, B. (2014). Buoyancy effect on MHD flow of nanofluid over a stretching sheet in the presence of thermal radiation. *Journal of Molecular Liquids*, 198, 234–238. DOI: 10.1016/j.molliq.2014.06.037

Reddy, B. M., Waqas, M., Khan, S. U., & Khan, M. I. (2021). Magnetic field effects in radiative MHD nanofluid flow considering prescribed heat flux and Hall current with cross-diffusion effects. *International Communications in Heat and Mass Transfer*, 125, 105311.

Revnic, C., Ghasemi, E., & Pop, I. (2019). Natural convection in a triangular cavity filled with a nanofluid under the effect of an inclined magnetic field using the Buongiorno model. *Applied Thermal Engineering*, 153, 175–184.

Revnic, C., Sheremet, M. A., & Pop, I. (2016). Double-diffusive natural convection in a square cavity filled with a nanofluid using Buongiorno's model. *Computers & Mathematics with Applications (Oxford, England)*, 71, 362–373.

Roshani, H., Jalili, P., Jalili, B., Ahmad, I., Hendy, A. S., Ali, M. R., & Ganji, D. D. (2024). The effect of magnetic field on the heat transfer in the porous medium octagonal cavity with Cassini oval barriers. *Case Studies in Thermal Engineering*, 56, 104194. DOI: 10.1016/j.csite.2024.104194

Seyyedi, S. M., Dayyan, M., Soleimani, S., & Ghasemi, E. (2015). Natural convection heat transfer under constant heat flux wall in a nanofluid-filled annulus enclosure. *Ain Shams Engineering Journal*, 6(1), 267–280. DOI: 10.1016/j.asej.2014.09.003

Shehzad, N., Hayat, T., & Rashidi, M. M. (2016). Convective heat transfer for Maxwell nanofluid with variable conductivity and thermal radiation. *International Journal of Heat and Mass Transfer*, 99, 1037–1046.

Sheremet, M. A., Ghasemi, E., & Ellahi, R. (2020). Double-diffusive natural convection in a nanofluid-saturated porous cavity under the influence of magnetic field. *Microsystem Technologies*, 26, 1415–1432.

Sheremet, M. A., & Pop, I. (2014). Natural convection in a porous cavity filled with a nanofluid under the influence of a magnetic field. *Transport in Porous Media*, 103, 419–436.

Sheremet, M. A., & Pop, I. (2014). MHD mixed convection in a square cavity filled with a nanofluid using the Buongiorno model. *Computers & Mathematics with Applications (Oxford, England)*, 67, 488–503.

Sheremet, M. A., & Pop, I. (2017). Free convection in a trapezoidal porous cavity filled with a nanofluid using Buongiorno's model with heat and mass transfer. *Transport in Porous Media*, 116, 325–344.

Sheremet, M. A., Revnic, C., & Pop, I. (2017). Free convection in a porous wavy cavity filled with a nanofluid using Buongiorno's mathematical model with thermal dispersion effect. *Applied Mathematics and Computation*, 299, 1–15. DOI: 10.1016/j.amc.2016.11.032

Waqas, M., Hayat, T., Shehzad, S. A., & Alsaedi, A. (2018). Transport of magnetohydrodynamic nanomaterial in a stratified medium considering gyrotactic microorganisms. *Physica B, Condensed Matter*, 529, 33–40. DOI: 10.1016/j.physb.2017.09.128

Waqas, M., Khan, M. I., Hayat, T., & Alsaedi, A. (2017). Numerical simulation for magneto Carreau nanofluid model with thermal radiation: A revised model. *Computer Methods in Applied Mechanics and Engineering*, 324, 640–653. DOI: 10.1016/j.cma.2017.06.012

Waqas, M., Muhammad, T., Shehzad, N., & Alghamdi, M. (2020). Magnetic dipole and thermal radiation in unsteady radiative nanofluid flow. *Journal of Thermal Analysis and Calorimetry*, 141, 2929–2943. DOI: 10.1007/s10973-020-09570-1

Zeeshan, A., Shehzad, N., Ellahi, R., & Alamri, S. Z. (2017). Convective Poiseuille flow of Al2O3-EG nanofluid in a porous wavy channel with thermal radiation. *Neural Computing & Applications*. Advance online publication. DOI: 10.1007/s00521-017-2924-9

Chapter 4
Analytical Approach to MHD

ABSTRACT

Due to the many applications of MHD in various industries, many studies have been conducted in this field. The Navier-Stokes equations for fluid dynamics and Maxwell's equations for electromagnetism are combined to provide the equations describing MHD. A thorough review of analytical techniques for solving MHD problems in linear and nonlinear equations is needed to advance engineering and applied sciences. One of the most important and challenging issues, especially in engineering and applied sciences, is the analysis of complex nonlinear systems. Considering the importance of solving the nonlinear equations governing the dynamics of magnetic fluids, this chapter presents new ideas to address practical problems using the most up-to-date analytical and semi-analytical techniques.

4.1. SQUEEZING FLOW OCCURS BETWEEN TWO FLAT EQUAL PLATES

Al_2O_3-Water nanofluid is under consideration. Figure 1 depicts the selection of a coordinate system. It is assumed that squeezing flow occurs between two flat equal plates located in an l (t) = L (1/ at) 1/2 unit distance (L denotes the plate's primary position, a > 0 indicates squeezing the plates until touching at t = 1/a and moving within the inverse course for a< 0). At (B_0 = the initial value), an even magnetic flux is used with density (t) = B_0/1. The governing equations are as follows under these conditions (Table 1) (Larestani, Aydin, & Mohamad, 20220 (Jalili *et al.*, 2024) (Kumar *et al.*, 2022) (Joens & Swan, 2022) (Moatimid, Zekry, & Ibrahim, 2023) (Berkan *et al.*, 2024a) (Zhang, Wang, & Wang, 2023) (Abdel-Aziz *et al.*, 2023) (Jalili *et al.*, 2024b) (Larestani *et al.*, 2022) (Dehghan *et al.*, 2024) (Barnoon *et al.*, 2022) (Mozafarifard *et al.*, 2022) (Shah *et al.*, 2022) (Lou *et al.*, 2022) (Jalili et al., 2024c):

DOI: 10.4018/979-8-3693-5335-6.ch004

$$\frac{\partial u}{\partial x} + \frac{\partial v}{\partial y} = 0$$

(4.1.1)

$$\rho\left(\frac{\partial u}{\partial t} + u\frac{\partial u}{\partial x} + v\frac{\partial u}{\partial y}\right) = \mu\left(\frac{\partial^2 u}{\partial x^2} + \frac{\partial^2 u}{\partial y^2}\right) - \sigma B^2(t)u$$

(4.1.2)

$$\rho\left(\frac{\partial v}{\partial t} + u\frac{\partial v}{\partial x} + v\frac{\partial v}{\partial y}\right) = \mu\left(\frac{\partial^2 v}{\partial x^2} + \frac{\partial^2 v}{\partial y^2}\right)$$

(4.1.3)

$$\frac{\partial T}{\partial t} + u\frac{\partial T}{\partial x} + v\frac{\partial T}{\partial y} = \frac{k}{\rho c_p}\left(\frac{\partial^2 T}{\partial x^2} + \frac{\partial^2 T}{\partial y^2}\right) + \frac{\mu}{\rho c_p}\left(2\left[\left(\frac{\partial u}{\partial x}\right)^2 + \left(\frac{\partial v}{\partial y}\right)^2\right] + \left(\frac{\partial u}{\partial y} + \frac{\partial v}{\partial x}\right)^2\right)$$
$$+\frac{1}{\rho c_p}\left(\frac{D_l}{T_W}\left(\left(\frac{\partial T}{\partial x}\right)^2 + \left(\frac{\partial T}{\partial y}\right)^2\right) + D_B\left(\frac{\partial C}{\partial x}\frac{\partial T}{\partial x} + \frac{\partial C}{\partial y}\frac{\partial T}{\partial y}\right)\right) - \frac{1}{\rho c_p}\frac{\partial q_r}{\partial y} + \frac{Q(T - T_0)}{\rho}$$

(4.1.4)

$$u\frac{\partial C}{\partial x} + v\frac{\partial C}{\partial y} + w\frac{\partial C}{\partial z} = D_B\left(\frac{\partial^2 C}{\partial x^2} + \frac{\partial^2 C}{\partial y^2} + \frac{\partial^2 C}{\partial z^2}\right) + \frac{D_T}{T_W}\left(\frac{\partial^2 T}{\partial x^2} + \frac{\partial^2 T}{\partial y^2} + \frac{\partial^2 T}{\partial z^2}\right)$$

(4.1.5)

Where q_r becomes: U is the velocity in the x direction, v is the velocity in the y direction, T is the temperature, c is the concentration parameter, σ is the electrical conductivity, B is the external magnetic effect, ρ is the density, κ is the permeability, μ is the dynamic viscosity, Cp is the specific heat at constant pressure, T_w is the temperature on below plate, and k is the thermal conductivity.

$$q_r = -\frac{4\sigma^*}{3k^*}\frac{\partial T^4}{\partial y} = -\frac{4\sigma^*}{3k}\frac{\partial\left(4T_0^{\,3}T - 3T_0^{\,4}\right)}{\partial y}$$

$\mu = \mu_{nf} = \mu_f(1+39.11\phi+533.9\phi^2)$ (4.1.6)

Where the radiation heat flux q_r is assumed concurring to Rosseland estimation such that q_r where σ^*, k^* are the Stefan–Boltzmann consistent and the cruel assimilation coefficient separately. The fluid-phase temperature contrasts inside the stream are accepted to be adequately little so that T^4 may be communicated as a direct work of temperature. This is often done by extending T^4 in a Taylor arrangement around the temperature T_0 and ignoring higher arranged terms to surrender, $T^4 = 4T_c^{\,3}T - 3T_c^{\,4}$. Table 1 presents the thermo-actual characteristics of nanoparticles and water, and the boundary layer for equations 1 to 5 is shown below. T_L is the temperature near the top plate, and C_L is the nanofluid concentration near the top plate.

$$u = 0; v = \frac{dl}{dt}; T = T_L; C = C_L \text{ at } y = l(t)$$

(4.1.7)

$$\frac{\partial u}{\partial y} = 0; v = 0; \frac{\partial T}{\partial y} = 0; \frac{\partial C}{\partial y} = 0 \text{ at } y = 0$$

(4.1.8)

Introducing non-dimensional variables:

$$\eta = \frac{y}{L(1-at)^{\frac{1}{2}}}, u = \frac{ax}{2(1-at)}f'(\eta), v = \frac{-aL}{(1-at)^{\frac{1}{2}}}f(\eta),$$

$$\theta(\eta) = \frac{T}{T_L}, C(\eta) = \frac{C}{C_L}, B(t) = \frac{B_0}{(1-at)^{\frac{1}{2}}},$$

$$\delta = \frac{L}{x}, \lambda = \frac{Q2L(T_L - T_0)(1-at)}{kT_L\rho}$$

(4.1.9)

And ϕ is the porosity, θ is the dimensionless temperature, f is the velocity parameter, M is the magnetic gauge, s is the squeeze parameter, Ec is the Eckert number, Nt is the thermo phoretic gauge, Pr is the Prandtl parameter, Nb is the Brownian motion gauge, and Sc is the Schmidt number.

$$f^{iv} - S(ff' + (-f+\eta)f''' + 3f'') - M^2f'' = 0$$

(4.1.10)

$$\theta'' + \Pr\left(S(f-\eta)\theta' + Ec(4\delta^2f^2 + f'^2)\right) + NbC'\theta' + Nt\theta'^2 + \theta\lambda = 0$$

(4.1.11)

Where:

$$\Pr = \frac{\mu C_p}{k}, M^2 = \frac{\sigma a B_0^2 L^2}{\rho v}, S = \frac{aL^2}{2v}$$

$$Ec = \frac{(ax)^2}{C_p T_L 4(1-at)^2}, \delta = \frac{L}{x}, Sc = \frac{\mu}{\rho_f D}, Nt = (\rho C)_p D_T(T_h^+)/(\rho C)_f \alpha T_c^+,$$

$$Nb = (\rho C)_P D_B(C_h)/(\rho C)_f \alpha$$

(4.1.12)

Subject to:

$$f = 0, f'' = 0, \theta' = 0, C' = 0 \text{ at } \eta = 0$$

(4.1.13)

$$f = 1, f' = 0, \theta = 1, C = 1 \text{ at } \eta = 1$$

(4.1.14)

Nusselt number and friction coefficient are defined as:

$$Nu = -\theta'$$

(4.1.15)

$$C_f = \frac{2\tau_s}{\rho u^2}$$

(4.1.16)

In this area, the elemental concept behind the HPM is clarified. To begin with, the common frame of a nonlinear differential condition and its boundary condition is considered: Where g (r) speaks to a known expository work, A(u) indicates the common differential administrator, ∂u/∂n speaks to separation along the standard drawn shapes from, and B is the space boundary and denotes a boundary administrator. The primary step in utilizing this strategy to illuminate an issue is to partition the equation's nonlinear and direct parts:

$$Y(x) - f(t) = 0, t \in \Psi \qquad (4.1.17)$$

Such that boundary condition:

$$B\left(x, \frac{\partial x}{\partial n}\right) = 0, t \in \Phi$$

(4.1.18)

Where could be a boundary administrator, a common differential administrator, the space boundary, a known expository work, and speaks to separation along the ordinary drawn outwards from. Commonly, the administrator is classified into straight and nonlinear parts. In this way, it is conceivable to modify Condition (4.1.18) as:

$$N(x) + L(x) = f(t) \tag{4.1.19}$$

If there is no "small parameter" in the nonlinear Eq. (18), the following Homotopy can be constructed:

$$H(\varepsilon,p) = L(\varepsilon) - L(x_0) + pL(x_0) + p(N(\varepsilon) - f(t)) = 0 \tag{4.1.20}$$

Where,

$$\varepsilon(t,p): \Psi\times[0,1]\rightarrow R \tag{4.1.21}$$

Eq. (4.1.21) indicates an inserting parameter, which speaks to the primary guess of assembly of the boundary condition. It is conceivable to assume that the arrangement to Eq. (4.1.21) is displayed as a control arrangement as takes after:

$$\varepsilon = \varepsilon_0 + p\varepsilon_1 + p^2\varepsilon_2 + \cdots \tag{4.1.22}$$

Also, the homotopy parameter is employed for expanding the square of an unknown angular frequency, which is shown below:

$$\gamma = \mu^2 - p\alpha_1 - p^2\alpha_2 - \cdots \tag{4.1.23}$$

Or

$$\mu^2 = \gamma + p\alpha_1 + p^2\alpha_2 + \cdots \tag{4.1.24}$$

This speaks to the Eq. (4.1.18) coefficient, and it is supplanted by the proper side of Eq. (4.1.24). There are subjective parameters that ought to be specified, too. The taking after is the most excellent estimation of an arrangement and standard recurrence μ:

$$x = \lim_{p \to 1} \varepsilon = \varepsilon_0 + \varepsilon_1 + \varepsilon_2 + \cdots \tag{4.1.25}$$

$$\mu^2 = 1 + \alpha_1 + \alpha_2 + \cdots \tag{4.1.26}$$

When Eqs. (4.1.18) And (4.1.26) become the approximate solution of Eq. (4.1.21), the best approximation of a solution and regular frequency is, (18).

Agreeing with the material science of the issue, expository strategies of non-linear and direct differential conditions require starting conditions and boundary conditions. As a result, any differential condition can be unraveled to any degree. To surpass what is shown in the present work. Two differential conditions have been opened to monitor the design of forms in this new way. The taking after presents the common frame of a differential condition, agreeing to the boundary conditions: The nonlinear differential condition of, as a work of, the parameter, as well as their subordinates, is considered as underneath:

$$p_k{:}f(y, y', y'', \ldots, y^{(n)}) = 0,\ y = y(x) \tag{4.1.27}$$

Boundary conditions:

$$\begin{cases} y(0) = y_0, & y'(0) = y_1, & y^{(n-1)}(0) = y_{n-1} \\ y(L) = y_L, & y'(L) = y_{L_1}, & y^{(n-1)}(L) = y_{L_{n-1}} \end{cases} \tag{4.1.28}$$

If x=0:

$$\begin{cases} y(0) = a_0 = y_0 \\ y'(0) = a_1 = y_1 \\ y''(0) = a_2 = y_2 \end{cases} \tag{4.1.29}$$

And when $x=L$:

$$\begin{cases} y(L) = a_0 + a_1 L + a_2 L^2 + \cdots + a_n L^n = yL_0 \\ y'(L) = a_1 + a_2 L + 3a_3 L^2 + \cdots + na_n L^{n-2} = yL_1 \\ y''(L) = 2a_2 + 6a_3 L + 12a_4 L^3 + \cdots + n(n-1)a_n L^{n-2} = yL_{m-1} \end{cases} \tag{4.1.30}$$

The primary differential condition in Eq. is unraveled with respect to the boundary conditions (4.1.28). The arrangement of letters should be considered within the arrange with constant coefficients that are the arrangement to the primary differential equation: Following the substitution of Eq. (4.1.31) into Eq. (4.1.27), it applies the boundary conditions on differential Eq. (4.1.27) as takes after: With respect to the determination of n sentences from Eq. (4.1.29), and for making a set of conditions comprised of questions and conditions, some additional questions are the coefficients of equation (4.1.29). To solve the issue, times ought to be determined from Eq. (4.1.27) since the extra questions within the already pronounced set of differential conditions and the boundary conditions of Eq. (4.1.28) are connected at that point.

$$\begin{aligned} p_0 &= f(y(0), y'(0), y''(0), \ldots, y^{(m)}(0)) \\ p_1 &= f(y(L), y'(L), y''(L), \ldots, y^{(m)}(L)) \end{aligned} \tag{4.1.31}$$

$$\begin{aligned} p_k^{'} &= f(y', y'', y''', \ldots, y^{(m+1)}) \\ p_k^{''} &= f(y'', y''', y^{(IV)}, \ldots, y^{(m+2)}) \end{aligned} \tag{4.1.32}$$

The boundary conditions are applied to the derivatives of the differential equations in Eq. (4.1.33) as follows:

$$p_k^{'}: \begin{cases} f(y'(0), y''(0), y'''(0), \ldots, y^{(m+1)}(0)) \\ f(y'(L), y''(L), y'''(L), \ldots, y^{(m+1)}(L)) \end{cases} \tag{4.1.33}$$

$$p_k^{''}: \begin{cases} f(y''(0), y'''(0), y^{(W)}(0), \ldots, y^{(m+2)}(0)) \\ f(y''(L), y'''(L), y^{(IV)}(L), \ldots, y^{(m+2)}(L)) \end{cases} \tag{4.1.34}$$

The following are the homotopy forms for the equations of the mentioned problem.

$$\text{Ueq} = (1-p)\left(\frac{d^4}{d\eta^4}f_0(\eta) + \left(\frac{d^4}{d\eta^4}f_1(\eta)\right)p - 3S\right.$$

$$\left.\left(\frac{d^2}{d\eta^2}f_0(\eta) + \left(\frac{d^2}{d\eta^2}f_1(\eta)\right)p\right) - M^2\left(\frac{d^2}{d\eta^2}f_0(\eta) + \left(\frac{d^2}{d\eta^2}f_1(\eta)\right)p\right)\right)$$

$$+p\left(\begin{array}{c} \frac{d^4}{d\eta^4}f_0(\eta) + \left(\frac{d^4}{d\eta^4}f_1(\eta)\right)p - S\left(\frac{d}{d\eta}f_0(\eta) + \left(\frac{d}{d\eta}f_1(\eta)\right)p\right) \\ \left(\frac{d^2}{d\eta^2}f_0(\eta) + \left(\frac{d^2}{d\eta^2}f_1(\eta)\right)p\right) + (-f_0(\eta) - f_1(\eta)p + \eta) \\ \left(\frac{d^3}{d\eta^3}f_0(\eta) + \left(\frac{d^3}{d\eta^3}f_1(\eta)\right)p\right) + 3\left(\frac{d^2}{d\eta^2}f_0(\eta)\right) + 3\left(\frac{d^2}{d\eta^2}f_1(\eta)\right) \\ p - M^2\left(\frac{d^2}{d\eta^2}f_0(\eta) + \left(\frac{d^2}{d\eta^2}f_1(\eta)\right)p\right) \end{array}\right) \quad (4.1.35)$$

$$eq = (1-p)\left(\frac{d^2}{d\eta^2}\theta_0(\eta) + \left(\frac{d^2}{d\eta^2}\theta_1(\eta)\right)p + 1.2\theta_0(\eta) + 1.2\theta_1(\eta)p\right)$$

$$+p\left(\begin{array}{c} \frac{d^2}{d\eta^2}\theta_0(\eta) + \left(\frac{d^2}{d\eta^2}\theta_1(\eta)\right)p + 0.045000\left(\frac{d}{d\eta}f_0(\eta) + p\left(\frac{d}{d\eta}f_0(\eta)\right)\right)^2 \\ +0.50\left(\frac{d}{d\eta}f_1(\eta) + p\left(\frac{d}{d\eta}f_1(\eta)\right)\right)^2 + 0.1\left(\frac{d}{d\eta}C_0(\eta) + p\left(\frac{d}{d\eta}C_0(\eta)\right)\right) \\ \left(\frac{d}{d\eta}\theta_0(\eta) + \left(\frac{d}{d\eta}\theta_1(\eta)\right)^2\right)^2 + 1.2\theta_0(\eta) + 1.2\theta_1(\eta)p \end{array}\right) \quad (4.1.36)$$

$$\text{Teq} = (1-p)$$

$$\left(0.1\left(\frac{d^2}{d\eta^2}C_0(\eta)\right) + 0.1\left(\frac{d^2}{d\eta^2}C_1(\eta)\right)p + 0.1\left(\frac{d^2}{d\eta^2}\theta_0(\eta)\right) + 0.1\left(\frac{d^2}{d\eta^2}\theta_1(\eta)\right)p\right)$$

$$+p\left(0.1\left(\frac{d^2}{d\eta^2}C_0(\eta)\right) + 0.1\left(\frac{d^2}{d\eta^2}C_1(\eta)\right)p + 0.1\left(\frac{d^2}{d\eta^2}\theta_0(\eta)\right) + 0.1\left(\frac{d^2}{d\eta^2}\theta_1(\eta)\right)p\right) \quad (4.1.37)$$

In the above mathematical expressions, U_{eq} means the f, P_{eq} means the θ, and T_{eq} means the C.

F, θ, and C are considered as follows:

$$f(\eta) = f_0(\eta) + f_1(\eta)p \quad (4.1.38)$$

$$\theta(\eta) = \theta_0(\eta) + \theta_1(\eta)p \quad (4.1.39)$$

$$C(\eta) = C_0(\eta) + C_1(\eta)p \quad (4.1.40)$$

Taking after that, Eqs. (4.1.16) And (4.1.17) will be substituted into Eqs. (4.1.12) to (4.1.14), as well as a few requesting and rearrangements based on the powers of perms, and P^0:

$$q_0 = \frac{d^2}{d\eta^2}\theta_0(\eta) + 1.2\theta_0(\eta)$$

(4.1.41)

$$s_0 = \frac{d^4}{d\eta^4}f_0(\eta) - 0.25\left(\frac{d^2}{d\eta^2}f_0(\eta)\right)$$

(4.1.42)

$$w_0 = 0.1\left(\frac{d^2}{d\eta^2}C_0(\eta)\right) - \frac{0.1200000000\cos\left(\frac{1}{5}\sqrt{30}\,\eta\right)}{\cos\left(\frac{1}{5}\sqrt{30}\right)}$$

(4.1.43)

Within the over-scientific expressions, q_0 implies the theta, S_0 implies the f, and W_0 implies the C. The boundary conditions are:

f[0](0)=0, f[0](1)=1, ((D^2)(f[0]))(0)=0, (D(f[0]))(1) =0 (4.1.44)

(D (theta[0]))(0) =0, theta[0](1)

=1)
(4.1.45)

(D(C[0]))(0) =0, C[0](1) =1) (4.1.46)

P^1:

$$s_1 = \frac{d^4}{d\eta^4}f_1(\eta) - 0.025\left(\frac{d^2}{d\eta^2}f_1(\eta)\right.$$

(4.1.47)

$$q_1 = \frac{d^2}{d\eta\text{Skipping malformed tag: msup}}\theta_1(\eta) + 1.2\theta_1(\eta) + 0.04500$$
$$\left(-\frac{e^{\frac{1}{2}} + e^{\frac{-1}{2}}}{e^{\frac{1}{2}} - 3e^{\frac{-1}{2}}} + \frac{e^{\frac{1}{2\eta}}}{e^{\frac{1}{2}} - 3e^{-\frac{1}{2}}} + \frac{e^{-\frac{1}{2\eta}}}{e^{\frac{1}{2}} - 3e^{-\frac{1}{2}}}\right)^2 + 0.50\left(\frac{1}{2}\frac{e^{\frac{1}{2\eta}}}{e^{\frac{1}{2}} - 3e^{-\frac{1}{2}}} - \frac{1}{2}\frac{e^{-\frac{1}{2\eta}}}{e^{\frac{1}{2}} - 3e^{-\frac{1}{2}}}\right)^2$$

(4.1.48)

$$w_1 = 0.1\left(\frac{d^2}{d\eta\text{Skipping malformed tag: msup}} C_1(\eta)\right)$$

$$+\frac{0.00000078\cos\left(\frac{1}{5\sqrt{30}\,\eta}\right)\left(807\,e^{\frac{5}{2}} - 2794\,e^{\frac{3}{2}} + 807\,e^{\frac{1}{2}}\right)e^{-\frac{1}{2}}}{\cos\left(\left(\frac{1}{5}\right)*\sqrt{30}\right)*\left(-\exp(2)+6*\exp(1)-9\right)}$$

$$+\frac{1}{-e^2+6e-9}\left(0.0000001361\left(36975\,e^{\frac{5}{2\eta}} - \frac{20097}{2}\,e^{\frac{3}{2\eta}} + 1479\,e^{\frac{1}{2\eta}} - 19008\,e^{2\eta+\frac{1}{2}}\right.\right.$$

$$\left.\left. -19008\,e^{2\eta-\frac{1}{2}} + \frac{25839}{4}\,e^{1+\frac{3}{2\eta}} + \frac{25839}{4}\,e^{1\mp\frac{3}{2\eta}} - 4752\,e^{\frac{1}{2}+\eta} - 4752\,e^{-\frac{1}{2}+\eta}\right)\right) \quad (4.1.49)$$

With the following boundary conditions:

$$f[1](0) = 0,\ f[1](1) = 1,\ ((D^2)(f[1]))(0) = 0,\ (D(f[1]))(1) = 0 \quad (4.1.50)$$

(D (theta[1]))(0) = 0, theta[1](1)

= 1)
(4.1.51)

$$(D(C[1]))(0) = 0,\ C[1](1) = 1) \quad (4.1.52)$$

By utilizing boundary conditions to fathom Eqs. (4.1.42)-(4.1.53), we get: When p = 1, the arrangements to these conditions are as takes after:

$$f(\eta) = \frac{-\left(e^{\frac{1}{2}} + e^{-\frac{1}{2}}\right)\eta}{e^{\frac{1}{2}} - 3\,e^{-\frac{1}{2}}} + \frac{2\,e^{\frac{1}{2\eta}}}{e^{\frac{1}{2}} - 3\,e^{-\frac{1}{2}}} - \frac{2\,e^{-\frac{1}{2\eta}}}{e^{\frac{1}{2}} - 3\,e^{-\frac{1}{2}}} \quad (4.1.53)$$

$$\theta(\eta) := \frac{\cos\left(\frac{1}{5}\sqrt{30}\,\eta\right)}{\cos\left(\frac{1}{5}\sqrt{30}\right)} - \frac{1}{15312}\,\frac{\text{os}\left(\frac{1}{5}\sqrt{30}\,\eta\right)\left(807\,e^{5/2} - 2794\,e^{3/2} + 807\,e^{1/2}\right)e^{-1/2}}{\cos\left(\frac{1}{5}\sqrt{30}\right)\left(-e^2+6e-9\right)} \quad (4.1.54)$$

$$+\frac{1}{76560(-e^2+6e-9)}\left(5916\,e^{5/2\eta} - 4466\,e^{3/2\eta} + 5916\,e^{1/2\eta} - 4752\,e^{2\eta+1/2} - 4752\,e^{2\eta-1/2}\right.$$

$$\left. +2871\,e^{-1+3/2\eta} + 2871\,e^{1+3/2\eta} - 4752\,e^{\eta+1/2} - 4752\,e^{\eta-1/2}\right)e^{1-3/2\eta}$$

$$C(\eta) = \frac{\cos\left(\frac{1}{5}\sqrt{30}\,\eta\right)}{\cos\left(\frac{1}{5}\sqrt{30}\right)} + 2 + \frac{1}{4\times10^{14}} \times \frac{1}{\cos\left(\frac{1}{5}\sqrt{30}\right)\left(-e^2+6e-9\right)}$$

$$\left(-6206896536192\cos(1/5\sqrt{30})\,e^{3/2-1/2}z = 6206896517184\cos(1/5\sqrt{30})\,e^{3/2+1/2}z1\right.$$

$$-6206896536192\cos(1/5\sqrt{30})e^{-1/2z_-^{1+1/2}} - 6206896517184\cos(1/5\sqrt{30})e^{1/2}z1 + 1/2$$

$$+(-87586206900896e + 25297805643888e^2 + 25297805643888)\cos(1/5\sqrt{30}z\_1)$$

$$22330(e - 9/14e^2 + 106583072058/77e^{--z1+1} - 532915359474/385e^{-z1+1} - 9/14)$$

$$-6206896536192\cos(1/5\sqrt{30})e^{-1/2z1+1/2} - 6206896517184\cos(1/5\sqrt{30})e^{1/2zz1+1/2}$$

$$\left(\int_0^1 + (-87586206900896e + 25297805643888e^2 + 25297805643888)\cos(1/5\sqrt{30}\_z1)\right)$$

$$22330(e - 9/14e^2 + 106583072058/77e^{--z1+1} - 532915359474/385e^{-z1+1} - 9/14)$$

$$\times\cos(1/530)d\text{-}z1\text{+-}4\times1015e2\text{+}24\times1014e\text{-}36\times1014\cos(1/530) \tag{4.1.55}$$

Given the most condition, the defenses advertised for the condition, and its boundary conditions, comprehend AGM and go over its application.

$$f^{iv} - S(ff' + (-f + \eta)f''' + 3f'') - M^2f'' = 0 \tag{4.1.56}$$

$$\theta'' + \Pr\left(S(f - \eta)\theta' + Ec(4\delta^2 f^2 + f''^2)\right) + NbC'\theta' + Nt\theta'^2 + \theta\lambda = 0 \tag{4.1.57}$$

$$NbC'' + Nt\theta'' + S.Nb \cdot Sc(fC' - \eta C'') = 0 \tag{4.1.58}$$

$$f(\eta) = a_5\eta^5 + a_4\eta^4 + a_3\eta^3 + a_2\eta^2 + a_1\eta + a_0 \tag{4.1.59}$$

$$\theta(\eta) = b_5\eta^5 + b_4\eta^4 + b_3\eta^3 + b_2\eta^2 + b_1\eta + b_0 \tag{4.1.60}$$

$$C(\eta) = c_5\eta^5 + c_4\eta^4 + c_3\eta^3 + c_2\eta^2 + c_1\eta + c_0$$

(4.1.61)

According to AGM, a test function is regarded as the equation's result.

$$f = 0, f'' = 0, \theta' = 0, C' = 0$$

at η=0 (4.1.62)

$$f = 1, f' = 0, \theta = 1, C = 1$$

at η=1 (4.1.63)

This is clarified by counting the boundary condition in the AGM condition. Here are the boundary conditions:

equ1=$f(0)$=0 equ1=a_0=0

equ2=$f(1)$=1
(4.1.64)

$$equ2 = a_5 + a_4 + a_3 + a_2 + a_1 + a_0 = 1$$

(4.1.65)

$$equ3 = f''(0) = 0 \ equ3 = 2a_2 = 0$$

(4.1.66)

$$equ4 = f'(1) = 0 \ equ4 = 5a_5 + 4a_4 + 3a_3 + 2a_2 + 1a_1 = 1$$

(4.1.67)

$$equ5 = \theta'(0) = 0 \ equ5 = b_1 = 0$$

(4.1.68)

$$equ6 = \theta(1) = 1 \text{ equ } 6 = \theta(\eta) = b_5 + b_4 + b_3 + b_2 + b_1 + b_0 = 1$$

(4.1.69)

$$equ7 = C'(0) = 0 \text{ equ } 7 = c_1 = 0$$

$$(4.1.70)$$

$$equ8 = C(1) = 1 \text{ equ } 8 = C(\eta) = c_5 + c_4 + c_3 + c_2 + c_1 + c_0 = 1$$

$$(4.1.71)$$

$$\text{equ9}=U\text{eq}(0) \text{ equ9}=24a_4 - 0.50a_2=0 \quad (4.1.72)$$

The solution is obtained by solving the primary problem using the boundary condition. It is followed by applying the boundary conditions to the aforementioned equation. Given the aforementioned information, 3 test functions are considered, which have 13 constant coefficients and 13 equations.

$$eeq = \{a_0 = 0, a_1 = 1.493902439, a_2 = 0,$$
$$a_3 = -.4878048780, a_4 = 0, a_5 = -0.6097560976\}$$
$$(4.1.73)$$

This chapter centered on the issue's geometry by visually investigating the temperature, concentration, and velocity profiles. Arrangements are created utilizing Computational strategies. The press stream depicts the outward improvement of a globule of matter, its extent of contact with the plate surfaces, and the impacts of interior and exterior components such as temperature, viscoelasticity, and matter heterogeneity. Figure 2 shows that as the crushing variable S rises, the velocity component f diminishes. Both numerical strategies are exceptionally near each other. Figure 3 compares temperature and heat exchange varieties to s parameter changes utilizing AGM and HPM scientific strategies. S has a negative effect on temperature, as shown in Figure 3. Usually levelheaded since kinematic thickness diminishes as crush stream increments, coming about in less friction between nanoparticles and plate surfaces. This sentence infers that the coefficient of surface contact includes a speaking relationship with the compressed stream of nanofluid. As the compression of the flow increases, the entirety of stickiness and consistency lessens. Figure 4 compares concentration varieties to s parameter changes utilizing AGM and HPM scientific strategies. Figure 4 shows that the concentration component increments as the crushing parameter press parameter increments. It is observed that there's an increment within the heat exchange prepared with expanding Pr. Besides, it appeared that with increasing Pr, there's a diminishment within the concentration. Figure 5 shows the Brownian impacts on concentration and temperature profiles. With a rise in Brownian movement, temperature characteristics diminish, and concentration increments. Brownian movement is known as the arbitrary development of a nanoparticle as a response to collisions with adjacent vaporous atoms. A concentration angle causes a collection of particles to emigrate in a handle called diffusiophoresis. This movement continuously moves from thickly populated zones to scantily populated places. Agreeing to Figure 6, by expanding the Brownian movement, the concentration impact increment. Figure 7 shows that the magnet parameter presents antagonistically influences temperature and velocity profiles. The attractive field, a vector field, depicts the attraction impact on electrical charges, electric streams, and nanomaterials. Thermophoresis (Nt), which is the particulate development in reaction to a bulk-gas heat slope towards the cold side, is imperative in numerous application zones, counting catalysis-driven plasmatic, adjusted chemical vapor statement, vaporized examining, microelectronic generation, and discussing decontamination. Figure 8 illustrates how taking a bigger esteem under consideration might lead to diminished concentration. The Schmidt number (Sc), known as v/Dm, is the relationship between kinematic thickness and atomic dissemination coefficient. Numerous field investigations have driven the advancement of experimental conditions that connect k to wind velocity. Figures 9 and 10 show a decrease in temperature and concentration patterns as Sc levels increment.

This ponder examines the effect of attraction on a 2D nanofluid stream through two equally dispersed sheets. The taking after measures is examined parametrically: Eckert number Ec, thermal era variable, Prandtl number Pr, radiation parameter Nr, nanofluid volume percent, attractive parameter M, and crush parameter S. Semi-analytical arrangements are inferred utilizing the HPM and the strategy displayed by Akbari Ganji, and the discoveries are examined utilizing charts, the numeric approach in an isolated chart at that point compares the resolutions of the proposed techniques. The oddity of this investigation is that it presents the HPM and the strategy proposed by Akbari Ganji to look at the heat stream, liquid stream, and nanofluid concentration by two rises to surfaces.

Figure 1. Schematic of the two parallel plates

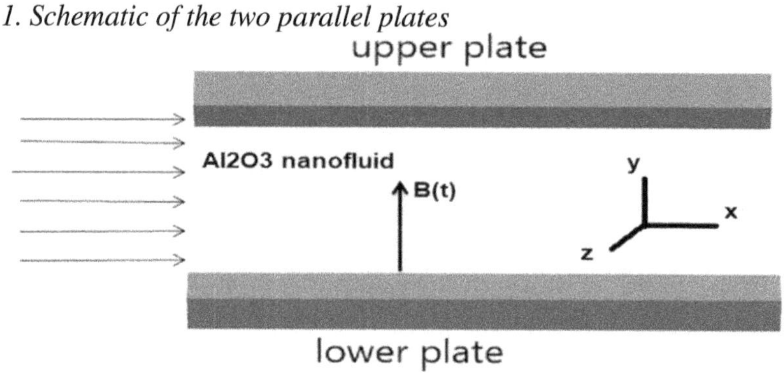

Table 1. Properties of water and Al_2O_3 nanofluid

	ρ(kg/m³)	K(w/m.k)	C_p (j/kg.k)	σ
Al_2O_3	3970	25	765	10^{-12}
H_2O	997	0.613	4179	0.05

Figure 2. Comparison of the velocity profile with different S parameter by AGM and HPM method at $\varphi = 0.02$, $M = 0.5$, $Nr = 1.0$, $Pr = 10.$, $Ec = 0.05$, $\delta = 0.15$, $\lambda = 1.2$, $Nb = 0.1$, $Nt = 0.1$, $Sc = 0.3$

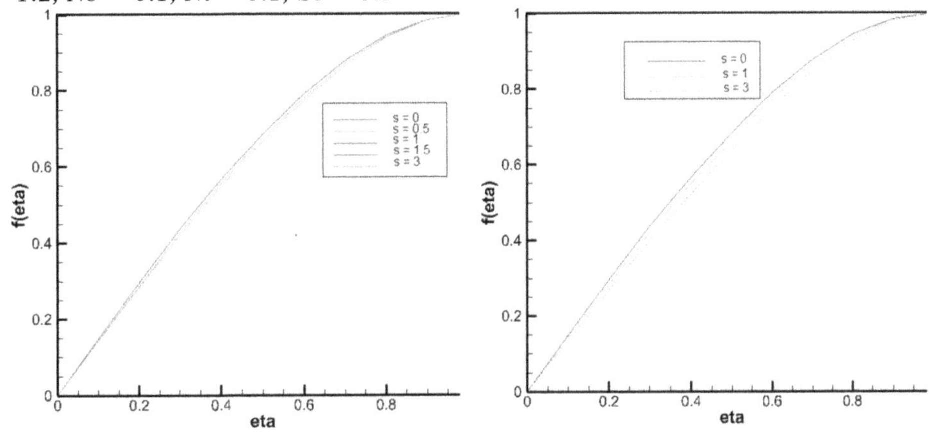

a) By AGM b) By HPM

Figure 3. Comparison of the temperature profile with different S parameter by AGM and HPM method at $\varphi = 0.02$, $M = 0.5$, $Nr = 1.0$, $Pr = 10$, $Ec = 0.05$, $\delta = 0.15$, $\lambda = 1.2$, $Nb = 0.1$, $Nt = 0.1$, $Sc = 0.3$

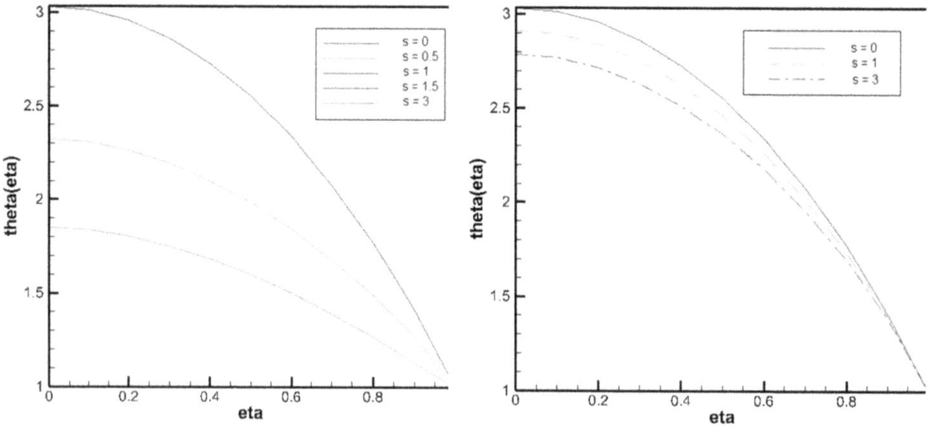

a) By AGM **b) By HPM**

Figure 4. Comparison of the concentration profile with different S parameter by HPM method at $\varphi = 0.02$, $M = 0.5$, $Nr = 1.0$, $Pr = 10$, $Ec = 0.05$, $\delta = 0.15$, $\lambda = 1.2$, $Nb = 0.1$, $Nt = 0.1$, $Sc = 0.3$

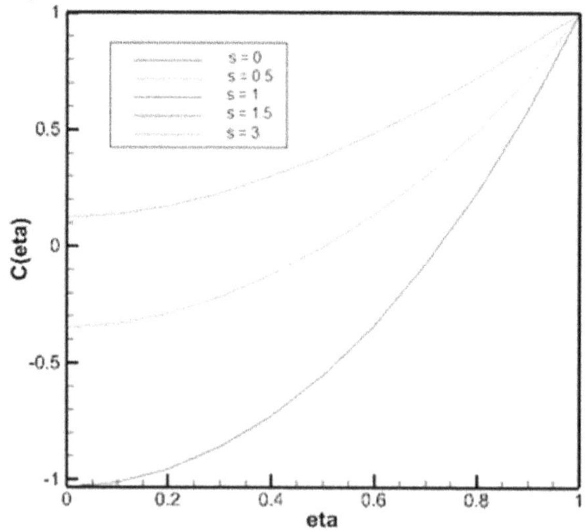

Figure 5. Comparison of the temperature profile with different Nb parameter by AGM and HPM method at $\varphi = 0.02$, $M = 0.5$, $Nr = 1.0$, $Pr = 10$, $Ec = 0.05$, $\delta = 0.15$, $\lambda = 1.2$, $s = 0.5$, $Nt = 0.1$, $Sc = 0.3$

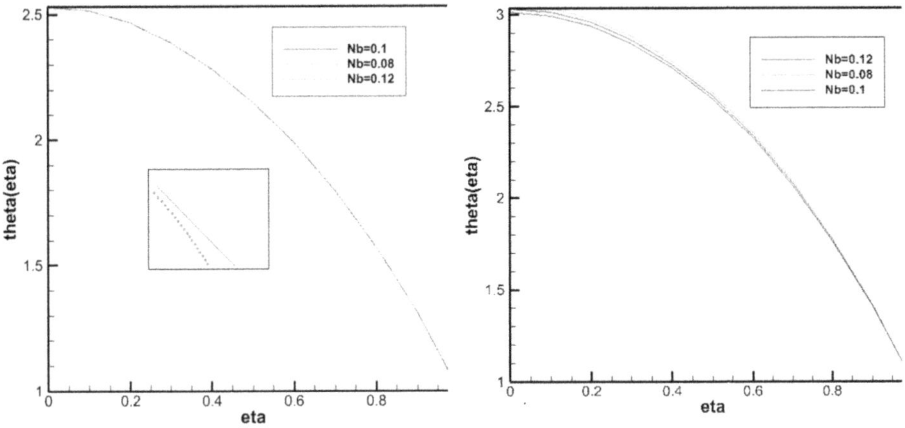

a) By AGM **b) By HPM**

Figure 6. Comparison of the concentration profile with different Nb parameter by HPM method at $\varphi = 0.02$, $M = 0.5$, $Nr = 1.0$, $Pr = 10$, $Ec = 0.05$, $\delta = 0.15$, $\lambda = 1.2$, $Nt = 0.1$, $Sc = 0.3$

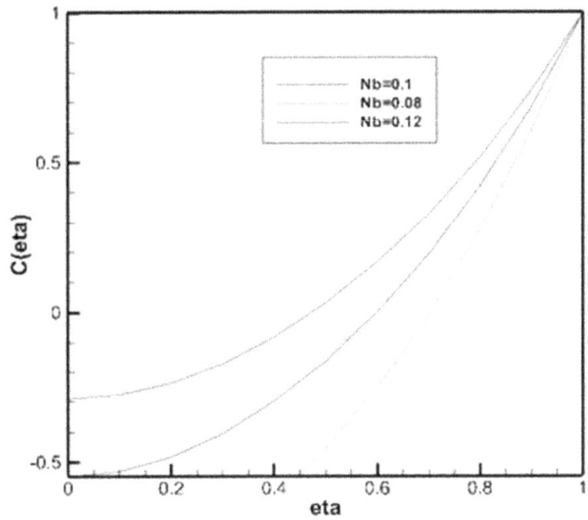

Figure 7. Comparison of the velocity profile with different M parameter by AGM method at $\varphi = 0.02$, $M = 0.5$, $Nr = 1.0$, $Pr = 10$, $Ec = 0.05$, $\delta = 0.15$, $\lambda = 1.2$, $Nb = 0.1$, $Nt = 0.1$, $Sc = 0.3$

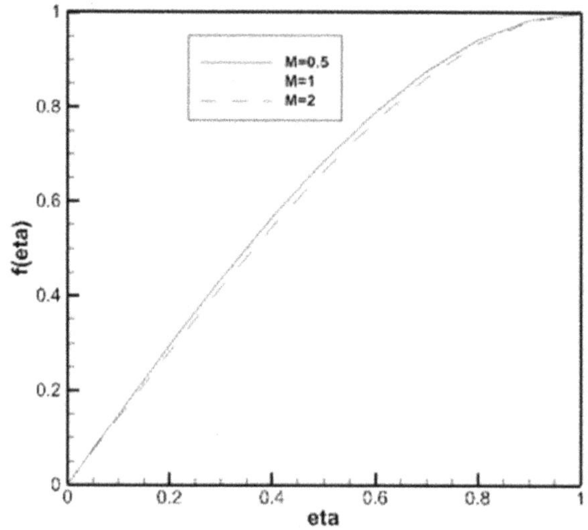

Figure 8. Comparison of the concentration profile with different Nt parameter by AGM method at $\varphi = 0.02$, $M = 0.5$, $Nr = 1.0$, $Pr = 10$, $Ec = 0.05$, $\delta = 0.15$, $\lambda = 1.2$, $Nb = 0.1$, $Sc = 0.3$

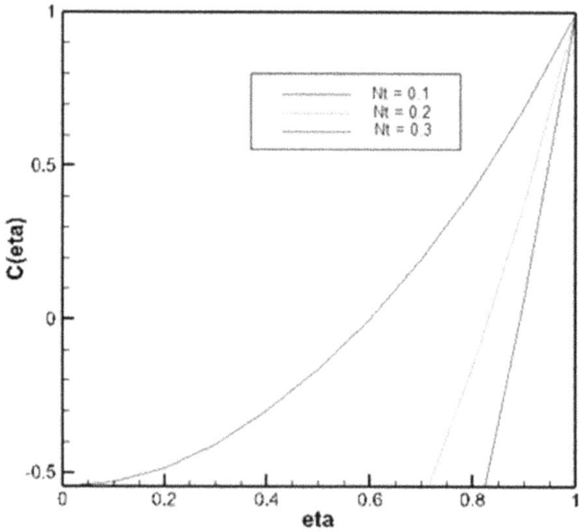

Figure 9. Comparison of the temperature profile with different Sc parameter by HPM method at φ = 0.02, M = 0.5, Nr = 1.0, Pr = 10, Ec = 0.05, δ =0.15, λ = 1.2, Nb = 0.1

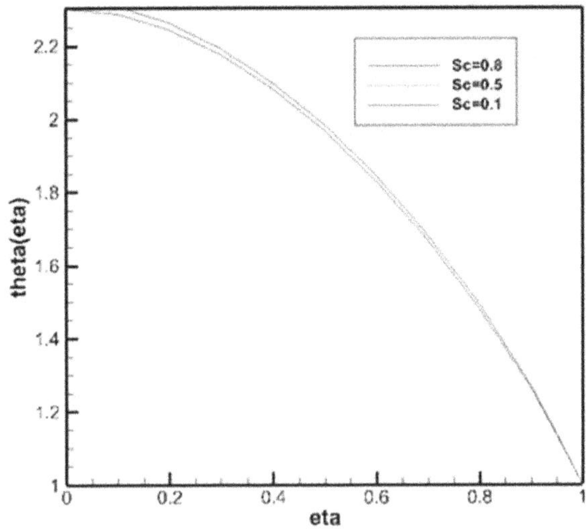

Figure 10. Comparison of the concentration profile with different Sc parameter by HPM method at φ = 0.02, M = 0.5, Nr = 1.0, Pr = 10, Ec = 0.05, δ =0.15, λ = 1.2, Nb = 0.1

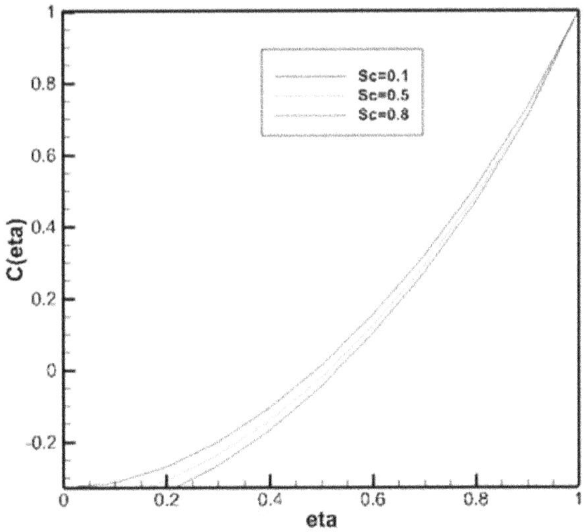

4.2. RAINER-RIVELIN FLUID FLOW BETWEEN TWO DISCS

Researchers named Rainer and Rivlin calculated the stress tensor formula as follows (Nadeem *et al.,* 2022) (Dey, 2018) (Sheykhi *et al.,* 2023) (Shadman *et al.,* 2023) (Fathollahi *et al.,* 2022) (Hafez, Abd-Alla, & Metwaly, 2023) (Jalili *et al.,* 2024d) (Shoukat *et al.,* 2023) (Hashemi & Domairry et al., 2011) (Abdollahi *et al.,* 2023)(Jalili *et al.,* 2023) (Jafaripournimchahi *et al.,* 2024) (Noreen *et al.,* 2020) (Ashraf *et al.,* 2021) (Al-Turki *et al.,* 2021) (Cheng *et al.,* 2023) (Li *et al.,* 2021):

$$\tau_{ij} = -p.\delta_{ij} + 2.\mu_{YN}.d_{ij} + \mu_{CV}.C_{ij} \tag{4.2.1}$$

Where d_{ij} and c_{ij} are:

$$d_{ij} = \frac{1}{2}\cdot(u_{ij} + u_{ji}), c_{ij} = d_{im}.d_{mj} \tag{4.2.2}$$

τ_{ij} is the stretch tensor, δ_{ij} is Kronecker's delta, μ_{nv} is the modulus of Newtonian viscosity and μ_{cv} is the modulus of cross viscosity, d_{ij} and c_{ij} are the physical details of the stress tensor.

In this chapter, Rainer-Rivelin fluid flow between two discs has been investigated. The fluid flow is continuous and remains constant over time. The lower disk is positioned at z=0, while the upper disk is positioned at z=d.

This analysis introduces a magnetic field B0, which is applied uniformly on a square shape. The lower disk undergoes rotational motion around the z axis at a constant angular velocity Ω, while the upper disk experiences radial stretching with a velocity ar. To facilitate our investigation, a cylindrical coordinate system with $\sqrt{r}$,θ,$\sqrt{z}$ coordinates, as shown in Figure 11 is adopted.

The velocity parameters exist in both the (r,θ,z) and (u,v,w) directions. The lower disk remains at a constant temperature of T_0, while the upper disk is subjected to a steady temperature of T_1. The equations that govern the momentum, continuity, and energy are as follows:

$$\frac{\partial u^+}{\partial r} + \frac{u^+}{r} + \frac{\partial w^+}{\partial z} = 0 \tag{4.2.3}$$

In a cylindrical framework, momentum is indeed conserved in the r direction:

$$\rho^{+}\left(u^{+}\frac{\partial u^{+}}{\partial r} - \frac{v^{2}}{r} + w\frac{\partial u^{+}}{\partial z}\right) = \frac{\partial \tau_{rr}}{\partial r} + \frac{\partial \tau_{rz}}{\partial z} + \frac{\tau_{rr} - \tau_{\theta\theta}}{r} - \frac{\sigma B_{0}^{2} u^{+}}{\rho}$$

(4.2.4)

Conservation of momentum in the θ direction in a cylindrical coordinate system:

$$\rho^{+}\left(u\frac{\partial v^{+}}{\partial r} - \frac{u^{+}v^{+}}{r} + w\frac{\partial v^{+}}{\partial z}\right) = \frac{\partial \tau_{r\theta}}{\partial r} + \frac{\partial \tau_{\theta z}}{\partial z} + \frac{2\tau_{r\theta}}{r} - \frac{\sigma B_{0}^{2} v^{+}}{\rho}$$

(4.2.5)

Conservation of momentum in the z direction in a cylindrical coordinate system:

$$\rho^{+}\left(u\frac{\partial w^{+}}{\partial r} + w^{+}\frac{\partial v^{+}}{\partial z}\right) = \frac{\partial \tau_{rz}}{\partial r} + \frac{\partial \tau_{zz}}{\partial z} + \frac{\tau_{rz}}{r}$$

(4.2.6)

Energy conservation in cylindrical coordinates:

$$\rho^{+}c_{v}\left(u^{+}\frac{\partial T}{\partial r} + w^{+}\frac{\partial T}{\partial z}\right) = K\left(\frac{\partial^{2} T}{\partial r^{2}} + \frac{1}{r}\frac{\partial T}{\partial r} + \frac{\partial^{2} T}{\partial z^{2}}\right) + \Phi$$

(4.2.7)

$$\Phi = \tau_{rr}.d_{rr} + \tau_{\theta\theta}.d_{\theta\theta} + \tau_{zz}.d_{zz} + 2\left(\tau_{r\theta}.d_{r\theta}.\tau_{rz}.d_{rz} + \tau_{\theta z}.d_{\theta z}\right)$$

(4.2.8)

Where Φ is the viscous-dissipation function, ρ is the density, σ is the conductivity, $\tau_{rr}, \tau_{zz}, \tau_{r\theta}, \tau_{z\theta}$, is the components of the stress tensor. k is the thermal conductivity, and T is the temperature.

The limitations or parameters that define the context of this current demonstration are as follows:

$$u^{+} = 0, = r., \ w^{+} = 0, \ T = T_{0} \rightarrow z = 0 u^{+} = ar,$$
$$v^{+} = 0, \ w^{+} = 0, \ T = T_{1} \rightarrow z = d$$

(4.2.9)

Now, to transform partial differential equations (PDEs) into ordinary differential equations (ODEs), the relevant variables proposed by von Karman have been used:

$$u^{+} = r.f(\xi),\ v^{+} = r.h(\xi),\ w^{+} = -2.df(\xi),\ T = \frac{T-T_0}{T_1} - T_0,$$

(4.2.10)

Where $\xi = z/d$is dimensionless variable.

The taking after temperature conveyance is proposed for Eqs. (7) and (10).

$$T = T_0 + \frac{v^{+}}{c_v}[\varnothing(\xi) + n^2\psi(\xi)]$$

(4.2.11)

Where the $\eta = r/d$is the non-dimensional variable.

By substituting the analogous variables from equation (10) into the PDEs mentioned in equations (4-6), as well as considering the temperature distribution given in equation (11) for the energy equation described in equation (7), A nonlinear ordinary differential equation is obtained:

$$f^4 + 2R_0(hh' + f\,.hh''') - 2R_0\tau cv(ff(4) + 3h'h'' + f'f'' = 0$$

(4.2.12)

$$h'' + 2R_0(f\,h' + fh) + 2R_0\tau_{cv}(f'h' + fh'') - M^2 h = 0$$

(4.2.13)

$$\psi'' - P_r R_0(2f\psi - 2f\psi' - 2f'' - 2h') - 3P_r R_0\tau_{cv}(2ff' + 2fh') = 0$$

(4.2.14)

$$\Phi'' + P_r R_0(2f\Phi' + 12f'2) + 4\psi - 24P_r R_0^2\tau_{cv}f3 = 0$$

(4.2.15)

Where $R_0 = \Omega\rho d2/\mu nv$ is the Reynolds number, $P_r = \mu nvcv/k$ is the Prandtl number, $\tau cv = \mu cv/\rho d2$ is the cross-viscosity parameter and $M = (B_0^2\sigma d2/\mu_1)1/2$is the magnetic parameter.

According to equation (11), the dimensionless temperature transport variable is expressed as.

$$T^{*} = \frac{T-T_0}{T_1-T_0} = E_r[\Phi(\xi) + n^2\psi(\xi)]$$

(4.2.16)

Where $Er=\mu_{cv}\Omega/(T_1 - T_0)$is the Eckert number.

According to the formula. (11) The dimensionless temperature distribution variable is:

$$f = 0, f = 0, h = 1, \phi = 0, \psi = 0, \; when \; \xi = 0$$

$$f = 0, f = a, h = 0, \phi = \frac{1}{E_r}, \psi = 0, \; when \; \xi = 01$$

(4.2.17)

Where $\alpha=a/\Omega$ is the dimensionless forced variable.

General nonlinear equations can be expressed in the form presented by:

$$L(u) + Ru + Nu = g(r) \qquad (4.2.18)$$

By applying the inverse operator L-1 to both sides of equation (49) and taking into account the provided conditions:

$$U=f(x) - L - 1(Ru) - L - 1(Nu) \qquad (4.2.19)$$

In the Adomian decomposition method for nonlinear differential equations, The nonlinear operator Nu=F(u) is considered:

$$F(u) = \sum_{m=0}^{\infty} A_m$$

(4.2.20)

The Adomian strategy characterizes the arrangement U(x) by means of a sequence:

$$u = \sum_{m=0}^{\infty} u_m$$

(4.2.21)

$$F(u) = F(u_0) + F'(u_0)(u - u_0) + F''(u_0)\frac{(u - u_0)}{2!} + F''(u_0)\frac{(u - u_0)}{3!} + \cdots$$

(4.2.22)

After dividing the terms evenly, the first few Adomian polynomials A0, A1, A2, etc. are:

$$A_0 = F(u_0)$$

$$A_1 = u_1 F'(u_0) \; A_2 = u_2 F'(u_0) + \frac{1}{2!} u_1 u_1 F''(u_0)$$

(4.2.23)

The finite element method (FEM) is a widely employed technique used to numerically solve differential equations commonly encountered in engineering and numerical modeling.

Some of the typical areas that people find interesting are the conventional fields of structural analysis, heat transfer, fluid flow, mass transfer, and electromagnetic potential. The finite element method offers engineers a valuable advantage by enabling them to simulate real-world phenomena. This reduces the reliance on physical prototypes and allows for component optimization during the project's design phase. The finite element method is commonly utilized in mechanical, aviation, automotive, civil engineering projects, and biomechanics.

By utilizing the Adomian Decomposition Method, the linear segment of the equation was isolated and equated to zero. Consequently, a differential equation with specific boundary conditions was successfully resolved.

$$\frac{d^4}{dn^4} f_0(\xi) = 0$$

(4.2.24)

$$\frac{d^2}{dn^2} h_0(\xi) = 0$$

(4.2.25)

$$\frac{d^2}{dn^2} \psi_0(\xi) = 0$$

(4.2.26)

$$\frac{d^2}{dn^2} \phi_0(\xi) = 0$$

(4.2.27)

$$f_0(\xi) = \xi^2 - 2\xi^2 + \xi \leftrightarrow \psi_0(\xi) = \xi + 1 \leftrightarrow \phi_0(\xi) = \xi + 1$$

(4.2.28)

Next, the nonlinear differential equation in Equation (12) is isolated.

$$A_0 = -R_0(3\xi^2 - 5\xi + 1)(6\xi - 5)$$

(4.2.29)

$$B_0 = R_0(\xi^2 - 2\xi)(6\xi - 5)$$

(4.2.30)

$$C_0 = -M(6\xi - 5)$$

(4.2.31)

$$A_1 = (-R_0(6\xi - 5) - 6R_0(3\xi^2 - 5\xi + 1)f_1(\xi)$$

(4.2.32)

$$B_1 = \left(R_0(3\xi^2 - 5\xi + 1)(6\xi - 5) + 6R_0(\xi^3 - 3\xi + \xi)\right)f_1(\xi)$$

(4.2.33)

$$C_1 = -6Mf_1(\xi)$$

(4.2.34)

For equation (13):

$$D_0 = -\tau cv R_0(\xi^3 - 2\xi + \xi)$$

(4.2.35)

$$D_1 = \tau cv R_0(3\xi^2 - 4\xi + 1)h_1(\xi)$$

(4.2.36)

For equation (14):

$$E_0 = -R_0 pr(\xi^3 - 2\xi^2 + \xi)$$

(4.2.37)

$$E_1 = R_0 pr(3\xi^2 - 4\xi + 1)\psi_1(\xi)$$

(4.2.38)

$$E_1 = -R_0 pr(3\xi^2 - 4\xi + 1)\psi_1(\xi)$$

(4.2.39)

According to the ADM strategy, the following values of the parameters in the equation were assumed: τcv=2, α=0.5, M=1, Er=1, Pr=20, η=1, and $R0$=0.1. The ADM method solution yielded the following functions:

$$f(\xi) = (0.01164396181)\xi^{12} - (0.04065039409)\xi^{11} + (0.11604434655)$$

$$\xi^{10} - (0.1317874340)\xi^{9} + (0.06445920287)\xi^{8} + (0.00145530639)$$

$$\xi^{7} - (0.04562227080)\xi^{6} + (0.1870027952)\xi^{5} - (0.1540473657)$$

$$\xi^{4} + (0.5911555227)\xi^{3} - (0.5439795898)\xi^{2},$$

(4.2.40)

$$f'(\xi) = (0.11142795417)\xi^{11} - (0.5552943350)\xi^{10} + (1.150642655)$$

$$\xi^{9} - (1.116086906)\xi^{8} + (0.6956736230)\xi^{7} + (0.05318714465)$$

$$\xi^{6} - (0.3637336248)\xi^{5} + (0.6650138760)\xi^{4} - (0.641894628)$$

$$\xi^{3} + (1.763666568)\xi^{2} - (0.9379591796\xi$$

(4.2.41)

In this section, our goal is to verify the precision of the ADM and FEM techniques by comparing them with the HPM, which was utilized for validation. Additionally, they will be compared with the corresponding HPM procedures.

Specifically, Tables 2-5 and Figures 13 to 16 will show how accurate the Adomian decomposition method is. By looking at graphs 2 and 3, it can be seen that the convergence of the three modes is precisely aligned, and there is no calculation error with minimal error. In simpler terms, when certain values like τ_{cv}, Pr, α, M,

Er, and η change, it will also affect the Reynolds number R_0. When the Reynolds number changes, it will independently impact the arrangement of spiral velocity, pivotal velocity, transverse velocity, and temperature profile. However, based on the four graphs in Figure 17, only the temperature profile will experience significant changes as the Reynolds number varies. The variations in other parameters, such as radial, axial, and transverse, are not significant. However, when the values of R_0, P, α, M, Er, and η change, it affects the cross-viscosity parameter τ_{cv}. Consequently, this leads to changes in the arrangements of outspread velocity, pivotal velocity, transverse velocity, and temperature profile in the ADM. By closely examining picture number (17), It can be seen that the heat transfer decreases as the Reynolds number increases. The four graphs in Figure 18 illustrate that there are no notable variations in radial velocity, axial velocity, transverse velocity, and temperature section when there is a change in the cross-viscosity variable. However, when the dimensionless forced parameter changes, specifically with $R_0= 0.2$, $Pr=10$, $\tau cv=2$, $M=2$, $Er=2$, $\eta=1$ as parameters, it will impact the ADM and FEM solutions for the radial and longitudinal velocity axes, lateral velocity, and the corresponding temperature profile. Upon further examination of the picture number (18), it becomes evident that as the stress tensor increases, the heat exchange rate peaks and the gradient of transverse velocity between different disks decreases.

Additionally, the four graphs in Figure 19 demonstrate that variations in the dimensionless forcing parameter substantially impact the axial velocity and temperature profile. When the parameters such as R_0, Pr, τcv, α, Er, and η change, it affects the magnetism parameter M. These changes then have an impact on the arrangements of spiral velocity, pivotal velocity, transverse velocity, and temperature profile in both the ADM and FEM. Figure 19 provides insights into how the dimensionless force parameter affects these arrangements. According to the results, as the dimensionless force parameter increases, the displacement of fluid velocity decreases, and the collision of particles also decreases. In simpler terms, the temperature gradient around the disks shows an opposite trend, causing more heat to be transferred between fluid particles compared to earlier. This is depicted in four charts, as opposed to Figure 20. It appears that when the Magnetic system changes, there are significant variations in the behavior of radial velocity, axial velocity, transverse velocity, and temperature profile.

This research paper introduces a novel technique called the Adomian Decomposition Method (ADM) along with Finite Element Methods. These methods are used to analyze the behavior of forced Reiner-Rivlin non-Newtonian Magnetohydrodynamic (MHD) fluid flow that is confined between two disks. Combining these techniques allows for a comprehensive study of the fluid motion in this specific scenario. This paper introduces a unique approach by combining both analytical and numerical methods, specifically the Adomian Decomposition Method (ADM) and Finite Element Method (FEM). The researchers employed this hybrid approach to solve a set of coupled linear differential equations. The advantage of this method is that it simplifies these equations, allowing for the calculation and analysis of parameters such as heat transfer and fluid velocity between the two disks.

Figure 11. Show of two disks

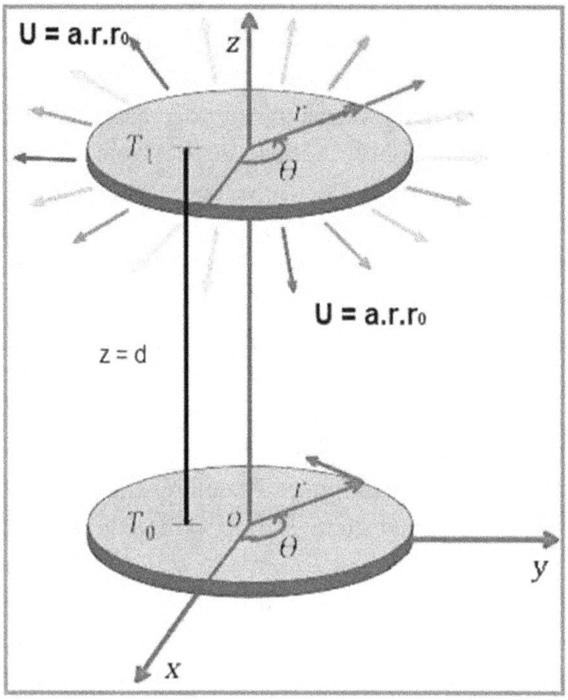

Figure 12. ADM techniques flow chart

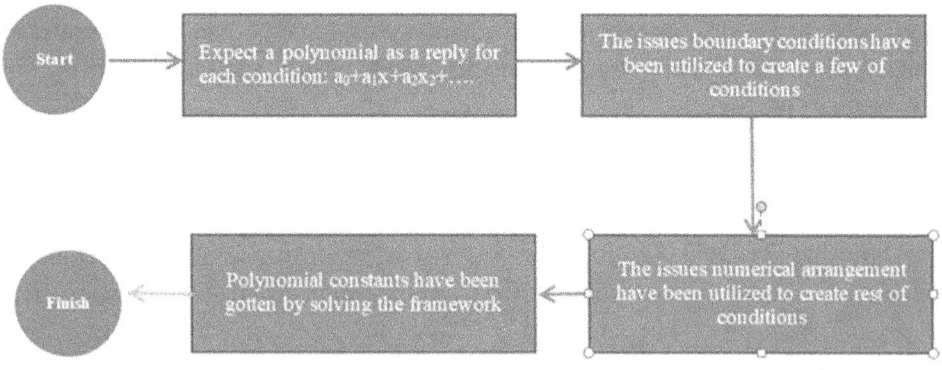

Figure 13. A comparison of the convergence process of the three mentioned methods

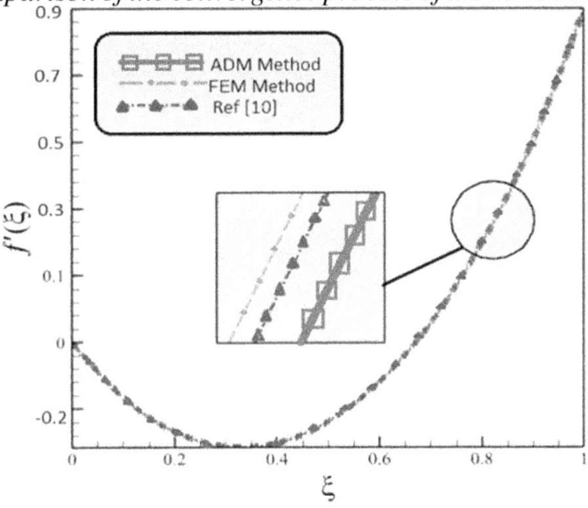

Figure 14. A comparison of the convergence process of the three mentioned methods for velocity

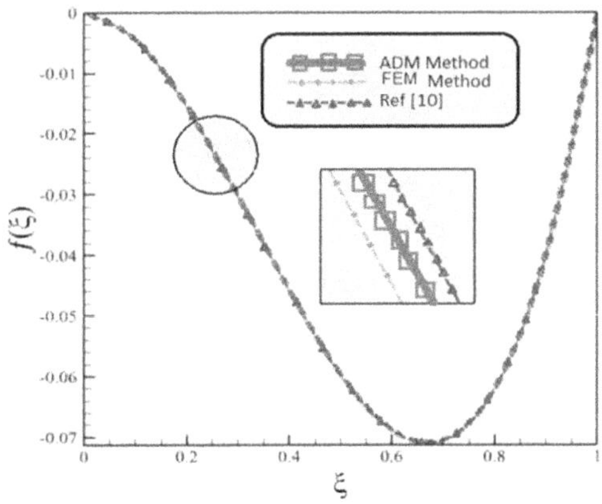

Figure 15. A comparison of the convergence process of the three mentioned methods for h (ε)

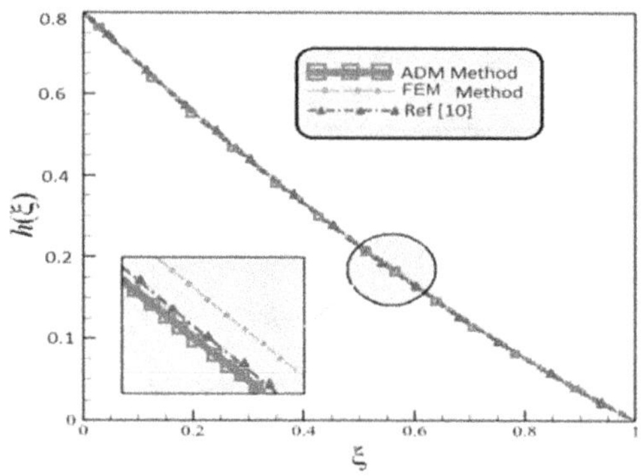

Figure 16. A comparison of the convergence process of the three mentioned methods for T (ε)*

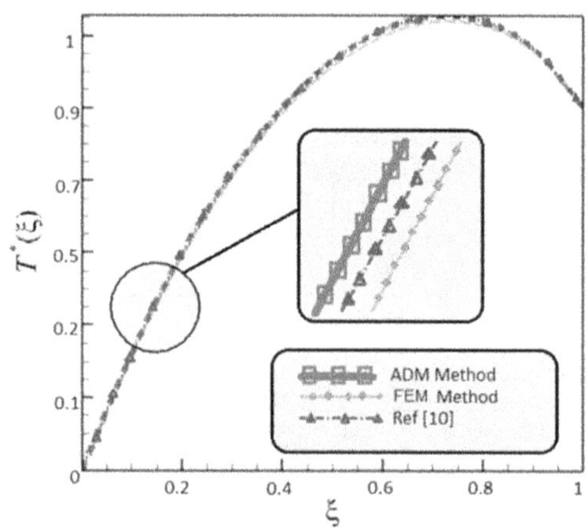

Figure 17. The effect of Reynolds number, R0, on f (ξ), (ξ), h(ξ) and T (ξ), when* τ_{ev} *= 2, Pr = 10, α = 0.6, M = 2, Er = 2, η = 1*

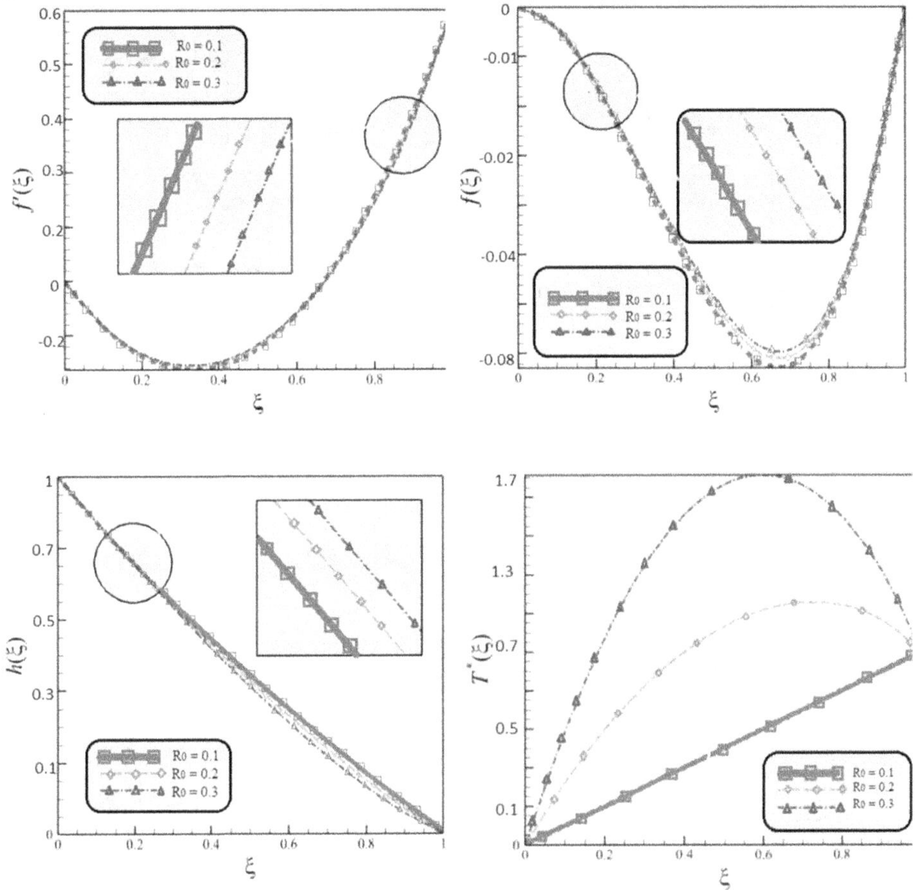

Figure 18. The effect of the cross-viscosity parameter, τcv, on f (ξ), (ξ), h (ξ) and T (ξ), when R_n = 0.2, Pr = 10, α = 0.6, M = 2, Er = 2, η = 1*

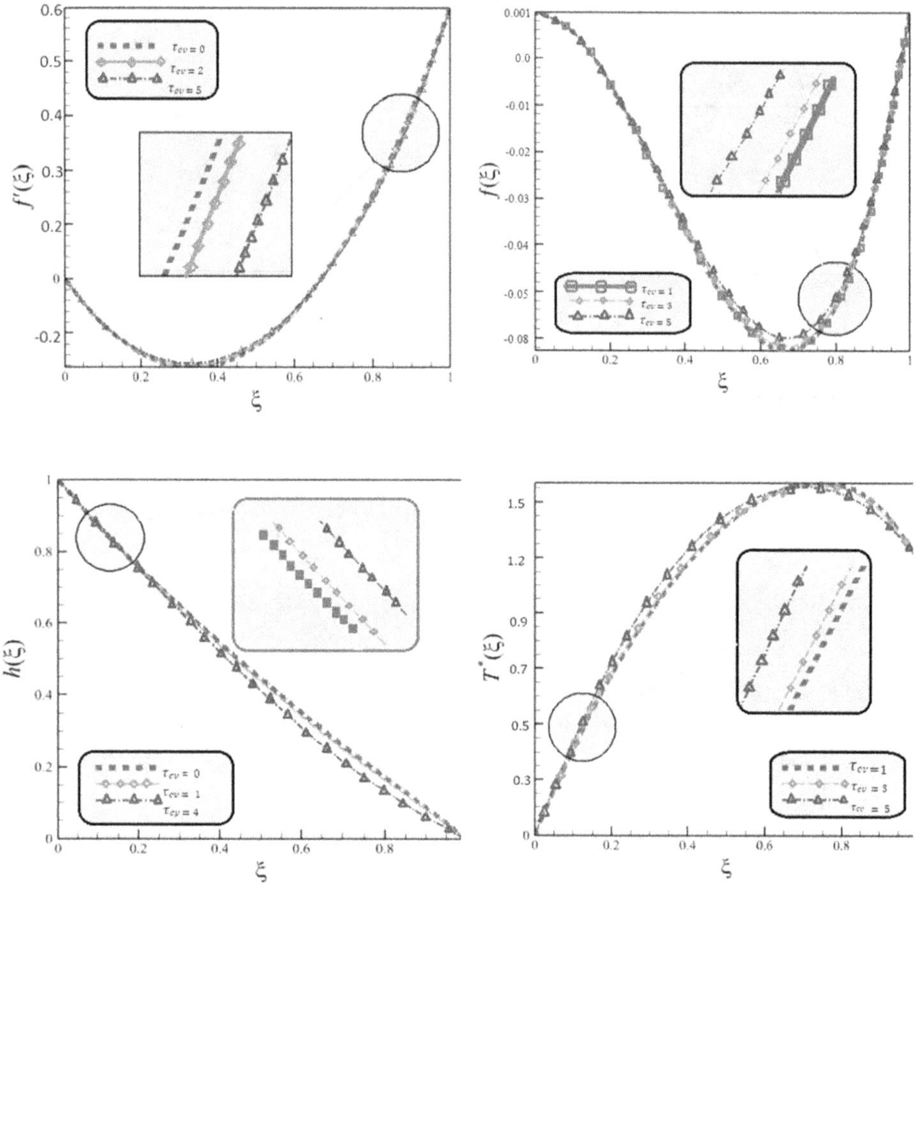

Figure 19. The effect of the dimensionless forced parameter, α, on f (ξ), (ξ), h(ξ), and T (ξ), when R_{n} = 0.2, Pr = 10, τ_{en} = 2, M = 2, Er = 2, η = 1*

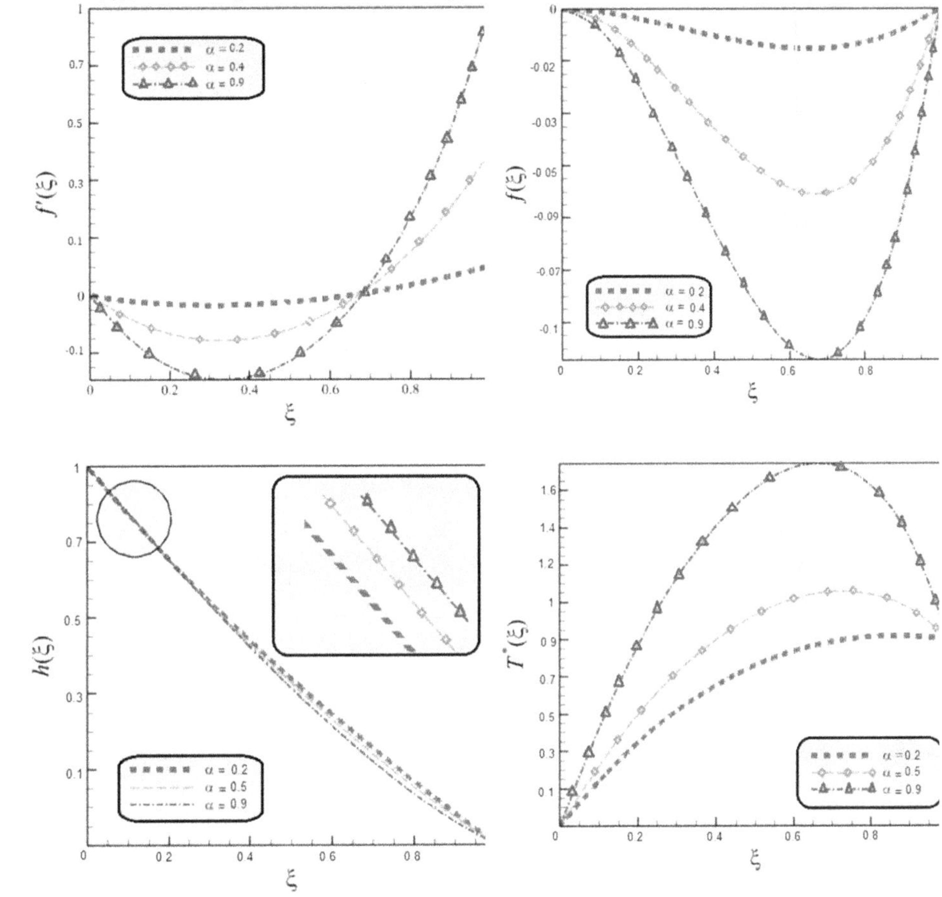

Figure 20. The effect of the magnetic variable, on f (ξ), (ξ), h(ξ) and T (ξ), when R0 = 0.2, Pr = 10, τ_{cn} = 2, M = 2, Er = 2, η = 1*

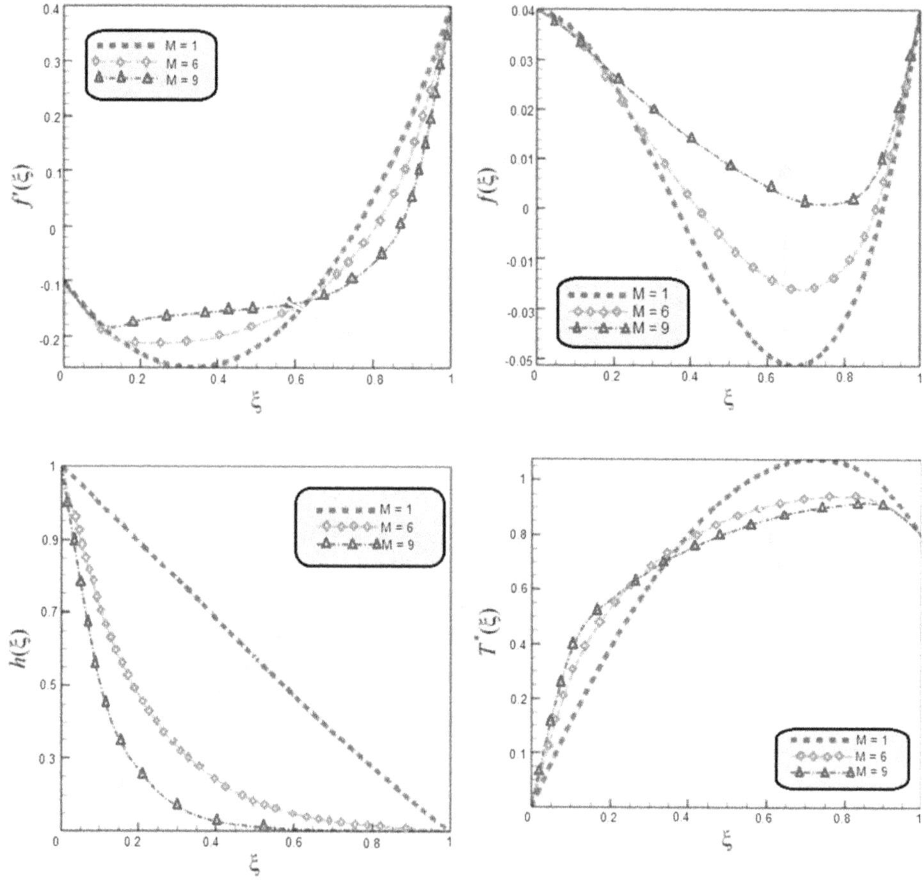

Table 2. Comparison of f (ξ) conclusions

ξ	ADM method	Ref	FEM method
0	0	0	0
0.1	-0.062262	-0.062261	-0.06227
0.2	-0.1636412	-0.1636532	-0.1636598
0.4	-0.191444	-0.191433	-0.191467
0.6	-0.039945	-0.0.03477	-0.0.03830
0.8	0.178021	0.178089	0.178670
1	0. 600000	0.6000000	0.6000000

Table 3. Comparison of f (ξ) conclusions

ξ	ADM method	Ref	FEM method
0	0	0	0
0.1	-0.002393	-0.002491	-0.002452
0.2	-0.025454	-0.0255454	-0.025454
0.4	-0.171417	-0.171457	-0.171457
0.6	-0.055667	-0.056667	-0.055732
0.8	-0.082112	-0.082244	-0.082031
1	0	0	0

Table 4. Comparison of h (ξ) conclusions

ξ	ADM method	Ref	FEM method
0	1	1	1
0.1	0.885088	0.886053	0.895083
0.2	0.735349	0.737141	0.735789
0.4	0.561990	0.564862	0.561901
0.6	0.350558	0.353458	0.350515
0.8	0.172401	0.173476	0.172407
1	0	0	0

Table 5. Comparison of T (ξ) conclusions*

ξ	ADM method	Ref	FEM method
0	0	0	0
0.1	0.313002	0.313470	0.329091
0.2	0.585394	0.585324	0.592344
0.4	0.979771	0.973621	0.984661
0.6	1.396277	1.386877	1.380912
0.8	1.271326	1.271344	1.281300
1	1	1	1

REFERENCES

Abdel-Aziz, M. H. (2023). One-dimensional ternary conducting polymers blend with 9.26% power conversion efficiency for photovoltaic devices applications. *Alexandria Engineering Journal*, 66, 475–488. DOI: 10.1016/j.aej.2022.11.013

Abdollahi, S. A., Alizadeh, A., Esfahani, C., Zarinfar, M., & Pasha, P. (2023). Investigating heat transfer and fluid flow betwixt parallel surfaces under the influence of hybrid nanofluid suction and injection with numerical analytical technique. *Alexandria Engineering Journal*, 70, 423–439. DOI: 10.1016/j.aej.2023.02.040

Al-Turki, Y. A. (2021). Numerical investigation of nanofluid flow and heat transfer in a pillow plate heat exchanger using a two-phase model: Effects of the shape of the welding points used in the pillow plate. *ZAMM-Journal of Applied Mathematics and Mechanics. Zeitschrift für Angewandte Mathematik und Mechanik.*

Ashraf, H., Siddiqui, A. M., Rana, M. A., & Gawo, G. A. (2021). Flow assessment of the shear rate dependent viscoelastic fluid: Application of biomechanics in growing human embryo transport. *Alexandria Engineering Journal*, 60(6), 5921–5934. DOI: 10.1016/j.aej.2021.04.055

Barnoon, P., Toghraie, D., Mehmandoust, B., Fazilati, M. A., & Eftekhari, S. A. (2022). Natural-forced cooling and Monte-Carlo multi-objective optimization of mechanical and thermal characteristics of a bipolar disk for use in a proton exchange membrane fuel cell. *Energy Reports*, 8, 2747–2761. DOI: 10.1016/j.egyr.2022.01.199

Berkan, S., Hosseini, S. R., Jalili, B., Ranjbar, A. A., & Ganji, D. D. (2024). A study on peristaltic flow and nanofluid in medication delivery systems considering heat transfer. *Discover Applied Sciences*, 6(4), 173. DOI: 10.1007/s42452-024-05847-9

Cheng, H.. (2023). The effect of temperature and external force on the thermal behavior of oil-based refrigerant inside a nanochannel using molecular dynamics simulation. *Journal of Molecular Liquids*, 369, 120893. DOI: 10.1016/j.molliq.2022.120893

Dehghan Afifi, M., Jalili, B., Mirzaei, A., Jalili, P., & Ganji, D. (2024). The effects of thermal radiation, thermal conductivity, and variable viscosity on ferrofluid in porous medium under magnetic field. *World Journal of Engineering*. Advance online publication. DOI: 10.1108/WJE-09-2023-0402

Dey, D. (2018). Viscoelastic fluid flow through an annulus with relaxation, retardation effects, and external heat source/sink. *Alexandria Engineering Journal*, 57(2), 995–1001. DOI: 10.1016/j.aej.2017.01.039

Fathollahi, R. (2022). Analyzing the effect of radiation on the unsteady 2D MHD Al2O3-water flow through parallel squeezing disks by AGM and HPM. *Alexandria Engineering Journal.*

Hafez, N. M., Abd-Alla, A. M., & Metwaly, T. M. N. (2023). Influences of rotation and mass and heat transfer on MHD peristaltic transport of casson fluid through inclined plane. *Alexandria Engineering Journal*, 68, 665–692. DOI: 10.1016/j.aej.2023.01.038

Hashemi Kachapi, S. H., & Domairry Ganji, D. (2011). Analysis of nonlinear equations in fluids. In Domairry Ganji, D. (Ed.), *Progress in Nonlinear Science* (Vol. 2).

Jafaripournimchahi, A., Shateri, A., Jalili, B., Jalili, P., & Ganji, D. D. (2024). The effects of magnetic field and thermal radiation on the mixed convection of Al2O3-Cu/water hybrid nanofluid over a permeable vertical flat plate. *Modern Physics Letters B*, 38(25), 2450242. DOI: 10.1142/S0217984924502427

Jalili, A., Sadinezhad, B., Fard, M., Khan, Y., Jalili, P., & Ganji, D. D. (2024). A novel technique to analyze the fractional model of Williamson and Casson non-Newtonian boundary layer flow. *Multidiscipline Modeling in Materials and Structures*, 20(2), 279–294. DOI: 10.1108/MMMS-09-2023-0309

Jalili, B., Emad, M., Malekshah, E. H., Jalili, P., Akgül, A., & Hassani, M. K. (2024). Investigating double-diffusive natural convection in a sloped dual-layered homogenous porous-fluid square cavity. *Scientific Reports*, 14(1), 7193. DOI: 10.1038/s41598-024-57395-2 PMID: 38531996

Jalili, B., Roshani, H., Jalili, P., Jalili, M., Pasha, P., & Ganji, D. D. (2023). The magnetohydrodynamic flow of viscous fluid and heat transfer examination between permeable disks by AGM and FEM. *Case Studies in Thermal Engineering*, 45, 102961. DOI: 10.1016/j.csite.2023.102961

Jalili, P., Alamdari, S. G., Jalili, B., Shateri, A., & Ganji, D. D. (2024). Analytical and numerical investigation of heat transfer of porous fin in a local thermal non-equilibrium state. *Heliyon*, 10(4). Advance online publication. DOI: 10.1016/j.heliyon.2024.e26424 PMID: 38420422

Jalili, P., Sharif Mousavi, S. M., Jalili, B., Pasha, P., & Ganji, D. D. (2024). Thermal evaluation of MHD Jeffrey fluid flow in the presence of a heat source and chemical reaction. *International Journal of Modern Physics B*, 38(08), 2450113. DOI: 10.1142/S0217979224501133

Joens, M. A., & Swan, J. W. (2022). Unsteady and lineal translation of a sphere through a viscoelastic fluid. *Physical Review Fluids*, 7(1), 013301. DOI: 10.1103/PhysRevFluids.7.013301

Kumar, V., Mukherjee, J., Sinha, S. K., & Ghosh, U. (2022). Combined electromechanically driven pulsating flow of nonlinear viscoelastic fluids in narrow confinements. *Journal of the Royal Society, Interface*, 19(189), 20210876. DOI: 10.1098/rsif.2021.0876 PMID: 35382577

Larestani, A., Aydin, M. J., & Mohamad, A. A. (2022). Predicting the surfactant-polymer flooding performance in chemical enhanced oil recovery: Cascade neural network and gradient boosting decision tree. *Alexandria Engineering Journal*, 61(10), 7715–7731. DOI: 10.1016/j.aej.2022.01.023

Larestani, A., Mousavi, S. P., Hadavimoghaddam, F., Ostadhassan, M., & Hemmati-Sarapardeh, A. (2022). Predicting the surfactant-polymer flooding performance in chemical enhanced oil recovery: Cascade neural network and gradient boosting decision tree. *Alexandria Engineering Journal*, 61(10), 7715–7731. DOI: 10.1016/j.aej.2022.01.023

Li, Y. M., Sedeh, S. N., Toghraie, D., & Alizadeh, A. (2021). Computational hemodynamics and thermal analysis of laminar blood flow for different types of hypertension. *Mathematics and Computers in Simulation*, 188, 330–341. DOI: 10.1016/j.matcom.2021.04.011

Lou, Q., Ali, B., Rehman, S. U., Habib, D., Abdal, S., Shah, N. A., & Chung, J. D. (2022). Micropolar dusty fluid: Coriolis force effects on dynamics of MHD rotating fluid when Lorentz force is significant. *Mathematics*, 10(15), 2630. DOI: 10.3390/math10152630

Moatimid, G. M., Zekry, M. H., & Ibrahim, D. A. (2023). Nonlinear EHD instability of two viscoelastic fluids under the influence of mass and heat transfer. *Scientific Reports*, 13(1), 357. DOI: 10.1038/s41598-023-27410-z PMID: 36611060

Mozafarifard, M., Azimi, A., Sobhani, H., Smaisim, G. F., Toghraie, D., & Rahmani, M. (2022). Numerical study of anomalous heat conduction in absorber disk of a solar collector using time-fractional single-phase-lag model. *Case Studies in Thermal Engineering*, 34, 102071. DOI: 10.1016/j.csite.2022.102071

Nadeem, S., Fuzhang, W., Alharbi, F. M., Sajid, F., Abbas, N., El-Shafay, A. S., & Al-Mubaddel, F. S. (2022). Numerical computations for Buongiorno nano fluid model on the boundary layer flow of viscoelastic fluid towards a nonlinear stretching sheet. *Alexandria Engineering Journal*, 61(2), 1769–1778. DOI: 10.1016/j.aej.2021.11.013

Noreen, S., Waheed, S., Lu, D., & Hussanan, A. (2020). Heat measures in performance of electro-osmotic flow of Williamson fluid in micro-channel. *Alexandria Engineering Journal*, 59(6), 4081–4100. DOI: 10.1016/j.aej.2020.07.013

Shadman, P., Parhizi, Z., Fathollahi, R., Zarinfar, M., Yu, A. E., & Pasha, P. (2023). Combined septum and chamfer fins on threaded stretching surface under the influence of nanofluid and the magnetic parameters for rotary seals in computer hardware. *Alexandria Engineering Journal*, 62, 489–507. DOI: 10.1016/j.aej.2022.07.044

Shah, S. A. A., Ahammad, N. A., Din, E. S. M. T. E., Gamaoun, F., Awan, A. U., & Ali, B. (2022). Bio-convection effects on prandtl hybrid nanofluid flow with chemical reaction and motile microorganism over a stretching disk. *Nanomaterials (Basel, Switzerland)*, 12(13), 2174. DOI: 10.3390/nano12132174 PMID: 35808009

Sheykhi, M., Eskandari, A., Ghafari, D., Ahmadi Arpanahi, R., Mohammadi, B., & Hosseini Hashemi, S. (2023). Investigation of fluid viscosity and density on vibration of nano beam submerged in fluid considering nonlocal elasticity theory. *Alexandria Engineering Journal*, 65, 607–614. DOI: 10.1016/j.aej.2022.10.016

Shoukat, Z., Hashir Zubair, M., Farman, M., Akgül, A., Sultan, M., Safarovich Sharipov, S., Botmart, T., Yahia, I. S., & Algarni, H. (2023). Impacts of joule heating with Cattaeno-Christove heat flux model in a MHD flow of Erying-Powell fluid on a Riga plate. *Alexandria Engineering Journal*, 64, 741–748. DOI: 10.1016/j.aej.2022.10.067

Zhang, A., Wang, Y., & Wang, H. (2023). Preparation of inorganic-polymer nanoemulsion inhibitor for corrosion resistance of steel reinforcement for concrete. *Alexandria Engineering Journal*, 66, 537–542. DOI: 10.1016/j.aej.2022.11.020

Section 3
Unsteady and Specialized MHD Flow

This section explores the unsteady behavior of MHD fluid flow and specialized topics such as magneto ferrofluid flow.

Chapter 5
Unsteady MHD Fluid Flow

ABSTRACT

This chapter introduces a thermal and flow analysis of an unsteady compressible nanofluid flow and heat transfer utilizing a nanofluid founded on the Brinkmann model in the presence of a varying magnetic field. An analytical approach is used to solve the problem's nonlinear differential equations. Nanofluid viscosity and effective thermal conductivity are calculated using the Brinkmann correlation. Nusselt number and skin friction coefficient are provided. Impacts of active parameters such as Hartmann number, pressure number, and heat source parameter are discussed. The results for temperature distribution and velocity profile, Nusselt number, and skin friction coefficient are presented employing the Galerkin technique.

5.1. TWO-DIMENSIONAL HEAT TRANSFER IN UNSTEADY SQUISH FLOWS

The matter to be considered is two-dimensional heat transfer in unsteady squish flows of incompressible nanofluids betwixt infinite, equal plates, as shown in Figure 1. The distance between the plates is 2. l (1 − αt) 1/ 2. Suppose α > 0, the two plates are pushed together until they touch at t = 1 / α. Two plates are separated. Suppose a time-varying magnetic field (B = Be, B = B_0 (1 - αt)) is applied in the y direction. As seen in Figure 1, a heat source (Q = Q0/ 1 - αt) is used between the two plates. Also, the electromagnetic force and current are defined by F and J, respectively. In this headache, the fluid is a water-based nanofluid containing copper oxide (CuO) nanoparticles. A nanofluid is a binary mixture with the following assumptions: no chemical reaction, negligible radiative heat transfer; the nano solid particles and the base fluid are in thermal equilibrium with no slippage betwixt them.

The governing equations of momentum and energy in the unsteady two-dimensional flow of nanofluids are (Alizadeh, Zekri, & Jafarmadar, 2020) (Niknejadi *et al.*, 2023) (Zhang *et al.*, 2023) (Li *et al.*, 2023) (Jalili et al., 2024) (Bai et al., 2022)

DOI: 10.4018/979-8-3693-5335-6.ch005

(Hai, Kadir, & Ghanbari, 2023) (Bahmani *et al.*, 2024) (Zhou, Talebizadehsardari, & Goharkhah, 2023) (Bazmi, Goharkhah, & Talebizadehsardari, 2019) (Qasemian, Goharkhah, & Goharkhah, 2021) (Salehi *et al.*, 2020) (Jalili *et al.*, 2024) (Zangooee, Hosseinzadeh, & Ganji, 2019) (Hussain & Xu, 2022) (Jalili *et al.*, 2024) (Anwar & Kumam, 2022) (Farahani *et al.*, 2021) (Kodi & Konduru, 2022):

$$u_x + v_y = 0, \tag{5.1.1}$$

$$\rho_{nf}(u_t + u u_x + v u_y) + P_x - \mu_{nf}(u_{xx} + u_{yy}) + \sigma_{nf} B_0^2 u = 0, \tag{5.1.2}$$

$$\rho_{nf}(v_t + u v_x + v v_y) + P_y - \mu_{nf}(v_{xx} + v_{yy}) \quad , \tag{5.1.3}$$

$$T_t + u T_x + v T_y - \frac{k_{nf}}{(\rho C_p)_{nf}}(T_{xx} + T_{yy}) - \frac{Q}{(\rho C_p)_{nf}} T = 0, \tag{5.1.4}$$

Pis the pressure, effective density ρ_{nf} and the effective heat capacity $(\rho C_p)_{nf}$ of the nanofluid, the effective electrical conductivity of nanofluid (σ_{nf}), and the effective thermal conductivity (k_{nf}) of the nanofluid are defined as:

$$\rho_{nf} = (1 - \varnothing)\rho_f + \varnothing \rho_s, \tag{5.1.5}$$

$$\mu_{nf} = \frac{\mu_f}{(1 - \varnothing)^{2.5}}, \tag{5.1.6}$$

$$(\rho c_p)_{nf} - (1 - \varnothing)(\rho c_p)_f - (\rho c_p)_s = 0, \tag{5.1.7}$$

$$\frac{\sigma_{nf}}{\sigma_f} - 1 - \frac{3\left(\frac{\sigma_s}{\sigma_f} - 1\right)\varnothing}{\left(\frac{\sigma_s}{\sigma_f} + 2\right) - \left(\frac{\sigma_s}{\sigma_f} - 1\right)\varnothing} = 0,$$

(5.1.8)

$$\frac{k_{nf}}{k_f} - \frac{k_s + 2k_f - 2\varnothing\,(k_f - k_s)}{k_s + 2k_f + 2\varnothing\,(k_f - k_s)} = 0,$$

(5.1.9)

The relevant boundary conditions are:

$$v = v_w = \frac{dh}{dt}, T = T_H \; at\, y = h(t), v = \frac{du}{dy} = \frac{dT}{dy} = 0 \; at\, y = 0\; v = v_w = \frac{dh}{dt},$$

$$T = T_H \; at\, y = h(t), v = \frac{du}{dy} = \frac{dT}{dy} = 0 \; at\, y = 0$$

(5.1.10)

The coming dimensionless group is introduced:

$$\eta = \frac{y}{1(1 - \alpha t)^{0.5}}, u = \frac{\alpha x}{2(1 - \alpha t)}f(\eta), v = -\frac{\alpha l}{2(1 - \alpha t)^{0.5}}, \theta = \frac{T}{T_H}$$

(5.1.11)

Use dimensionless parameters and substitute expressions. (14) Removing the pressure gradient from (2)-(4) in Eq. and the resulting equation yields the ordinary derivative equation:

$$f^{iv} - s\left(\frac{A_1}{A^4}\right)(\eta f''' + 3f'' + ff'' - ff''') - Ha^2\left(\frac{A_s}{A_4}\right)f'' = 0,$$

(5.1.12)

$$\theta'' + prs\left(\frac{A_2}{A_1}\right)(f\theta' - \eta\theta') + \frac{Hs}{A_3}\theta = 0,$$

(5.1.13)

And the boundary conditions (10) become:

$$f(0) = 0, f''(0) = 0, f(1) = 1, f'(1) = 0, \theta'(0) = 0, \theta(1) = 1$$

(5.1.14)

An important consideration is to arrange some background ability of the mathematical approach used. Therefore, this section introduces some basic connections and theories about the Galerkin method (GM). The many advantages of WRM, especially GM, compared to other analytical methods make them more important and motivate researchers to use them to solve problems. Some of these benefits are indexed below.

a) WRM solves the equations directly, so no simplifications are needed.
b) No perturbation, linearization, or small parameters are required compared to homotopy perturbation (HPM) and parametric perturbation (PPM).
c) Compared to the homotopy analysis method (HAM), there is no need to determine ancillary parameters and ancillary functions.

It is assumed that the differential operator D is applied to the function u to obtain the function p.

$D\ (u\ (x)) = p(x)$ (5.1.15)

u is approximated by a function $\tilde{u}$, which is a linear combination of basic functions chosen from a linearly in dependent set. That is,

$$u = \tilde{u} = \sum c_i \varphi_i$$

(5.1.16)

The most popular thought of the Galerkin strategy is to constrain the leftover to zero in a few normal senses over space. That's

$\int x R(x) W_i(x)=0,\ i=1,2,\ldots,n$ (5.1.17)

Where the number of weight functions Wi is exactly equal to the number of unknown constants ci in $\tilde{u}$ function. The result is a set of n algebraic equations for the unknown constants c_i.

$$W_i = \frac{\partial u}{\partial g_i},\ i = 1,2,\ldots n$$

(5.1.18)

Presently GM is connected to the Eqs. (5.1.12) and (5.1.13). To begin with, consider the trial work as:

$$f(\eta) = \eta^3 + c_1(\eta^3 - \eta) + c_2(\eta^3 - \eta^5), \tag{5.1.19}$$

$$\theta(\eta) = 1 + c_4(1 - \eta^3) + c_5(1 - \eta^2), \tag{5.1.20}$$

Which fulfills the beginning condition in Eqs. (5.1.12) and (5.1.13). Utilizing Eq. (5.1.22), weight capacities will be gotten as:

$$W_1 = \eta^3 - \eta,\ W_2 = \eta^3 - \eta^5, W_3 = \eta^3 - \eta^7 \tag{5.1.21}$$

$$W_4 = 1 - \eta^3, W_5 = 1 - \eta^5, W_6 = 1 - \eta^6, \tag{5.1.22}$$

Applying Eq. (5.1.21), a set of algebraic equations is defined that show in Appendix . Finally f(η) and θ(η) functions for CuO–water nanofluid when d p = 29 nm and φ= 0.04 will be calculated as following:

$$f(\eta) = -0.35\eta^3 + 1.4\eta - 0.06\eta^5, \tag{5.1.23}$$

$$\theta(\eta) = 0.76 + 0.44\eta^3 - 0.305\eta^5, \tag{5.1.24}$$

In this consideration, after examining the heat exchange within the shaky crushing stream of an incompressible nanofluid between the unbounded parallel plates. The Brinkman relationship must calculate the thickness of nanofluid and successful heat conductivity. Galerkin Strategy (GM) is utilized to unravel this issue. A comparison of the current work with that of Mustafa et al. appears in Figure 2 and is a great assertion. Figure 3 shows two nanofluids (CuO and Al_2O_3) comparing Nusselt numbers and skin grinding coefficients. As can be seen from this figure, the Nusselt number and skin contact coefficient values of CuO are prevalent to those of Al_2O_3. The development of all the taking after figures 4 to (5.1.10), showing the impacts of a few vital parameters on the velocity profile, temperature dissemination, Nusselt number, and skin grinding coefficient, is based on CuO water. Figure 4 illustrates the impact of volume division (ϕ) of CuO–water nanofluid on Nusselt number and skin contact co-proficiency. This figure concurs with nanofluid volume division in wrinkles, Nusselt number increments, and skin grinding coefficient de-wrinkles. Also, Figure 5 affirms the impact of volume division and Hartman number on the Nusselt number and skin contact coefficient. The impact of the Hartmann number (Ha) on velocity and temperature profiles appears in Figure 6. As seen in this figure, an increase in the Hartmann number results in an increase in velocity and temperature profiles. Figure 7 outlines the impact of the crush number (S) on velocity and temperature profiles. It is very self-evident that when two plates move together, the heat boundary layer thickens and increments as the supreme greatness of the press number upgrades. The temperature is generally high when the plates are moving towards each other. Moreover, an increase in crush number can be related to the diminish within the velocity.

In this work, a heat and stream investigation of an insecure pressing nanofluid stream and heat exchange utilizing nanofluid based on Brinkman show the nearness of variable attractive field. Galerkin's strategy (GM) is connected to fathom the nonlinear differential conditions overseeing the issue. As seen in the results, the values of Nusselt number and skin contact coefficient for CuO are superior to Al_2O_3's. Moreover, according to figures, as the nanofluid volume division increases, the Nusselt number increments, and the skin friction coefficient diminishes. An increment within the Hartman number comes about in an increment in velocity and temperature profiles, and an increment in the press number can be associated with the diminishment within the velocity.

Figure 1. Geometry of the problem

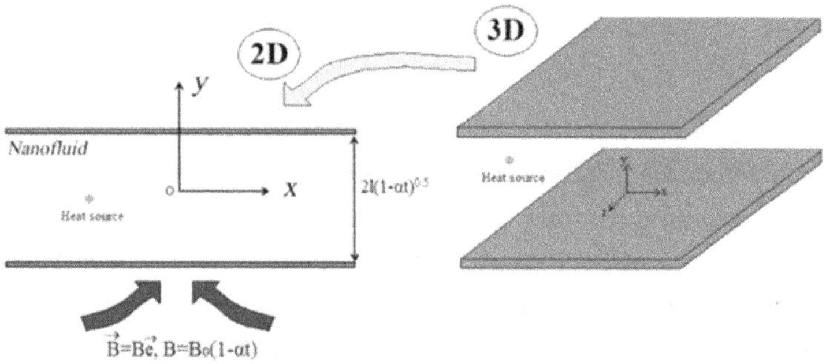

Figure 2. Comparison between present work and Mustafa et al. CuO–water nanofluid for (a) velocity profile, (b) temperate profile when ϕ= 0.04, d p = 29 nm, Ha = 1, S = 1 and Hs = −1

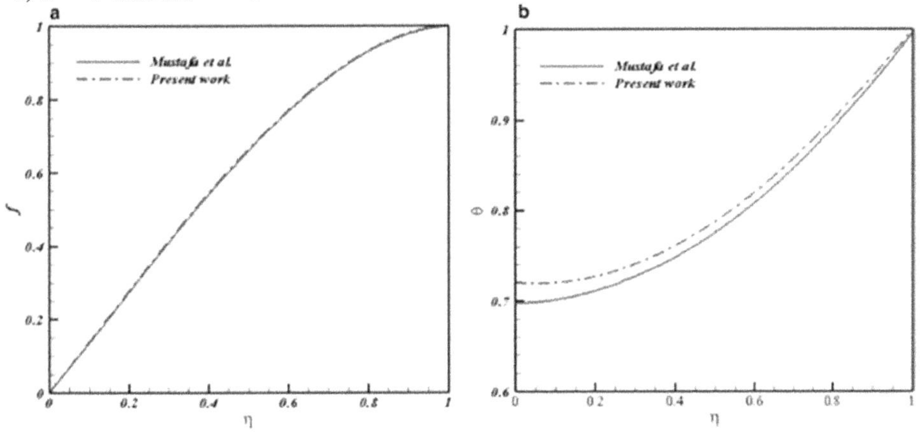

Figure 3. Comparison between CuO–water and Al 2 O 3 –water nanofluids (a) Nusselt number, (b) skin friction coefficient when ϕ= 0.04, d p = 29 nm, Ha = 1, S = 1 and Hs = −1

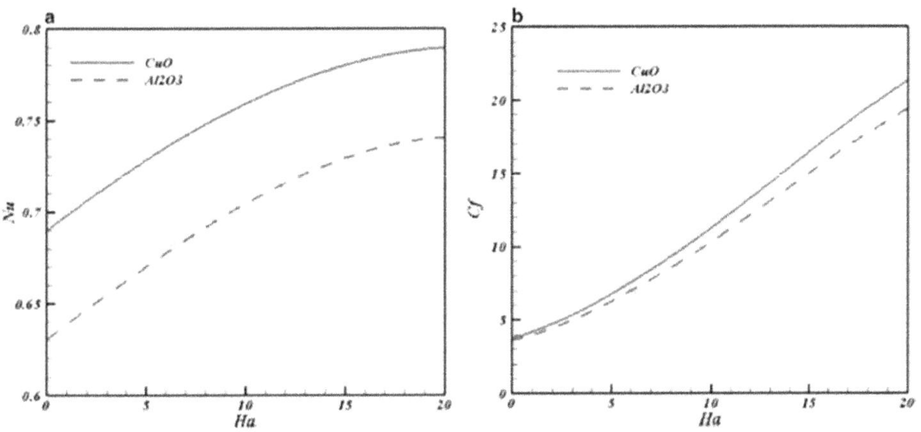

Figure 4. Effect of volume fraction of CuO–water nanofluid on (a) Nusselt number, (b) skin friction coefficient when Pr = 6.2, Hs = −1, S = 1

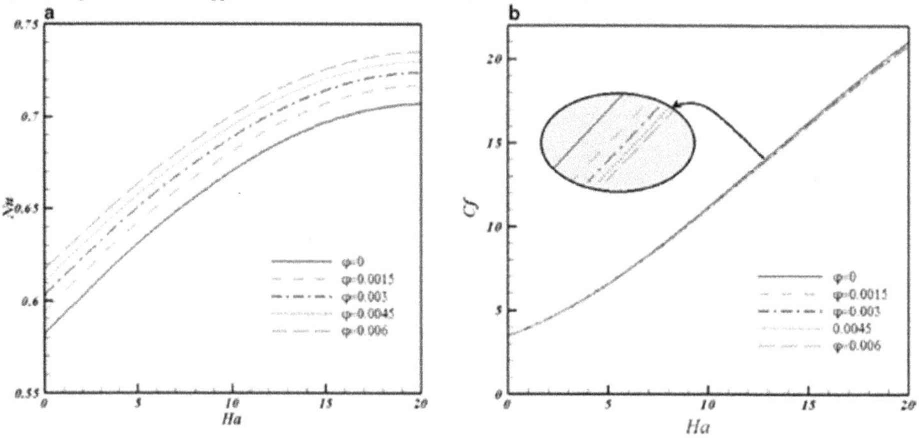

Figure 5. Contours for effects of volume fraction and Hartman number on (a) Nusselt number, (b) skin friction coefficient when S = 1, Hs = −1 and d p = 29 nm (CuO–water)

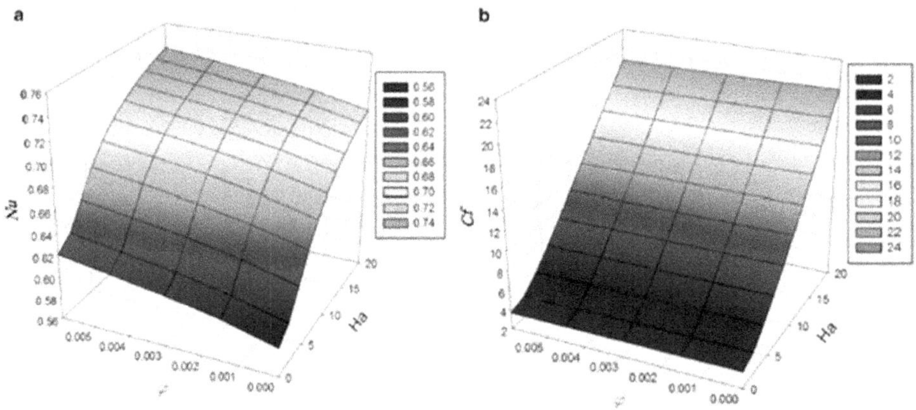

Figure 6. Effect of the Hartman number on velocity and temperature profiles when Pr = 6.2, Hs = −1, S = 1, (CuO–water)

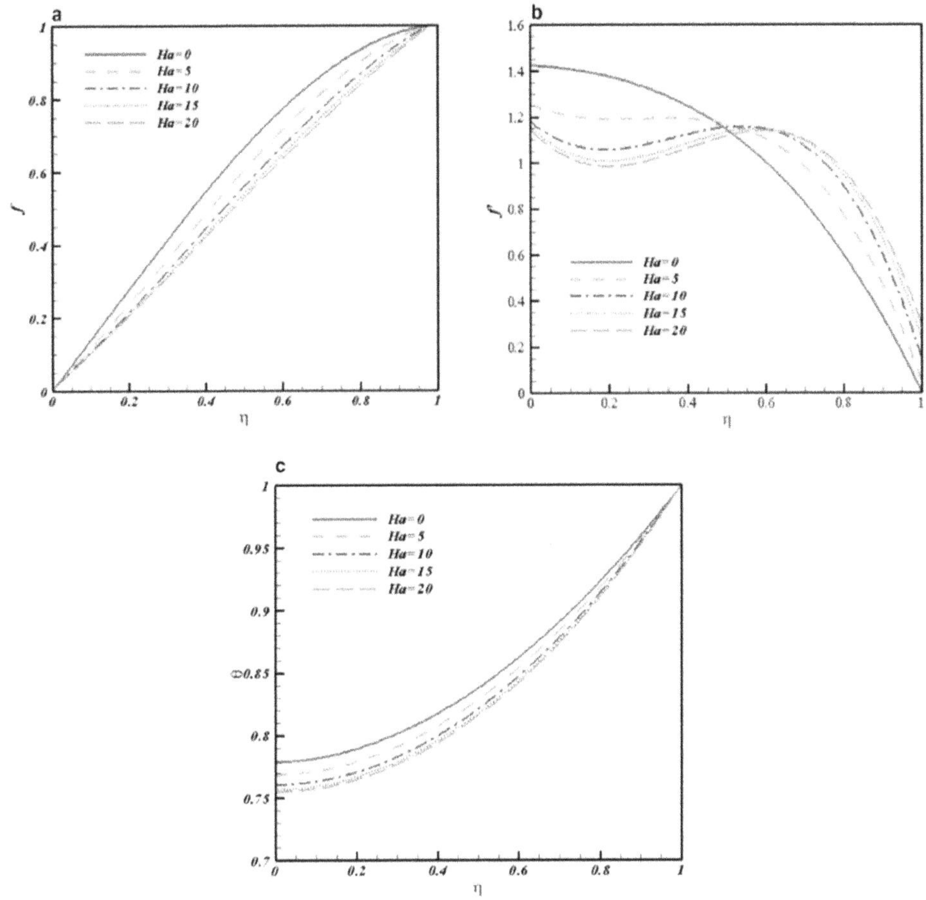

Figure 7. Effect of the squeeze number on velocity and temperature profiles when Pr = 6.2, Hs = −1, S = 1, (CuO–water)

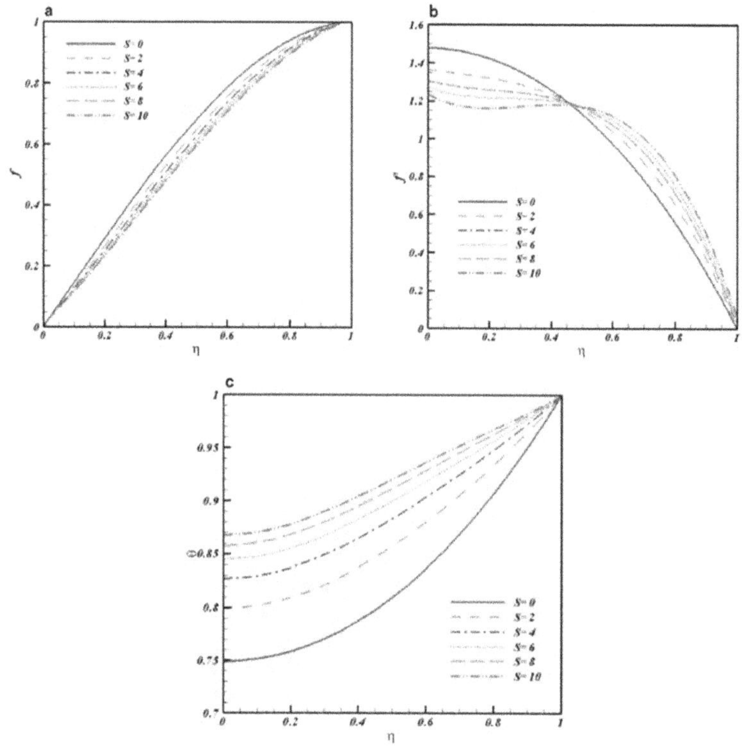

5.2. SQUEEZING FLOW TAKING PLACE BETWEEN TWO HORIZONTAL PARALLEL PLATES

The study focuses on Al_2O_3-Water nanofluid. Figure 8 illustrates the choice of a coordinate system. The analysis assumes a squeezing flow taking place between two horizontal parallel plates initially positioned l(t) = L(1 - at) (1/2) units apart. Here, L represents the plates' initial separation, a > 0 implies plates are squeezed together until they touch at t = 1/a, and for a < 0, the plates move in the opposite direction. At (B_0 is the initial value), a uniform magnetic flux with density (t) = $B_0/1$ is applied. Under these conditions, the governing equations are (Ghosh, Meena, & Das, 2021) (Jalili et al., 2019) (Mittal, 2022) (Ojjela, Raju, & Kambhatla, 2017) (Hosseinzadeh & Ganji, 2022) (Raju, 2019) (Ahmadi *et al.*, 2023) (Nouar *et al.*,

2021) (Saeed *et al.*, 2022) (Jalili *et al.*, 2023) (Raza & Ahmad, 2020) (Reza *et al.*, 2021) (Yazdi, Saedodin, & Shaterzadeh, 2020):

$$\frac{\partial u}{\partial x} + \frac{\partial v}{\partial y} = 0$$

(5.2.1)

$$\rho\left(\frac{\partial v}{\partial t} + u\frac{\partial u}{\partial x} + v\frac{\partial u}{\partial y}\right) = \mu\left(\frac{\partial^2 u}{\partial x^2} + \frac{\partial^2 u}{\partial y^2} - \sigma B^2(t)u\right)$$

$$\rho\left(\frac{\partial v}{\partial t} + u\frac{\partial v}{\partial x} + v\frac{\partial v}{\partial y}\right) = \mu\left(\frac{\partial^2 v}{\partial x^2} + \frac{\partial^2 v}{\partial y^2}\right)$$

(5.2.2)

$$\frac{\partial T}{\partial t} + u\frac{\partial T}{\partial x} + v\frac{\partial T}{\partial y} = \frac{k}{\rho c_\rho}\left(\frac{\partial^2 T}{\partial x^2} + \frac{\partial^2 T}{\partial y^2}\right) + \frac{\mu}{\rho c_\rho}\left(2\left[\left(\frac{\partial u}{\partial x}\right)^2 + \left(\frac{\partial v}{\partial y}\right)^2\right] + \left(\frac{\partial u}{\partial y} + \frac{\partial v}{\partial x}\right)^2\right)$$

$$+\frac{1}{\rho c_\rho}\left(\frac{D_T}{T_W}\left(\left(\frac{\partial T}{\partial x}\right)^2 + \left(\frac{\partial T}{\partial y}\right)^2\right) + D_B\left(\frac{\partial C}{\partial x}\frac{\partial T}{\partial x} + \frac{\partial C}{\partial y}\frac{\partial T}{\partial y}\right)\right)$$

(5.2.3)

$$-\frac{1}{\rho c_\rho}\frac{\partial q_r}{\partial y} + \frac{Q(T - T_0)}{\rho}$$

(5.2.4)

$$u\frac{\partial C}{\partial x} + v\frac{\partial C}{\partial y} + w\frac{\partial C}{\partial z} = D_B\left(\frac{\partial^2 C}{\partial x^2} + \frac{\partial^2 C}{\partial y^2} + \frac{\partial^2 C}{\partial z^2}\right) + \frac{D_T}{T_W}\left(\frac{\partial^2 T}{\partial x^2} + \frac{\partial^2 T}{\partial y^2} + \frac{\partial^2 T}{\partial z^2}\right)$$

(5.2.5)

Where q_r becomes:

$$q_r = -\frac{4\sigma^*}{3k^*}\frac{\partial T^4}{\partial y} = -\frac{4\sigma^*}{3k^*}\frac{\partial(4T_0^3 T - 3T_0^4)}{\partial y}$$

(5.2.6)

The thermo-actual properties of water and nanoparticles are displayed in Table 1:

$$u = 0; v = \frac{dl}{dt}; T = T_L; C = C_L \; at \; y = 1(t)$$

(5.2.7)

$$\frac{\partial u}{\partial y} = 0; v = 0; \frac{\partial T}{\partial y} = 0; \frac{\partial C}{\partial y} = 0 \, at \, y = 0$$

(5.2.8)

Introducing non-dimensional variables:

$$n = \frac{y}{L(1-at)^{\frac{1}{2}}}; u = \frac{ax}{2(1-at)} f'(n), \; v = \frac{-aL}{(1-at)^{\frac{1}{2}}} f(n), \theta(n) = \frac{T}{T_L},$$

$$C(n) = \frac{C'}{C_L} B(t) = \frac{B_0}{(1-at)^{1/2}}, \; \delta = \frac{L}{x}, \; \lambda = \frac{Q2L(T_L - T_0)(1-at)}{kT_L\rho}$$

(5.2.9)

Hence, the dimensionless form of the governing momentum and energy equations for this situation is provided as follows:

$$f^{iv} - S(ff' + (-f + n)f''' + 3f) \text{-} M^2 f = 0$$

(5.2.10)

$$\theta'' + \Pr\left(S(f-n)\theta' + Ec(4\delta^2 f^2 + f'^2)\right) + NbC'\theta' + Nt\theta'^2 + \theta\lambda = 0$$

(5.2.11)

$$NbC + Nt\theta + S.Nb.Sc(fC' - nC'') = 0$$

(5.2.12)

Where:

$$Pr = \frac{\mu C_p}{k}, \; M^2 = \frac{\sigma a B_0^2 L^2}{pv}, \; S = \frac{aL^2}{2v}, \; Ec = \frac{(ax)^2}{C_p T_L 4(1-at)^2}, \; Nr = \frac{16\sigma^* T_0^3}{3k^* k}$$

(5.2.13)

Subject to

$$f = 0, f'' = 0, \theta' = 0, C' = 0 \, at \, n = 0$$

(5.2.14)

$$f = 1, f' = 0, \theta = 1, C = 1 \; at \; n = 1$$

(5.2.15)

The fundamental concept of the homotopy perturbation method is elucidated in this section. Initially, let's examine the generic structure of a nonlinear differential equation along with its boundary condition: Here, A(u) represents the general differential operator, g® is a known analytic function, B is a boundary operator, and ∂u/∂n signifies differentiation along the outward normal direction. To apply this technique to problem-solving, the initial step involves breaking down the equation into its linear and nonlinear components:

$$Y(x) - f(t)=0, t \in \psi \qquad (5.2.16)$$

Such that boundary condition:

$$B\left(x, \frac{\partial x}{\partial n}\right) = 0, t \in \phi$$

(5.2.17)

The elements involved here include a boundary operator, a general differential operator, the domain boundary, a known analytic function, and differentiation along the outward normal direction. Typically, the operator can be split into two segments: one representing the nonlinear component and the other the linear component. Equation (5.2.16) can consequently be restated in the following manner:

$$N(x)+L(x)=f(t) \qquad (5.2.18)$$

If there is no "small parameter" in the nonlinear Eq. (5.2.16), construct the following Homotopy:

$$H(\varepsilon,p) = L(\varepsilon) - L(x_0)_{pL(x_0)} + p(N(\varepsilon) - f(t)) = 0$$

(5.2.19)

Where,

$$\varepsilon(t,p): \psi \times [0,1] \rightarrow R \qquad (5.2.20)$$

In Equation (5.2.19), ξ stands as an embedding parameter, and it signifies the initial approximation that satisfies the boundary condition. It is plausible to hypothesize that the solution to Equation (5.2.19) can be represented as a power series, as demonstrated below:

$$\varepsilon = \varepsilon_0 + p\varepsilon_1 + p^2\varepsilon_2 + \cdots$$

(5.2.21)

The homotopy parameter is also used to expand the square of an unknown angular frequency, as shown below:

$$\gamma = \mu^2 - p\alpha_1 - p^2\alpha_2 - \cdots$$

(5.2.22)

or

$$\mu^2 = \gamma + p\alpha_1 + p^2\alpha_2 + \cdots \text{ 5.2.23)}$$

The coefficient of "α" in Equation (5.2.16) is replaced by the right-hand side of Equation (5.2.22), along with some arbitrary parameters that need to be determined. The most suitable approximation for the solution and the regular frequency is as follows:

$$x = \lim_{p \to 1} \varepsilon = \varepsilon_0 + \varepsilon_1 + \varepsilon_2 + \cdots$$

(5.2.24)

$$\mu^2 = 1 + \alpha_1 + \alpha_2 + \cdots$$

(5.2.25)

When Eqs. (5.2.16) And (5.2.24) become the approximate solution of Eq. (5.2.19), the best approximation for the solution and regular frequency is:

In this physics problem, boundary and initial conditions are indispensable for applying analytical techniques to linear and nonlinear differential equations. As a result, differential equations of any degree with suitable conditions have been dealt with.

To gain a deeper comprehension of the content in this paper, this innovative approach will be used to solve two differential equations that govern engineering processes. The general form of a differential equation, considering boundary conditions, is outlined as follows:

The nonlinear differential equation, which depends on a parameter and its derivatives, is expressed as follows:

$$P_k{:}f(y,y',y'',\ldots,y^{(n)}) = 0,\ y = y(x) \tag{5.2.26}$$

Boundary conditions:

$$\begin{cases} y(0) = y_0,\ y'(0) = y_1,\ y^{(n-1)}(0) = y_{n-1} \\ y(L) = y_L,\ y'(L) = y_{L_1},\ y^{(n-1)}(L) = y_{L_{n-1}} \end{cases} \tag{5.2.27}$$

In line with the principles of this physics problem, linear and nonlinear differential equations necessitate boundary conditions and initial conditions for analytical approaches. This enables us to address differential equations of any degree. This innovative method has been used to solve two differential equations governing engineering processes to increase the understanding of the article's content. The general form of a differential equation, considering boundary conditions, is articulated as follows:

The nonlinear differential equation of, which is a function of, the parameter, which is a function of, and their derivatives are taken into account as follows:

If $x=0$:

$$\begin{cases} y(0) = a_0 = y_0 \\ y'(0) = a_1 = y_1 \\ y''(0) = a_2 = y_2 \end{cases} \tag{5.2.28}$$

and when $x=L$:

$$\begin{cases} y(L) = a_0 + a_1 L + a_2 L^2 + \cdots + a_n L^n = yL_0 \\ y'(L) = a_1 + a_2 L + 3a_3 L^2 + \cdots + n a_n L^{n-2} = yL_1 \\ y''(L) = 2a_2 + 6a_3 L + 12a_4 L^3 + \cdots + n(-1)a_n L^{n-2} = yL_{m-1} \end{cases} \tag{5.2.29}$$

Obtain the solution to the first differential equation in equation (5.2.26) by applying the given boundary conditions and considering a series of letters with constant coefficients in the prescribed order. After substituting Equation (5.2.29) into Equation (5.2.25), the boundary conditions for Equation (5.2.25) are implemented in the following manner. To form a set of equations with both equations and unknowns, n sentences are selected from Equation (5.2.27). However, this selection introduces additional unknowns, essentially the coefficients of Equation (5.2.27). To solve this problem, the equation (5.2.25) has been differentiated several times to these additional unknowns identified in the set above of differential equations. Subsequently, the boundary conditions specified in Equation (5.2.26) are applied to these differentiated equations.

$$\begin{aligned} p_0 &= f(y(0), y'(0), y''(0), \ldots, y^{(m)}(0)) \\ p_1 &= f(y(L), y'(L), y''(L), \ldots, y^{(m)}(L)) \end{aligned} \tag{5.2.30}$$

$$\begin{cases} p_k' = f(y', y'', y''', \ldots y^{(m+1)} \\ p_k'' = f(y'', y''', y^{(IV)}, \ldots, y^{(m+2)} \end{cases} \tag{5.2.31}$$

Application of the boundary conditions on the derivatives of the differential equations in Eq. (5.2.31) is done in the form of:

$$p_k' : \begin{cases} f(y'(0), y''(0), y'''(0), \ldots, y^{(m+1)}(0)) \\ f(y'(L), y''(L), y'''(L), \ldots, y^{(m+1)}(L)) \end{cases} \tag{5.2.32}$$

$$p_k'' : \begin{cases} f(y''(0), y'''(0), y^{(IV)}(0), \ldots, y^{(m+2)}(0)) \\ f(y''(L), y'''(L), y^{(IV)}(L), \ldots, y^{(m+2)}(L)) \end{cases} \tag{5.2.33}$$

The following are the homotopy forms for the equations of the mentioned problem.

$$\begin{aligned} Ueq &= (1-p)\Big(\frac{d^4}{dn^4}f_0(n) + \Big(\frac{d^4}{dn^4}f_1(n)\Big)p \\ &-3S\Big(\frac{d^2}{dn^2}f_0(n) + \Big(\frac{d^2}{dn^2}f_{1(n)}\Big)p\Big) - M^2\Big(\frac{d^2}{dn^2}f_0(n) + \Big(\frac{d^2}{dn^2}f_1(n)\Big)p\Big)\Big) \end{aligned}$$

$$+p\left(-S\left(\begin{array}{c}\frac{d^4}{dn^4}f_0(n)+\left(\frac{d^4}{dn^4}f_1(n)\right)p\\ \left(\frac{d}{dn}f_0(n)+\left(\frac{d}{dn}f_1(n)\right)p\right)\left(\frac{d^2}{dn^2}f_0(n)+\left(\frac{d^2}{dn^2}f_1(n)\right)p\right)\\ +(-f_0(n)-f_1(n)p+n)\left(\frac{d^3}{dn^3}f_0(n)+\left(\frac{d^3}{dn^3}f_1(n)\right)p\right)\\ +3\left(\frac{d^2}{dn^2}f_0(n)\right)+3\left(\frac{d^2}{dn^2}f_1(n)\right)p-M^2\left(\frac{d^2}{dn^2}f_0(n)+\left(\frac{d^2}{dn^2}f_1(n)\right)p\right)\end{array}\right)\right) \quad (5.2.34)$$

$$Peq = (1-p)\left(\frac{d^2}{dn^2}\theta_0(n)+\left(\frac{d^2}{dn^2}\theta_1(n)\right)p+1.2\theta_0(n)+1.2\theta_1(n)p\right)$$

$$+p\left(\begin{array}{c}\frac{d^2}{dn^2}\theta_0(n)+\left(\frac{d^2}{dn^2}\theta_1(n)\right)p+0.045000\left(\frac{d}{dn}f_0(n)+p\left(\frac{d}{dn}f_0(n)\right)\right)^2\\ +0.50\left(\frac{d}{dn}f_1(n)+p\left(\frac{d}{dn}f_1(n)\right)\right)^2\\ +0.1\left(\frac{d}{dn}C_0(n)+p\left(\frac{d}{dn}C_0(n)\right)\right)\left(\frac{d}{dn}\theta_0(n)+\left(\frac{d}{dn}\theta_1(n)\right)p\right)^2\\ +1.2\theta_0(n)+1.2\theta_1(n)p\end{array}\right) \quad (5.2.35)$$

$$Teq = (1-p)\left(0.1\left(\frac{d^2}{dn^2}C_0(n)\right)+0.1\left(\frac{d^2}{dn^2}C_1(n)\right)p+0.1\left(\frac{d^2}{dn^2}\theta_0(n)\right)+0.1\left(\frac{d^2}{dn^2}\theta_1(n)\right)p\right)$$
$$+p\left(0.1\left(\frac{d^2}{dn^2}C_0(n)\right)+0.1\left(\frac{d^2}{dn^2}C_1(n)\right)p+0.1\left(\frac{d^2}{dn^2}\theta_0(n)\right)+0.1\left(\frac{d^2}{dn^2}\theta_1(n)\right)p\right)$$

(5.2.36)

$$f(\eta) = f_0(\eta)+f_1(\eta)\mathrm{p}$$

(5.2.37)

$$\theta(\eta) = \theta_0(\eta)+\theta_1(\eta)\mathrm{p}$$

(5.2.38)

$$C(\eta) = C_0(\eta)+C_1(\eta)\mathrm{p}$$

(5.2.39)

Subsequently, after substituting Equations (5.2.14) and (5.2.15) into Equations (5.2.10) to (5.2.12) and simplifying them while organizing them based on the powers of permutations, the resulting equations will be as follows:

P^0:

$$q_0 = \frac{d^2}{dn^2}\theta_0(n) + 1.2\theta_0(n)$$

(5.2.40)

$$S_0 = \frac{d^4}{dn^4}f_0(n) - 0.25\left(\frac{d^2}{dn^2}f_0(n)\right)$$

(5.2.41)

$$w_0 = 0.1\left(\frac{d^2}{d\eta^2}C_0(\eta)\right) - \frac{0.1200000000\cos\left(\frac{1}{5}\sqrt{30}\,\eta\right)}{\cos\left(\frac{1}{5}\sqrt{30}\right)}$$

(5.2.42)

In the aforementioned mathematical expressions, q0 corresponds to theta, S_0 corresponds to f, and W_0 corresponds to C. Additionally, the boundary conditions are as follows:

f[0](0) = 0, f[0](1) = 1, ((D²)(f[0]))(0) = 0, (D(f[0]))(1) = 0 (5.2.43)

(D (theta[0]))(0) = 0, theta[0](1)

= 1
(5.2.44)

(D(C[0]))(0) = 0, C[0](1) = 1 (5.2.45)

P^1:

$$s_1 = \frac{d^4}{d\eta^4}f_1(\eta) - 0.025\left(\frac{d^2}{d\eta^2}f_1(\eta)\right)$$

(5.2.46)

$$q_1 = \frac{d^2}{d\eta^2}\theta_1(\eta) + 1.2\theta_1(\eta) + 0.04500$$

$$\left(-\frac{e^{\frac{1}{2}} + e^{\frac{-1}{2}}}{e^{\frac{1}{2}} - 3e^{\frac{-1}{2}}} + \frac{e^{\frac{1}{2\eta}}}{e^{\frac{1}{2}} - 3e^{-\frac{1}{2}}} + \frac{e^{-\frac{1}{2\eta}}}{e^{\frac{1}{2}} - 3e^{-\frac{1}{2}}}\right)^2 + 0.50\left(\frac{1}{2}\frac{e^{\frac{1}{2\eta}}}{e^{\frac{1}{2}} - 3e^{-\frac{1}{2}}} - \frac{1}{2}\frac{e^{-\frac{1}{2\eta}}}{e^{\frac{1}{2}} - 3e^{-\frac{1}{2}}}\right)^2$$

(5.2.47)

$$w_1 = 0.1\left(\frac{d^2}{d\eta^2} C_1(\eta)\right)$$

$$+\frac{0.00000078\cos\left(\frac{1}{5\sqrt{30}\,\eta}\right)\left(807\,e^{\frac{5}{2}} - 2794\,e^{\frac{3}{2}} + 807\,e^{\frac{1}{2}}\right)e^{-\frac{1}{2}}}{\cos\left(\left(\frac{1}{5}\right)*\sqrt{30}\right)*(-\exp(2)+6*\exp(1)-9)}$$

$$+\frac{1}{-e^2+6e-9}(0.0000001361\left(36975\,e^{\frac{5}{2\eta}} - \frac{20097}{2}e^{\frac{3}{2\eta}} + 1479\,e^{\frac{1}{2\eta}} - 19008\,e^{2\eta+\frac{1}{2}}\right.$$

$$\left.-19008\,e^{2\eta-\frac{1}{2}} + \frac{25839}{4}e^{1+\frac{3}{2\eta}} + \frac{25839}{4}e^{1\mp\frac{3}{2\eta}} - 4752\,e^{\frac{1}{2}+\eta} - 4752\,e^{-\frac{1}{2}+\eta}\right)$$

(5.2.48)

With the following boundary conditions:

f[1](0) = 0, f[1](1) = 1, ((D²)(f[1]))(0) = 0, (D(f[1]))(1) = 0 (5.2.49)

(D (theta[1]))(0) = 0, theta[1](1)

= 1
(5.2.50)

(D(C[1]))(0) = 0, C[1](1) = 1 (5.2.51)

By using boundary conditions to solve Eqs. (40)-(51): When p = 1, the solutions to these equations are as follows:

$$f(n) = \frac{-(e^{1/2}+e^{-1/2})n}{e^{1/2}-3e^{-1/2}} + \frac{2e^{1/2n}}{e^{1/2}-3e^{-1/2}} - \frac{2e^{-1/2n}}{e^{1/2}-3e^{-1/2}}$$

(5.2.52)

$$\theta(n) := \frac{\cos\left(\frac{1}{5}\sqrt{30}\,n\right)}{\cos\left(\frac{1}{5}\sqrt{30}\right)}$$

$$-\frac{1}{15312}\frac{os\left(\frac{1}{5}\sqrt{30}\,n\right)\left(807\,e^{\frac{5}{2}} - 2794\,e^{\frac{3}{2}} + 807\,e^{\frac{1}{2}}\right)e^{-\frac{1}{2}}}{\cos\left(\frac{1}{5}\sqrt{30}\right)(-e^2+6e-9)} + \frac{1}{76560(-e^2+6e-9)}$$

$$\left(5916\,e^{\frac{5}{2n}} - 4466\,e^{\frac{3}{2n}} + 5916\,e^{\frac{1}{2n}} - 4752\,e^{2n+\frac{1}{2}} - 4752\,e^{2n-\frac{1}{2}}\right.$$
$$\left.+2871\,e^{-1+\frac{3}{2n}} + 2871\,e^{1+\frac{3}{2n}} - 4752\,e^{n+\frac{1}{2}} - 4752\,e^{n-1/2}\right)e^{1-3/2n}$$

(5.2.53)

$$
\begin{aligned}
C(n) &= \frac{\cos\left(\frac{1}{5}\sqrt{30}\,n\right)}{\cos\left(\frac{1}{5}\sqrt{30}\right)} + 2 + \frac{1}{4\times 10^{14}} \times \frac{1}{\cos\left(\frac{1}{5}\sqrt{30}\right)(-e^2+6e-9)} \\
&\Bigg(-6026896536192\cos\left(\frac{1}{5\sqrt{30}}\right)e^{\frac{3}{2}-\frac{1}{2}z1} - 6206896517184\cos\left(\frac{1}{5\sqrt{30}}\right)e^{\frac{3}{2}+\frac{1}{2}z1} \\
&-6206896536192\cos\left(\frac{1}{5\sqrt{30}}\right)e^{-\frac{1}{2}z1+\frac{1}{2}} - 6206896517184\cos\left(\frac{1}{5\sqrt{30}}\right)e^{\frac{1}{2}z1+\frac{1}{2}} \\
&+(-87586206900896e + 25297805643888e^2 + 25297805643888) \\
&\cos\left(\frac{1}{5\sqrt{30}_{z1}}\right)22330\left(e - \frac{9}{14e^2} + \frac{106583072058}{77e^{-z1+1}} - \frac{532915359474}{385e^{z1=1}} - \frac{9}{14}\right)n\_z1 \\
&\times\cos\left(\frac{1}{5\sqrt{30}}\right)d\_z1d\_z1\Bigg) + (n-1) \\
&\times\left(-\int_1\int_1\left(-6206896536912\cos\left(\frac{1}{5\sqrt{30}}\right)e^{\frac{3}{2}-\frac{1}{2}z1} - 6206896517184\cos\left(\frac{1}{5\sqrt{30}}\right)e^{\frac{3}{2}+\frac{1}{2}z1}\right)\right. \\
&-620689656192\cos\left(\frac{1}{5\sqrt{30}}\right)e^{-\frac{1}{2}z1+\frac{1}{2}} - 620689617184\cos(1/5\sqrt{30})e^{\frac{1}{2}z1+1/2} \\
&\left(\int_0^1 +(-87586206900896e + 252978 - 5643888e^2 + 25297805643888)\right. \\
&\cos(1/5\sqrt{30\_z1})22330\left(e - \frac{9}{14e^2} + 1 - \frac{6583072058}{77e^{-z1+1}} - 9/14\right) \\
&\times\cos(1/5\sqrt{30})d\_z1) + (-4\times 10^{15}e^2 + 24\times 10^{14}e - 36\times 10^{14})\cos(1/5\sqrt{30}))
\end{aligned}
$$

(5/2/54)

The main focus of this chapter is to analyze the geometry of the problem by presenting visual representations of the velocity, temperature, and concentration profiles. Computational methods are used to obtain solutions. Figure 9 illustrates that as the squeezing parameter S increases, the velocity component f decreases. The numerical methods used in this study produced very similar results. Figure 10 compares temperature and heat transfer variations concerning alterations in the s parameter using AGM and HPM mathematical methods. As evidenced by Figure 10, s has an adverse impact on temperature. This is reasonable since an increase in S reduces kinematic viscosity, leading to reduced friction between nanoparticles and plate surfaces. Figure 11 compares concentration variations with alterations in the s parameter using AGM and HPM mathematical methods. The concentration component is observed to increase as the squeezing parameter S increases, as depicted in Figure 11. Furthermore, the heat transfer process is found to increase with an increase in the Prandtl number (Pr). Conversely, it is shown that as Pr increases, the concentration decreases. In Figure 12, the impacts of Nb on temperature and concentration profiles are displayed. It can be observed that with an increase in Nb, the temperature characteristics decrease while the concentration increases. The motion of nanoparticles resulting from collisions with surrounding gas molecules is referred to as Brownian motion. When there is a concentration gradient, particles undergo migration known as diffusiophoresis, moving from densely populated areas to sparsely populated regions. As depicted in Figure 13, an increase in Brownian motion leads to an amplified concentration effect. Additionally, Figure 14 demonstrates that the magnetic parameter has an adverse effect on velocity and temperature profiles. The impact of magnetism on electrical charges, electric currents, and nanomaterials is elucidated by the presence of a magnetic field, which is a vector field. When a charge moves through a magnetized region, it experiences a force perpendicular to its own velocity and the magnetic field. Thermophoresis (Nt) refers to the movement of particles towards the colder side in response to a thermal gradient in the surrounding gas. This phenomenon plays a crucial role in various applications, such as catalysis-driven plasmas, modified chemical vapor deposition, aerosol sampling, microelectronic production, and air purification. Figure 15 illustrates that considering a higher value may result in a decrease in concentration. The Schmidt number (Sc), defined as the ratio between kinematic viscosity and molecular diffusion coefficient, is a crucial parameter in fluid dynamics. Several field studies have resulted in empirical equations that relate k to wind velocity. Figures 16 and 17 demonstrate that an increase in Sc leads to a decrease in temperature and concentration trends.

This research examines the influence of magnetism on the flow of two-dimensional nanofluids through two parallel sheets. Several parameters are investigated, including the heat generation variable, Eckert number (Ec), Prandtl number (Pr), nanofluid volume percent, radiation parameter (Nr), squeeze parameter (S), and magnetic parameter (M). The study utilizes semi-analytical solutions obtained through homotopy perturbation and Akbari Ganji's methods. The results are analyzed through diagrams and subsequently compared with numerical solutions in a separate chart to evaluate the effectiveness of these methods. This research is unique in that it employs the Homotopy perturbation Method (HPM) and Akbari Ganji's Method to investigate nanofluids' heat transfer, fluid dynamics, and concentration behavior flowing between two identical surfaces.

Table 1. Properties of water and Al_2O_3 nanofluid

	ρ(kg/m³)	**K(w/m.k)**	**C_p (j/kg.k)**	**σ**
Al_2O_3	3970	25	765	10^{-12}
H_2O	997	0.613	4179	0.05

Figure 8. Schematic of the two parallel plates

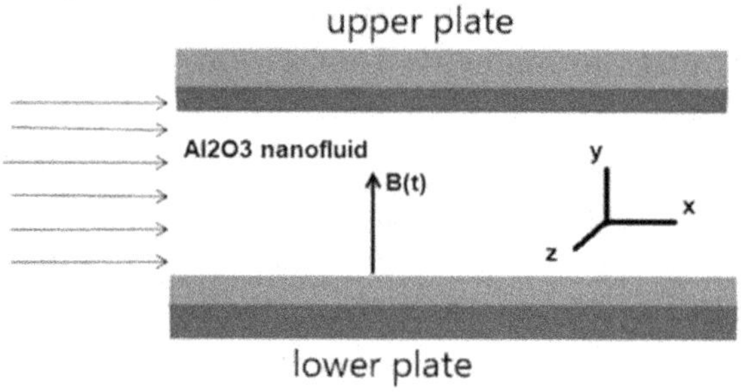

Figure 9. Comparison of the velocity profile with different S parameter by AGM and HPM method at $\varphi = 0.02$, $M = 0.5$, $Nr = 1.0$, $Pr = 10.0$, $Ec = 0.05$, $\delta = 0.15$, $\lambda = 1.2$, $Nb = 0.1$, $Nt = 0.1$, $Sc = 0.3$

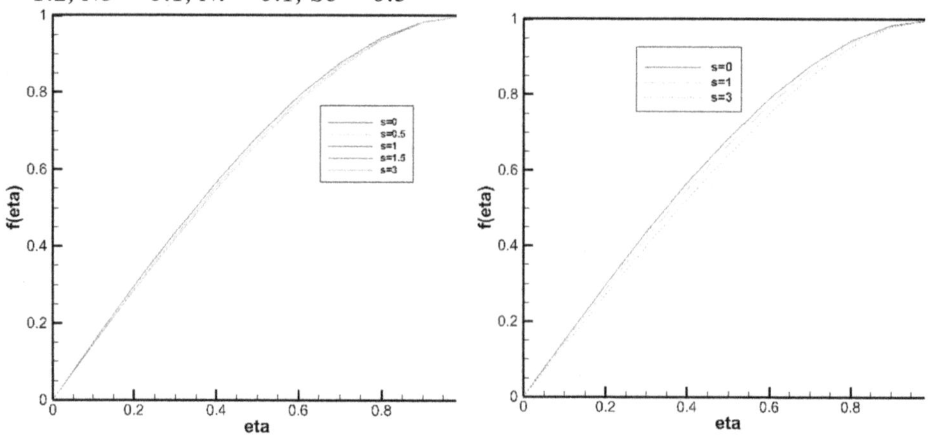

a) By AGM **b) By HPM**

Figure 10. Comparison of the temperature profile with different S parameter by AGM and HPM method at $\varphi = 0.02$, $M = 0.5$, $Nr = 1.0$, $Pr = 10.0$, $Ec = 0.05$, $\delta = 0.15$, $\lambda = 1.2$, $Nb = 0.1$, $Nt = 0.1$, $Sc = 0.3$

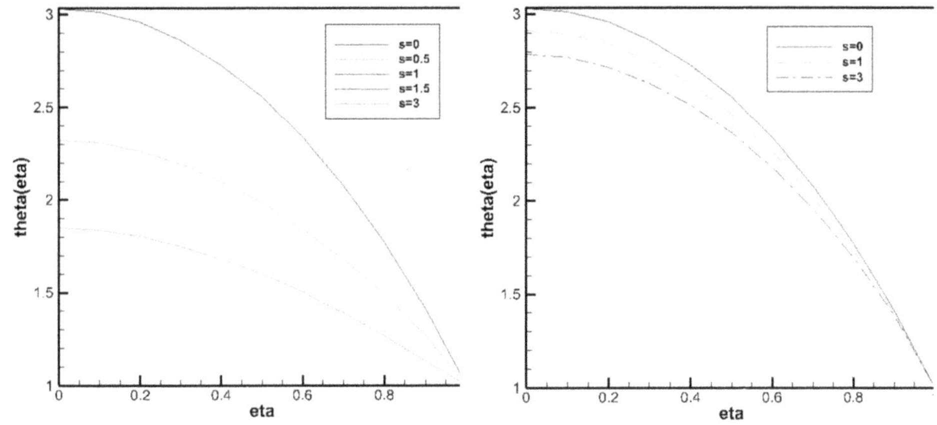

a) By AGM

b) By HPM

Figure 11. Comparison of the concentration profile with different S parameters by HPM method at $\varphi = 0.02$, $M = 0.5$, $Nr = 1.0$, $Pr = 10.0$, $Ec = 0.05$, $\delta = 0.15$, $\lambda = 1.2$, $Nb = 0.1$, $Nt = 0.1$, $Sc = 0.3$

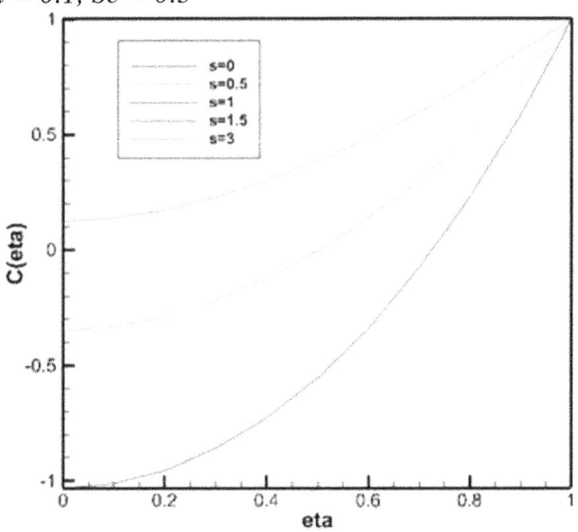

Figure 12. Comparison of the temperature profile with different Nb parameters by AGM and HPM method at $\varphi = 0.02$, $M = 0.5$, $Nr = 1.0$, $Pr = 10.0$, $Ec = 0.05$, $\delta = 0.15$, $\lambda = 1.2$, $s = 0.5$, $Nt = 0.1$, $Sc = 0.3$

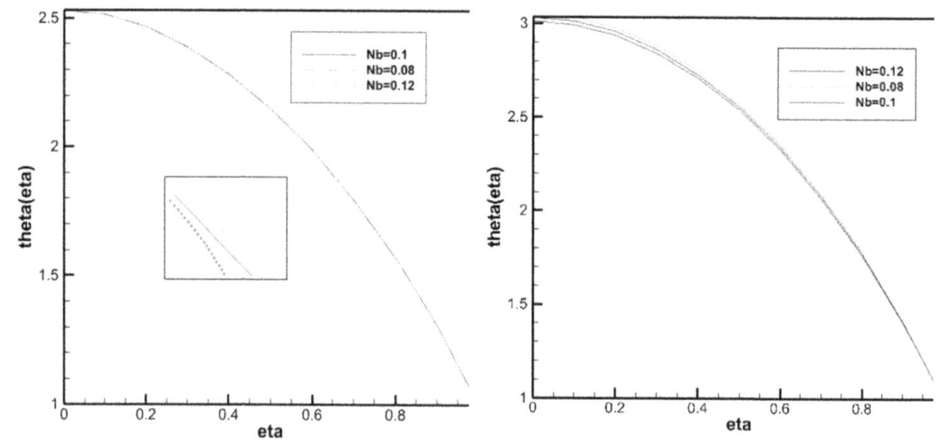

a) By AGM

b) By HPM

Figure 13. Comparison of the velocity profile with different M parameter by HPM method at $\varphi = 0.02$, $M = 0.5$, $Nr = 1.0$, $Pr = 10.0$, $Ec = 0.05$, $\delta = 0.15$, $\lambda = 1.2$, $Nt = 0.1$, $Sc = 0.3$

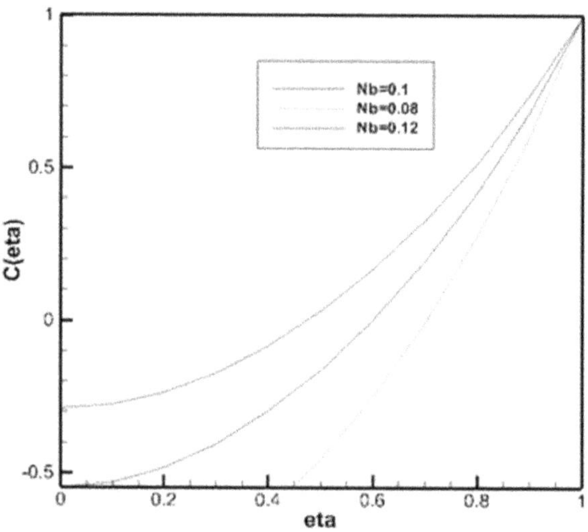

Figure 14. Comparison of the velocity profile with different M parameter by AGM method at $\varphi = 0.02$, $M = 0.5$, $Nr = 1.0$, $Pr = 10.0$, $Ec = 0.05$, $\delta = 0.15$, $\lambda = 1.2$, $Nb = 0.1$, $Nt = 0.1$, $Sc = 0.3$

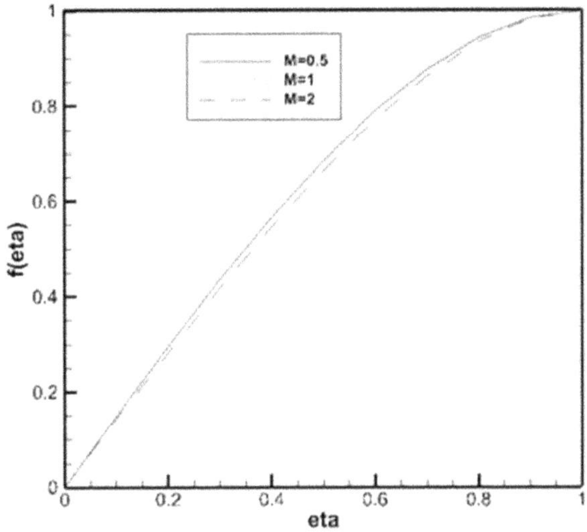

Figure 15. Comparison of the concentration profile with different Nt parameter by AGM method at $\varphi = 0.02$, $M = 0.5$, $Nr = 1.0$, $Pr = 10.0$, $Ec = 0.05$, $\delta = 0.15$, $\lambda = 1.2$, $Nb = 0.1$, $Sc = 0.3$

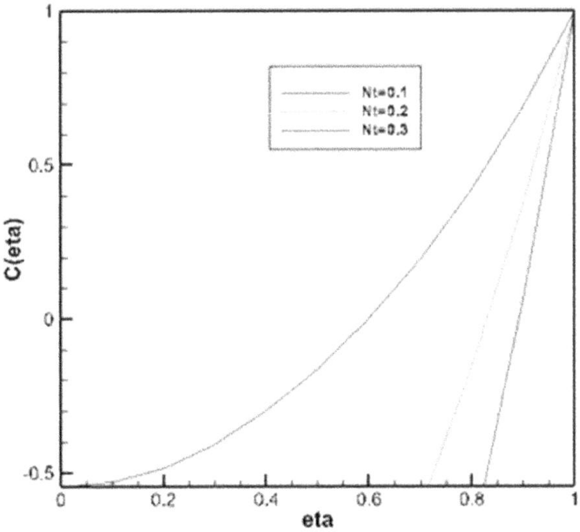

Figure 16. Comparison of the temperature profile with different Sc parameter by HPM method at φ = 0.02, M = 0.5, Nr = 1.0, Pr = 10.0, Ec = 0.05, δ =0.15, λ = 1.2, Nb = 0.1

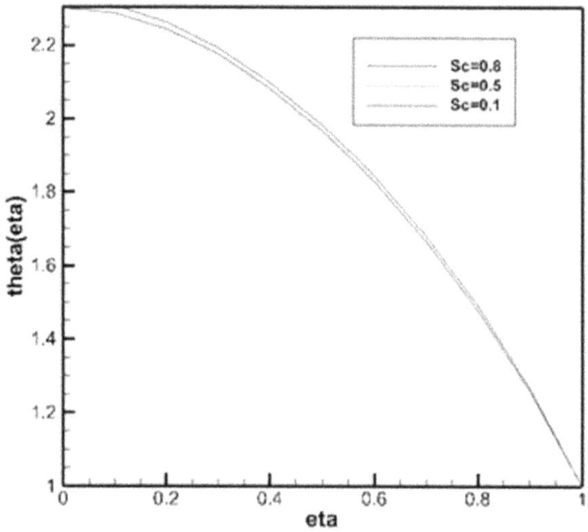

Figure 17. Comparison of the concentration profile with different Sc parameter by HPM method at φ = 0.02, M = 0.5, Nr = 1.0, Pr = 10.0, Ec = 0.05, δ =0.15, λ = 1.2, Nb = 0.1

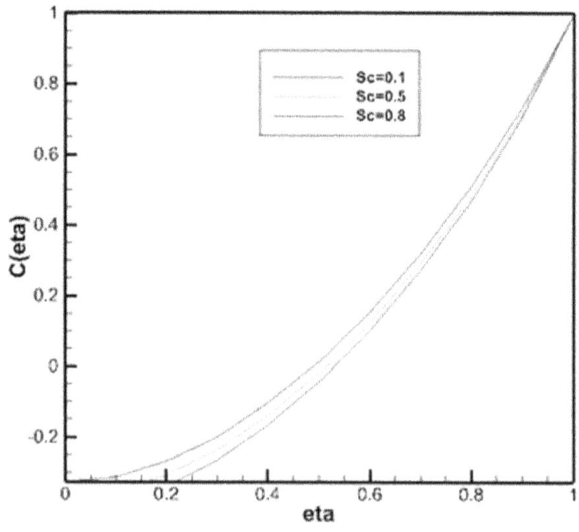

REFERENCES

Ahmadi Azar, A., Jalili, B., Jalili, P., & Domiri Ganji, D. (2023). Investigating the effect of structural changes of two stretching disks on the dynamics of the MHD model. *Scientific Reports*, 13(1), 21833. DOI: 10.1038/s41598-023-48988-4 PMID: 38071256

Alizadeh, A., Zekri, H. J., & Jafarmadar, S. (2020). Numerical simulation of the formation of vortices around rigid cylinders as a issue of fluid-structure interaction using immersed interface method. *Mechanics*, 26(1), 18–24. DOI: 10.5755/j01.mech.26.1.23186

Anwar, T., & Kumam, P. (2022). A fractal fractional model for thermal analysis of GO– NaAlg– Gr hybrid nanofluid flow in a channel considering shape effects. *Case Studies in Thermal Engineering*, 31, 101828. DOI: 10.1016/j.csite.2022.101828

Bahmani, M., Jalili, B., Jalili, P., Mirzaei, A., & Ganji, D. D. (2024). Effect of variations hollow of octagon porous media on heat and mass transfer. *International Journal of Thermofluids*, 21, 100576. DOI: 10.1016/j.ijft.2024.100576

Bai, J., Talebizadehsardari, P., Yekani Motlagh, N., & Goharkhah, M. (2022). Numerical analysis and two-phase modeling of water Graphene Oxide nanofluid flow in the riser condensing tubes of the solar collector heat exchanger. *Sustainable Energy Technologies and Assessments*, 53, 102408. DOI: 10.1016/j.seta.2022.102408

Bazmi, M., Goharkhah, M., & Talebizadehsardari, P. (2019). Nitrogen-doped carbon nanotubes for heat transfer applications: Enhancement of conduction and convection properties of water/N-CNT nanofluid. *Journal of Thermal Analysis and Calorimetry*, 138(1), 69–79. DOI: 10.1007/s10973-019-08024-y

Farahani, S. D., Farahani, A. D., Tayebzadeh, F., & Mosavi, A. H. (2021). Melting of non-Newtonian phase change material in a finned triple-tube: Efficacy of non-uniform magnetic field. *Case Studies in Thermal Engineering*, 28, 101543. DOI: 10.1016/j.csite.2021.101543

Ghosh, D., Meena, P. R., & Das, P. K. (2021). Heat transfer from a ferrofluid during generalized Couette flow through parallel plates in the presence of an orthogonal magnetic field. *International Journal of Thermal Sciences*, 164, 106895. DOI: 10.1016/j.ijthermalsci.2021.106895

Hai, T., Kadir, D. H., & Ghanbari, A. (2023). Modeling the emission characteristics of the hydrogen-enriched natural gas engines by multi-output least-squares support vector regression: Comprehensive statistical and operating analyses. *Energy*, 276, 127515. DOI: 10.1016/j.energy.2023.127515

Hosseinzadeh, S., & Ganji, D. D. (2022). A novel approach for assessment of MHD mixed fluid around two parallel plates by consideration hybrid nanoparticles and shape factor. *Alexandria Engineering Journal*, 61(12), 9779–9793. DOI: 10.1016/j.aej.2022.03.031

Hussain, T., & Xu, H. (2022). Time-dependent squeezing bio-thermal MHD convection flow of a micropolar nanofluid between two parallel disks with multiple slip effects. *Case Studies in Thermal Engineering*, 31, 101850. DOI: 10.1016/j.csite.2022.101850

Jalili, B., Azar, A. A., Jalili, P., & Ganji, D. D. (2023). impact of variable viscosity on asymmetric fluid flow through the expanding/contracting porous channel: A thermal analysis. *Case Studies in Thermal Engineering*, 52, 103672. DOI: 10.1016/j.csite.2023.103672

Jalili, B., Sadighi, S., Jalili, P., & Ganji, D. D. (2019). Characteristics of ferrofluid flow over a stretching sheet with suction and injection. *Case Studies in Thermal Engineering*, 14, 100470. DOI: 10.1016/j.csite.2019.100470

Jalili, H., Ahmadi Azar, A., Esmaeili, K., Liu, D., Jalili, P., & Domiri Ganji, D. (2024). A novel approach to micropolar fluid flow between a non-porous disk and a porous disk with slip. *Zhongguo Wuli Xuekan*, 87, 118–137. DOI: 10.1016/j.cjph.2023.11.023

Jalili, P., Jalili, B., Ahmad, I., Hendy, A. S., Ali, M. R., & Ganji, D. D. (2024). Python approach for using homotopy perturbation method to investigate heat transfer problems. *Case Studies in Thermal Engineering*, 54, 104049. DOI: 10.1016/j.csite.2024.104049

Jalili, P., Mahboob, M., Shateri, A., Jalili, B., & Domiri Ganji, D. (2024). A Comparative Study of Hybrid Analytical and Laplace Transform Approaches for Solving Partial Differential Equations in Python. *International Journal of Engineering*, 37(2), 352–364. DOI: 10.5829/IJE.2024.37.02B.11

Kodi, R., & Konduru, V. (2022). Heat and mass transfer on MHD convective unsteady flow of a Jeffrey fluid past an inclined vertical porous plate with thermal diffusion Soret and aligned magnetic field. *Materials Today: Proceedings*, 50, 2128–2134. DOI: 10.1016/j.matpr.2021.09.436

Li, H., Chen, H., Fan, H., & Zang, W. (2023). A comprehensive investigation of thermal conductivity in of monolayer graphene, helical graphene with different percentages of hydrogen atom: A molecular dynamics approach. *Colloids and Surfaces. A, Physicochemical and Engineering Aspects*, 656, 130324. DOI: 10.1016/j.colsurfa.2022.130324

Mittal, A. S. (2022). Study of radiation effects on unsteady 2D MHD Al2O3-water flow through parallel squeezing plates. *International Journal of Ambient Energy*, 43(1), 653–660. DOI: 10.1080/01430750.2019.1662843

Niknejadi, M., Goharkhah, M., Talebizadehsardari, P., & Zanganeh, H. (2023). Numerical simulation of the thermal-hydraulic performance of solar collector equipped with vector generators filled with two-phase hybrid nanofluid Cu-TiO2/H2O. *Engineering Analysis with Boundary Elements*, 151, 670–685. DOI: 10.1016/j.enganabound.2023.03.035

Nouar, A., Dib, A., Kezzar, M., Sari, M. R., & Eid, M. R. (2021). Numerical treatment of squeezing unsteady nanofluid flow using optimized stochastic algorithm. *Zeitschrift für Naturforschung. Section A. Physical Sciences*, 76(10), 933–946. DOI: 10.1515/zna-2021-0163

Ojjela, O., Raju, A., & Kambhatla, P. K. (2017). Influence of thermophoresis and induced magnetic field on chemically reacting mixed convective flow of Jeffrey fluid between porous parallel plates. *Journal of Molecular Liquids*, 232, 195–206. DOI: 10.1016/j.molliq.2017.02.061

Paradis, P. Y., & Champliaud, H. (2021). New correlations for heat transfer in parallel-plate ducts at low Peclet number. *International Journal of Heat and Mass Transfer*, 178, 121595. DOI: 10.1016/j.ijheatmasstransfer.2021.121595

Qasemian, A., Goharkhah, M., & Goharkhah, S. (2021). Hydraulic and thermal analysis of automatic transmission fluid in the presence of nanoparticles and twisted tape: An experimental and numerical study. *Journal of Central South University*, 28(11), 3404–3417. DOI: 10.1007/s11771-021-4864-x

Raju, T. L. (2019). MHD heat transfer two-ionized fluids flow between parallel plates with Hall currents. *Results in Engineering*, 4, 100043. DOI: 10.1016/j.rineng.2019.100043

Raza, M., & Ahmad, S. (2020). Entropy analysis of MHD and radiation effects on squeezing flow in porous medium with variable fluid properties. *Results in Physics*, 18, 103209.

Reza, A., Houshmand, A., Nazari, F., & Ganji, D. D. (2021). A study of radiative and magnetic fields on squeezing nanofluid flow with heat source and sink using analytical method. *Results in Physics*, 20, 103708.

Saeed, A., Shah, R. A., Khan, M. S., Fernandez-Gamiz, U., Bani-Fwaz, M. Z., Noeiaghdam, S., & Galal, A. M. (2022). Theoretical analysis of unsteady squeezing nanofluid flow with physical properties. *Mathematical Biosciences and Engineering*, 19(10), 10176–10191. DOI: 10.3934/mbe.2022477 PMID: 36031990

Salehi, S., Nori, A., Hosseinzadeh, K., & Ganji, D. D. (2020). Hydrothermal analysis of MHD squeezing mixture fluid suspended by hybrid nanoparticles between two parallel plates. *Case Studies in Thermal Engineering*, 21, 100650. DOI: 10.1016/j.csite.2020.100650

Yazdi, M., Saedodin, S., & Shaterzadeh, P. (2020). Numerical analysis of the thermophysical properties and convective heat transfer enhancement of the two-phase nanofluid in a squeezing flow between two parallel plates. *Results in Physics*, 18, 103160.

Zangooee, M. R., Hosseinzadeh, K., & Ganji, D. D. (2019). Hydrothermal analysis of MHD nanofluid (TiO2-GO) flow between two radiative stretchable rotating disks using AGM. *Case Studies in Thermal Engineering*, 14, 100460. DOI: 10.1016/j.csite.2019.100460

Zhang, Y. X., Gao, Y., Luo, Z., Song, Z., Huang, Q., & Zhou, W. (2023). Investigating the effect of size and number of layers of iron nanochannel on the thermal behavior and phase change process of calcium chloride/sodium sulfate hexa-hydrate with molecular dynamics simulation. *Journal of Energy Storage*, 62, 106762. DOI: 10.1016/j.est.2023.106762

Zhou, Y., Talebizadehsardari, P., & Goharkhah, M. (2023). Computational fluid dynamics and multi-objective response surface methodology optimization of perforated-finned heat sinks. *Journal of the Taiwan Institute of Chemical Engineers*, 145, 104823. DOI: 10.1016/j.jtice.2023.104823

Chapter 6
Magneto Ferrofluid Flow

ABSTRACT

Ferrofluids are useful in numerous industries and sciences, like mechanical and electrical engineering. Recent improvements in nanoelectromechanical systems (NEMS), microfluid, nanofluid, and Ferrofluid indicate the importance of this new scientific field. Various and complex factors reduce the heat transfer process. The low heat transfer coefficient has become a significant concern in many areas of heat transfer, such as small systems. In this chapter, the characteristics of Ferrofluid over a shrinking sheet with an effective thermal conductivity model are studied using the homotopy perturbation method (HPM) and Akbari-Ganji's method (AGM). Also, the Finite Element Method (FEM) has been applied to numerical answers. The governing equations were formulated using the Tiwari-Das model. The influence of related micro-rotation and dimensionless velocity parameters have been taken for suction and injection of mass transfer parameters. Results indicate that the magnetic and boundary parameters, in contrast to the micro-rotation parameter, have the same effect on velocity.

6.1. MICROPOLAR FERROFLUIDS

Micropolar ferrofluids are a fascinating topic for researchers. This article assumes a two-dimensional equilibrium flow of a slightly polar ferrofluid across a foil contracting in a normal magnetic field. The sheet can only change in the x direction, and the velocity follows. The velocity should be negative if the sheet is contracting, and the y direction is vertical.

The governing equations of this problem are (Gorla, 1988) (Gorla, Pender, & Eppich, 1983) (Arafa & Gorla, 1992) (Mirzaei *et al.*, 2023) (Mohamed & Abo-Dahab, 2009) (Jalili *et al.*, 2023a) (Rashidi *et al.*, 2011) (Jalili *et al.*, 2023b) (Turkyilmazoglu, 2016) (Mirgolbabaee, Ledari, & Ganji, 2017) (Jalili *et al.*, 2023c) (Mehmood *et al.*, 2019) (Jalili *et al.*, 2023d) (Agarwal *et al.*, 2021) (Jalili *et al.*, 2023e) (Ahmad,

DOI: 10.4018/979-8-3693-5335-6.ch006

Ashraf, & Ali, 2019) (Jalili *et al.,* 2023f) (Fatunmi & Adeniyan, 2018) (Krishna, Swarnalathamma, & Chamkha, 2018) (Ahmad, Ashraf, & Ali, 2020) (Jalili *et al.,* 2023g)(Abd-Alla *et al.*, 2022) (Agarwal *et al.,* 2022):

$$\frac{d}{dt}(\rho_{nf}) = \nabla.(\rho_{nf}V) \tag{6.1.1}$$

$$\rho_{nf}\left(\frac{dN}{dt}\right) = -\nabla p + \left(2\mu_{nf} + k\right)\nabla(\nabla.V) - \left(\mu_{nf} + K\right)\nabla\times(\nabla\times V) + K(\nabla\times N) + J\times B + \rho_{nf}g\Big) \tag{6.1.2}$$

$$\rho_{nf}j\left(\frac{dN}{dt}\right) = \left(\varphi + \lambda + \gamma_{nf}\right)\nabla.(\rho_{nf}\bullet N) - \gamma_{nf}\nabla N\Big) + K\left(\nabla\times V\right) - 2_K N + \rho_{nf}I \tag{6.1.3}$$

Brinkman Bourantas and Loukopoulos also introduce effective density, spin gradient viscosity, and effective kinematic viscosity.

$$\rho_{nf} = (1-\varphi)\rho_f + \phi\rho_s \tag{6.1.4}$$

$$\mu_{nf} = \frac{\mu_f}{(1-\varphi)^{2.5}} \tag{6.1.5}$$

$$\gamma_{nf} = \left(\mu_{nf} + \frac{K}{2}\right)j \tag{6.1.6}$$

Also, there are no external forces, and simplified equations are as follows:

$$\rho_{nf}\left(\frac{dN}{dt} + (\boldsymbol{V}.\nabla)\boldsymbol{V}\right) = \left(\mu_{nf} + K\right)\nabla^2\boldsymbol{V} + K(\nabla\times\boldsymbol{N}) + \mathbf{J}\times\mathbf{B} \tag{6.1.7}$$

$$\rho_{nf} j\left(\frac{d\mathbf{N}}{dt} + (\mathbf{N} \bullet \nabla)\mathbf{N}\right) = \gamma_{nf} \nabla^2 \mathbf{N} + \mathbf{K}(\nabla \times \mathbf{V}) - 2_K \mathbf{N}$$

(6.1.8)

Electrical conductivity and bulk force of ferrofluids represented by Sheikholeslami and Hussanan et al.:

$$\sigma_{nf} = \left[1 + \frac{3(\sigma - 1)\phi}{(\sigma + 2) - (\sigma - 1)\phi}\right] \sigma_f$$

(6.1.9)

$$\mathbf{J} \times \mathbf{B} = -\sigma_{nf} B_0^2 \mathbf{V}$$

(6.1.10)

The 2-D energy equation is:

$$u\frac{\partial T}{\partial x} + v\frac{\partial T}{\partial y} = \frac{k_{nf}}{\left(\rho C_p\right)_{nf}} \frac{\partial^2 T}{\partial y^2} - \frac{\partial q_r}{\partial y}$$

(6.1.11)

The effective thermal conductivity and the specific heat capacity proposed by Mohyud-Din et al and Khan et al.

$$\frac{k_{nf}}{k_f} = \frac{k_s + 2k_f - 2\phi\left(k_f - k_s\right)}{k_s + 2k_f + \phi\left(k_f - k_s\right)}$$

(6.1.12)

$$C_{p,nf} = \frac{\phi\left(\rho C_p\right)_s + (1 - \phi)\left(\rho C_p\right)_f}{\rho_{nf}}$$

(6.1.13)

The boundary conditions are expressed as:

$$u = \alpha u_w(x)\,,\; v = v_w$$

$$@\, y = 0,\, u \to 0,\, y \to \infty$$

(6.1.14)

$$N = -\delta \frac{\alpha u}{\alpha y}$$

(6.1.15)

$$T = T_W$$

$$@\, y = 0, T \rightarrow T_w, y \rightarrow \infty$$

(6.1.16)

In these equations, u and v are the velocity components in the x and y directions. Injection and aspiration are related to negative and positive quantities of μ, dynamic and eddy viscosities. δ is a constant parameter between [0,1], j is the micro-inertia per unit mass, and γ is the spin gradient viscosity. Use a similarity variable like this:

$$n = y\sqrt{\frac{a}{v_f}}, u = axF'(n), v = -\sqrt{a v_f}F(n), N = ax\sqrt{\frac{a}{v_f}}G(n), \theta(n) = \frac{T - T_\infty}{T_w - T_\infty}$$

(6.1.17)

The non-dimensional variables are expressed as:

$$M^2 = \frac{\sigma_f B_0^2}{a\rho_f}, K = \frac{K}{\mu_f}, j = \frac{v_f}{a}, Pr = \frac{(\rho C_p)_f v_f}{k_f}, R = \frac{16\sigma^* T_\infty^3}{3k^* k_f}, S = -\frac{v_w}{\sqrt{a v_f}} \qquad (61.18)$$

The equations turned to a system of ODEs such as:

$$\left(\frac{\mu_{nf}}{\mu_f} + K\right)F'''(n)$$

$$+(\rho_{nf})F(n)F''(n) - (\rho_{nf})F'^2(n) - M\left(\frac{\sigma_{nf}}{\sigma_f}\right)F'(n) + KG'(n) = 0$$

(6.1.19)

$$\left(\frac{\mu_{nf}}{\mu_f} + \frac{K}{2}\right)G''(n) + (\rho_{nf})F(n)G'(n)$$

$$-(\rho_{nf})F'(n)G(n) - K(2G(n) + F''(n)) = 0$$

(6.1.20)

$$\frac{1}{Pr}\left(\frac{k_{nf}}{k_f}+R\right)\theta''\left(n\right)+\left(1-\phi+\phi\frac{\left(\rho C_p\right)_s}{\left(\rho C_p\right)_f}\right)F\left(n\right)\theta'\left(n\right)=0$$

(6.1.21)

The boundary conditions of the problem are:

$$F(n)=S,F'(n)=a,G(n)=-\delta F''(n),\theta(n)=1 \ @ \ n=0$$

$$F(n)=S,F'(n)=a,G(n)=-\delta F''(n),\theta(n)=1 \ @ \ n=0$$

(6.1.22)

$$F'(n)\rightarrow 0,G(n)\rightarrow 0,\theta(n)\rightarrow 0 \ @ \ n\rightarrow\infty$$

(6.1.23)

To demonstrate the procedure of this method, first, the formal state and boundary conditions of the equations are:

$$A(u)-f(r)=0 \ r\in\Omega$$

(6.1.24)

$$B\left(u,\frac{\partial u}{\partial n}\right)=0 \ r\in\Gamma$$

(6.1.25)

A and B are general differential and boundary operators, respectively. f is the analytic function, the boundary of the domain. Of course, A can contain L and N. where L is linear, and N is the nonlinear term in the equation. Hence the formula. (6.1.25) takes the form

$$L(u)+N(u)-f(r)=0, r\in\Omega \qquad (6.1.26)$$

Then, after the apply method to the equation, the general formation should be like this:

$$H(v,p)=(1-p)\left[L(v)-L(u_0)\right]+p\left[A(v)-f(r)\right]=0$$

(6.1.27)

Where:

$v(r,p)$: $\Omega\times[0,1]\rightarrow R$ (6.1.28)

Eq. (6.1.28) is an embedding parameter and the first approximation that satisfies the boundary condition. The solution of Eq. (6.1.28) is a power series in p, as follows:

$$v = v_0 + pv_1 + p^2 v_2 + \cdots$$

(6.1.29)

Eventually, the best approximation for the solution is:

$$u = \lim_{p\rightarrow 1} v = v_0 + v_1 + v_2 + \cdots$$

Conforming to section 3, HPM is applied to the equation as follows:

$$\begin{aligned}
H(F,p) = {} & (1-p)\Big(\Big(\frac{1}{(1-0.03)^{2.5}}+2\Big)F'''(n) + \Big(1-0.03+0.03\frac{5180}{997.1}\Big)F''(n) \\
& -10\Big(1+\frac{3(499999)0.03}{500002-(499999)0.03}\Big)F'(n) - \Big(\frac{1}{(1-0.03)^{2.5}}+2\Big)F_0''(n) \\
& -\Big(1-0.03+0.03\frac{5180}{997.1}\Big)F_0''(n) - 10\Big(1+\frac{3(499999)0.03}{500002-(499999)0.03}\Big)F_0'(n) \\
& +p\Big(\Big(\frac{1}{(1-0.03)^{2.5}}+2\Big)F'''(n) + \Big(1-0.03+0.03\frac{5180}{999.7}\Big)F(n)F''(n) \\
& -\Big(1-0.03+0.03\frac{5180}{999.7}\Big)(F'(n))^2 - 10\Big(1+\frac{3(499999)0.03}{500002-(499999)0.03)}F'(n) + 2G'(n)\Big)\Big)
\end{aligned}$$

(6.1.30)

$$\begin{aligned}
H(F,p) = {} & (1-p)\Big(\Big(\frac{1}{(1-0.03)^{2.5}}+1\Big)G''(n) + \Big(1-0.03+0.03\frac{5180}{997.1}\Big)G'(n) \\
& -\Big(1-0.03+0.03\frac{5180}{997.1}+4\Big)G(n) - \Big(\frac{1}{(1-0.03)^{2.5}}+1\Big)G_0''(n) \\
& +\Big(1-0.03+0.03\frac{5180}{997.1}\Big)G_0'(n) - \Big(1-0.03+0.03\frac{5180}{997.1}+4\Big)G_0(n)\Big) \\
& +p\Big(\Big(\frac{1}{(1-0.03)^{2.5}}+1\Big)G''(n) + \Big(1-0.03+0.03\frac{5180}{997.1}\Big)F(n)G'(n) \\
& -\Big(1-0.03+0.03\frac{5180}{997.1}\Big)F'(n)G(n) - 4(2G(n)+F''(n))\Big)
\end{aligned}$$

(6.1.31)

F and G polynomials are as follows:

$$F(n) = F_0(n) + F_1(n) + \ldots = \sum_{i=0}^{n} F_i\left(n\right)$$

(6.1.32)

$$G(n) = G_0(n) + G_1(n) + \ldots = \sum_{i=0}^{n} G_i\left(n\right)$$

(6.1.33)

After replacing F from the formula. (6.1.32) in (6.1.34) to G (6.1.35) in (6.1.33). Some simplifications and rearrangements based on powers of the p terms yield

$$3.079122293\,F_0'''(n) + 1.12585197\,F_o''(n) - 10.92782931\,F_0'(n) = 0$$

(6.1.34)

$$2.079122293\,G_o''(n) + 1.125851971\,G_0'(n) - 5.125851971\,G_0(n) = 0$$

(6.1.35)

For the boundary conditions:

$$F_0(0) = -1, F_0'(0) = -2, F_0'(10) = 0$$

(6.1.36)

$$G_0(0) = -1.9, G_0(10) = 0$$

(6.1.37)

By these boundary conditions, the solutions of Eq. (6.1.36) and (6.1.37) will be like these:

$$F(n) = -1.963599355 + 3.612833932 \times 10^{(-8)} e^{17.09910858-2{,}075551408n}$$

$$+ 4.385387999 \times 10^{(-8)} e^{-20.75551408+1.709910858n}$$

(6.1.28)

According to section 5, the polynomials are assumed as follows:

$$f(n) = \sum_{k=0}^{n} a_k(-e^{-1.2kn})$$

$$= -a_0 - a_1 e^{-1.2n} - a_2 e^{-2.4n} - a_3 e^{-3.6n} - a_4 e^{-4.8n} - a_5 e^{-6n} - a_6 e^{-7.2n}$$

(6.1.39)

According to this method, 12 equations and 12 unknowns are needed to solve ODEs. Therefore, some of the equations are pointed out below:

$$f(0) = -1 \rightarrow -a_0 - a_1 - a_2 - a_3 - a_4 - a_5 - a_6 = -1$$

(6.1.40)

$$f'(0) = -2 \rightarrow 1.2a_1 + 2.4a_2 + 3.6a_3 + 4.8a_4 + 6a_5 + 7.2a_6 = -2$$

(6.1.41)

$$g(0) = -1.9 \rightarrow b_0 + b_1 + b_2 + b_3 + b_4 = -1.9$$

(6.1.42)

After doing all the above steps, these proposed constants could be easily calculated.

$$a_0 = 7.281809116, a_1 = 2.5017098960 \times 10^{-42}, a_2 = -50.75623470,$$

$$a_3 = 106.6007624, a_4 = -98.77556994, a_5 = 44.74960348, a_6 = -8.100370343,$$

$$b_0 = -2.098282999 \times 10^{-43}, b_1 = -2.641415483 \times 10^{-32}, b_2 = -3.206675023,$$

$$b_3 = 1.561579050, b_4 = -.2549040276$$

(6.1.43)

By substituting the calculated constants into Eqs. (6.1.48) and (6.1.49), the series solution of this method will be as follows:

$$f(n) = -7.281809116 - 2.501798960 \times 10^{-42} e^{-1.2n} + 50.75623470 e^{-2.4n}$$

$$-106.6007624 e^{-3.6n} + 98.77556994 e^{-4.8n} - 44.74960348 e^{-6n} + 8.100370343 e^{-7.2n}$$

(6.1.44)

$$g(n) = -2.098282999 \times 10^{-43} - 2.64145483 \times 10^{-32} e^{-1.2n}$$

$$-3.206675023 e^{-2.4n} + 1.561579050 e^{-3.6n} - .2549040276 e^{-4.8n}$$

(6.1.45)

FEM is a numerical technique for solving partial differential and integral equations. In this study, FEM is applied based on the FlexPDE trade code. FlexPDE is powerful software that uses finite element methods to solve equations in minimal time. Some of the advantages of this software are that it can solve simultaneous partial differential equations, display 3D output, and take a short time to solve the equations. It belongs to the OSS (Open Source Software) category that allows arbitrary modification of the underlying equations. The mesh grid can be controlled with MESH_SPACING and MESH_DENSITY. ERRLIM, which is set to 0.2% by default, is the tolerance-bound formula. If any grid cell exceeds the tolerance, it is split, and the reconstruction process restarts. Flex PDE is, therefore, very reliable. This software has never been widely used in the field of mechanics, but its high accuracy and reliability will make it widely used in the future.

In the current manuscript, the Tiwari-Das model has been applied to study the interface of micropolar ferrofluids on foils that contract under thermal irradiation by a transverse magnetic field. To evaluate these semi-analytical methods, HPM, AGM, FEM, and comparison with the graphical results of Hussanan et al. The dimensionless velocity F'(η) and the dimensionless microrotational velocity G(η) were studied for magnetic, microrotational, and boundary parameter variations. A graphical and tabular comparison of these results demonstrates the reliability and accuracy of the semi-analytical method. Figures 2(a) and (b) show the effect of the magnetic parameter on the dimensionless velocity, showing that F'(η) increases with increasing magnetic parameters. Additionally, Tables 2 and 3 show the accuracy of the solution. Figure 3 (a) and (b) show how F'(η) behaves at different values of micro rotational velocity for aspiration and injection mass transfer. In suction mode, the thickness of F'(η) is smaller than in injection mode. Tables 4 and 5 show that the values of F'(η) in the absence of microrotation parameters are very low. From Figure 4 (a) and (b), The limiting parameters are inversely related to the dimensionless velocity for both suction and injection modes. As evident in the suction mode, the dimensionless velocity reaches sustenance faster than in the injection mode, and the injection mold has a thicker boundary layer than in the suction mode. Tables 6 and 7 are prepared for the absence of boundary parameters. Also, Figure 5 (a), (b), and (c) show the exclusive behavior of the micro-rotation velocity concerning the variation of the magnetic parameters. G(η) for suction and jet tuning swap passes with η=1. This means that it initially has the highest value at M=10 and the lowest value after η=1. There is no significant change in the boundary layer thickness compared to suction and injection moods. Figure 5 (c) shows a detailed and concise jet tuning showing the accuracy and compatibility of the solution. Tables 8 and 9 set the dimensionless fine rotation velocity values for the highest magnetic parameters.

Figure 1. Shrinking sheet (α < 0)

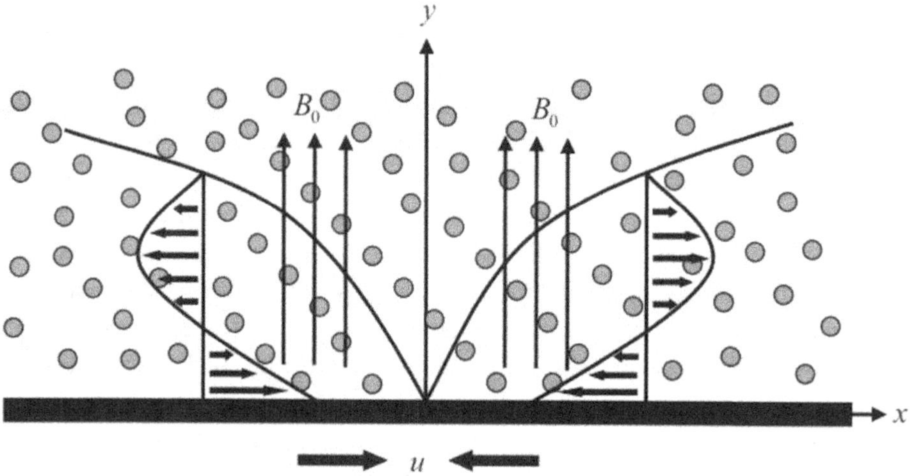

Table 1. Thermophysical properties of Fe_3O_4 and H_2O

	ρ(kg/m³)	C_p (J/kg K)	k (W/m k)	σ (Ω m)⁻¹
H_2O	997.1	4179	0.613	0.05
Fe_3O_4	5180	670	9.7	25000

Figure 2. Graphical comparison for the variation of the magnetic parameter between the result of HPM, AGM, FEM and Hussanan et al. for K=2, δ=0.5, Ø=0.03 (a) suction (S>0) and (b) injection (S<0)

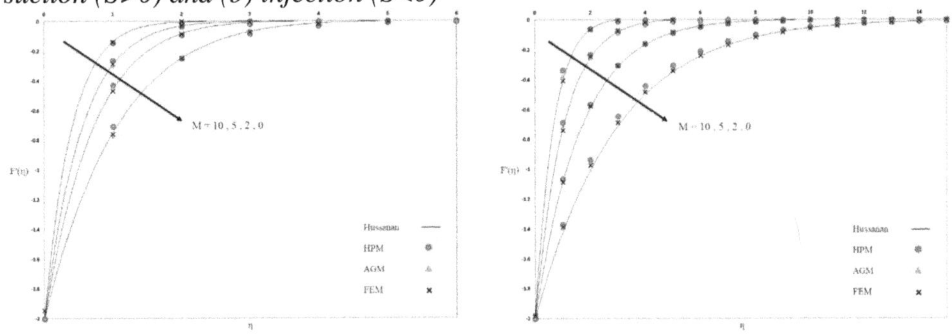

Table 2. Comparison for F (η) with S > 0, α < 0, Ø = 0.03, δ = 0.5, K = 2 & M=

η	Hussanan et al.	HPM	AGM
0	-2.00000000000000	-2.000000000	-2.000000003
1	-0.134837743679157	-0.1433646206	-0.1397245666
2	-0.00373289830387080	-0.01027670742e	-0.005098730486
3	0.000623499807710272	-0.0007366581527	-0.0001516284790
4	0.000136969133976153	-0.00005280535829	-0.000004240220158
5	0.0000168412066207519	-0.000003785210154	-0.0000001158451435
6	0.00000151736766844738	-0.0000002713326121	-0.000000003090808141

Table 3. Comparison for F (η) with S < 0, α < 0, Ø = 0.03, δ = 0.5, K = 2 & M=10

η	Hussanan et al.	HPM	AGM
0	-2.00000000000000	-2.000000001	-2.000000000
1	-0.414123789756412	-0.3370861682	-0.3755054253
2	-0.0636786979467218	-0.05681354297	-0.07050216295
3	0.000196140464266996	-0.009575529753	-0.01323697247
4	0.00623927388269486	-0.001613889332	-0.002485277540
5	0.00387455550545896	-0.0002720098854	-0.0004666176001
6	0.00181984492399330	-0.00004584538544	-0.00008760872099

continued on following page

Table 3. Continued

η	Hussanan et al.	HPM	AGM
7	0.000764434369589246	-0.000007726922578	-0.00001644877519
8	0.000302714510190792	-0.000001302319387	-0.000003088302227
9	0.000115689376412867	-0.0000002194969260	-0.0000005798371207
10	0.0000432032594190610	-0.00000003699468885	-0.0000001088659934
11	0.0000158812869313168	-0.000000006235198954	-0.00000002043988578
12	0.00000577322223020775	-0.000000001050899673	-0.000000003837644063
13	0.00000208189235919662	-0.0000000001771218715	-0.0000000007205280890
14	0.000000746335432178264	-2.985266660×10^ (-11)	-0.0000000001352811038
15	2.6638171373821×10^ (-7)	-5.031460155×10^ (-12)	-2.539938987×10^ (-11)

Figure 3. Graphical comparison for the variation of micro rotation parameter between the result of HPM, AGM, FEM and Hussanan et al. for M=5, δ=0.5, Ø=0.03 (a) suction (S>0) and (b) injection (S<0)

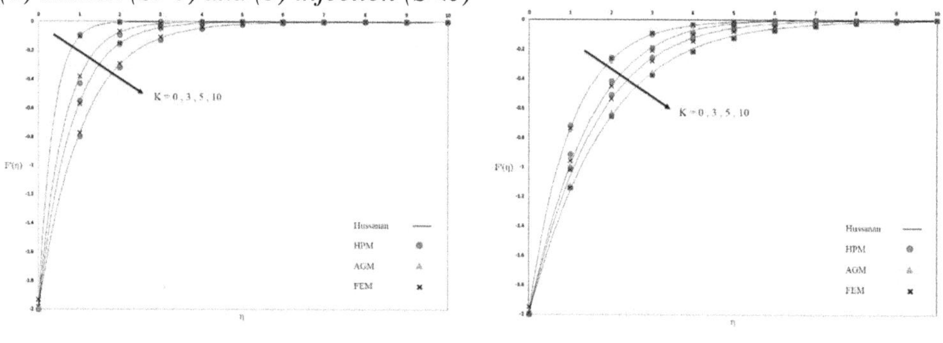

Table 4. Comparison for F (η) with S > 0, α < 0, Ø = 0.03, δ = 0.5, M = 5 & K = 0

η	Hussanan et al.	HPM	AGM
0	-2.00000000000000	-2.000000000	-2.000000002
1	-0.102333609351827	-0.09085689531	-0.1056737693
2	-0.00472524126910642	-0.004127487795	-0.005583472757
3	-0.000211299294752186	-0.0001875053632	-0.0002950133116
4	-9.344588354554×10^ (-6)	-0.000008518077579	-0.00001558758431
5	-4.116263362015×10^ (-7)	-0.0000003869630451	-0.0000008235993933

continued on following page

Table 4. Continued

η	Hussanan et al.	HPM	AGM
6	-1.810540435449×10^ (-8)	-0.00000001757913027	-0.00000004351642656
7	-7.95755412531×10^ (-10)	-7.985926063×10^ (-10)	-2.299272408×10^ (-9)
8	-3.49324815378×10^ (-11)	-3.627882279×10^ (-11)	-1.214863922×10^ (-10)
9	-1.48849027828×10^ (-12)	-1.648090585×10^ (-12)	-6.418962483×10^ (-12)
10	0	-7.487019758×10^ (-14)	-3.391579836×10^ (-13)

Table 5. Comparison for $F(\eta)$ with $S < 0$, $\alpha < 0$, $\emptyset = 0.03$, $\delta = 0.5$, $M = 5$ & $K = 0$

η	Hussanan et al.	HPM	AGM
0	-2.00000000000000	-2.000000000	-2.000000000
1	-0.883420212761216	-0.7163178409	-0.7530204800
2	-0.313431202840136	-0.2565555867	-0.2835197936
3	-0.0777170103009256	-0.09188757889	-0.1067477600
4	-0.0000535012803330599	-0.03291010632	-0.04019083832
5	0.0159923681370134	-0.01178633953	-0.01513055069
6	0.0133748983606653	-0.004219459597	-0.005693107311
7	0.00788558141944817	-0.001506342386	-0.002135425488
8	0.00383898328649040	-0.0005276650103	-0.0007860492343
9	0.00158718883783134	-0.0001612240546	-0.0002555530363
10	0.000537614831461316	0.000006264553393	-0.000004636463283

Figure 4. Graphical comparison for the variation of boundary parameter between the result of HPM, AGM, FEM and Hussanan et al. for M=5, K=2, Ø=0.03 (a) suction (S>0) and (b) injection (S<0)

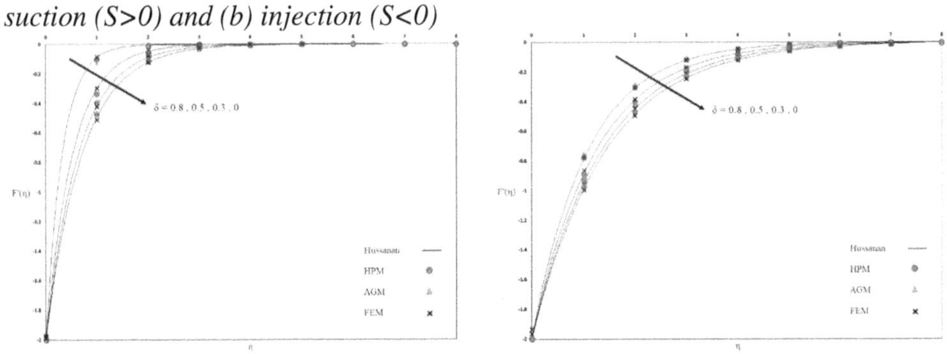

Table 6. Comparison for F (η) with S > 0, α < 0, Ø = 0.03, M = 5, K = 2 & δ = 0

η	Hussanan et al.	HPM	AGM
0	-2.00000000000000	-2.000000000	-2.000000001
1	-0.491245580743495	-0.4735392108	-0.4853585471
2	-0.121042808107032	-0.1121196920	-0.1177864595
3	-0.0298248356273956	-0.02654653523	-0.02858433270
4	-0.00734867387895637	-0.006285412736	-0.006936824953
5	-0.00181047670349067	-0.001488194678	-0.001683423655
6	-0.000445765469784635	-0.0003523592702	-0.0004085320254
7	-0.000109359907502846	-0.00008342796532	-0.00009914225411
8	-0.0000262702138934811	-0.00001975320641	-0.00002405977043

Table 7. Comparison for F (η) with S < 0, α < 0, Ø = 0.03, K = 2, M = 5 & δ = 0

η	Hussanan et al.	HPM	AGM
0	-2.000000002	-2.000000001	-1.999999999
1	-0.985711121	-0.9727407368	-0.9943924912
2	-0.485730705	-0.4731069249	-0.4944007836
3	-0.239198124	-0.2300926030	-0.2457968505
4	-0.117496977	-0.1118854088	-0.1221754719
5	-0.057154389	-0.05437062083	-0.06068201555
6	-0.0026761097	-0.02635552994	-0.03005400709
7	-0.00105651338	-0.01265185508	-0.01472712559
8	0.000000000000366	-0.005840608346	-0.006924790995

Figure 5. Graphical comparison for the variation of the magnetic parameter between the result of HPM, AGM, FEM and Hussanan et al. for M=5, K=2, Ø=0.03 (a) suction (S>0) and (b) injection (S<0) and (c) illustrate the precision of the results of (b) M=5

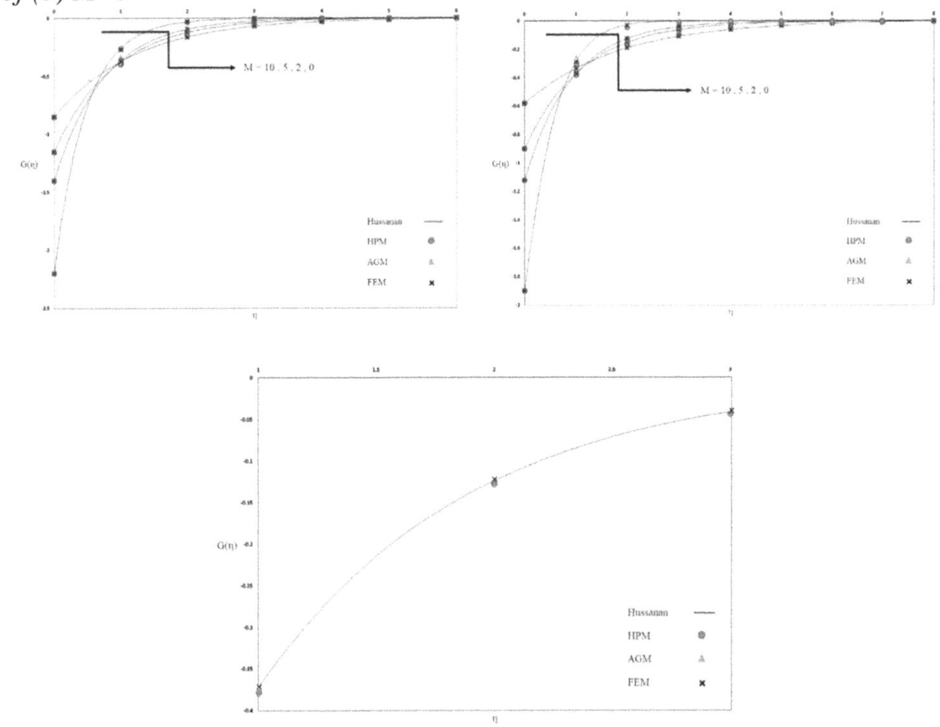

Table 8. Comparison for $G(\eta)$ with $S > 0$, $\alpha < 0$, $\emptyset = 0.03$, $\delta = 0.5$, $K = 2$ & $M=10$

η	Hussanan et al.	HPM	AGM
0	-2.20000000000000	-2.200000001	-2.200000000
1	-0.241907293990170	-0.2583874828	-0.2425628858
2	-0.0162473197020185	-0.03034731420	-0.02674397915
3	-0.000123939760672729	-0.003564257331	-0.002948680374
4	0.000153483826141641	-0.0004186179453	-0.0003251092821
5	0.0000270761703532344	-0.00004916619818	-0.00003584520256
6	0.00000289647030379403	-0.000005774513720	-0.000003952143573

Table 9. Comparison for $G(\eta)$ with $S < 0$, $\alpha < 0$, $\emptyset = 0.03$, $\delta = 0.5$, $K = 2$ & $M=10$

η	Hussanan et al.	HPM	AGM
0	-1.90000000000000	-1.900000001	-1.900000001
1	-0.282597729558912	-0.2945733285	-0.2503326381
2	-0.0263333169716698	-0.04567023493	-0.02524153588
3	-0.00122815380067409	-0.007080649030	-0.002362344708
4	0.000105284559385470	-0.001097773868	-0.0002163148078
5	0.0000341815454877478	-0.0001701972977	-0.00001967871909
6	0.00000485122299566926	-0.00002638707620	-0.000001786720019
7	0.000000450107702633713	-0.000004090738711	-0.0000001621287726
8	0.0000000209376760261659	-0.0000006331847208	-0.00000001470911598

REFERENCES

Abd-Alla, A., Abbas, I. A., Abo-Dhab, S. M., Elmhedy, Y., Sapoor, H., & Abdel-hafez, M. A. (2022). Effect of magnetic field and heat transfer on peristaltic flow of a micropolar fluid through a porous medium. *Waves in Random and Complex Media*, 1–12. DOI: 10.1080/17455030.2022.2058111

Agarwal, V., Singh, B., Kumari, A., Jamshed, W., Nisar, K. S., Almaliki, A. H., & Zahran, H. Y. (2021). Steady magnetohydrodynamic micropolar fluid flow and heat and mass transfer in permeable channel with thermal radiation. *Coatings*, 12(1), 11. DOI: 10.3390/coatings12010011

Agarwal, V., Singh, B., Kumari, A., Jamshed, W., Nisar, K. S., Almaliki, A. H., & Zahran, H. Y. (2022). Steady magnetohydrodynamic micropolar fluid flow and heat and mass transfer in permeable channel with thermal radiation. *Coatings*, 12(1), 11. DOI: 10.3390/coatings12010011

Ahmad, S., Ashraf, M., & Ali, K. (2019). Numerical simulation of viscous dissipation in a micropolar fluid flow through a porous medium. *Journal of Applied Mechanics and Technical Physics*, 60(6), 996–1004. DOI: 10.1134/S0021894419060038

Ahmad, S., Ashraf, M., & Ali, K. (2020). Micropolar fluid flow with heat generation through a porous medium. *Punjab University Journal of Mathematics, 52*(4).

Arafa, A. A., & Gorla, R. S. R. (1992). Mixed convection boundary layer flow of a micropolar fluid along vertical cylinders and needles. *International Journal of Engineering Science*, 30(12), 1745–1751. DOI: 10.1016/0020-7225(92)90092-U

Fakour, M., Vahabzadeh, A., Ganji, D. D., & Hatami, M. (2015). Analytical study of micropolar fluid flow and heat transfer in a channel with permeable walls. *Journal of Molecular Liquids*, 204, 198–204. DOI: 10.1016/j.molliq.2015.01.040

Fatunmbi, E., & Adeniyan, A. (2018). MHD stagnation point-flow of micropolar fluids past a permeable stretching plate in porous media with thermal radiation, chemical reaction, and viscous dissipation. *Journal of Advances in Mathematics and Computer Science*, 26(1), 1–19. DOI: 10.9734/JAMCS/2018/38595

Gorla, R. S. R. (1988). Buoyancy effects on the boundary layer flow of a micropolar fluid along a vertical cylinder. *International Journal of Engineering Science*, 26(8), 883–892. DOI: 10.1016/0020-7225(88)90039-0

Jalili, B., Alamdari, S. G., Jalili, P., & Gani, D. D. (2023). Analysis of bio-nanofluid flow over a stretching sheet with slip boundaries. *Results in Physics*, 54, 107083. DOI: 10.1016/j.rinp.2023.107083

Jalili, B., Azar, A. A., Jalili, P., & Ganji, D. D. (2023). Analytical approach for micropolar fluid flow in a channel with porous walls. *Alexandria Engineering Journal*, 79, 196–226. DOI: 10.1016/j.aej.2023.08.015

Jalili, B., Ganji, A. M., Shateri, A., Jalili, P., & Ganji, D. D. (2023). Thermal analysis of non-Newtonian visco-inelastic fluid MHD flow between rotating disks. *Case Studies in Thermal Engineering*, 49, 103333. DOI: 10.1016/j.csite.2023.103333

Jalili, B., Jalili, P., Rostami, E., & Ganji, D. D. (2023). *A Semi-Analytical Approach to Nonlinear Mechanical Engineering Issues*. Cambridge Scholars Publishing.

Jalili, B., Roshani, H., Jalili, P., & Ganji, D. D. (2023). Numerical study of Newtonian fluid flow characteristics in triangle enclosure considering various obstacle geometries. *International Journal of Thermofluids*, 20, 100515. DOI: 10.1016/j.ijft.2023.100515

Jalili, B., Shateri, A., Akgül, A., Bariq, A., Asadi, Z., Jalili, P., & Ganji, D. D. (2023). An investigation into a semi-porous channel's forced convection of nano fluid in the presence of a magnetic field as a result of heat radiation. *Scientific Reports*, 13(1), 18505. DOI: 10.1038/s41598-023-44275-4 PMID: 37898603

Jalili, P., & Afifi, , M., Jalili, B., Mirzaei, A. M., & Ganji, D. (2023). Numerical study and comparison of two-dimensional ferrofluid flow in semi-porous channel under magnetic field. *International Journal of Engineering*, 36(11), 2087–2101. DOI: 10.5829/IJE.2023.36.11B.13

Jalili, P., Shateri, A., Ganji, A. M., Jalili, B., & Ganji, D. D. (2023). Analytical analyzing mixed convection flow of nanofluid in a vertical channel using python approach. *Results in Physics*, 52, 106908. DOI: 10.1016/j.rinp.2023.106908

Krishna, M. V., Swarnalathamma, B., & Chamkha, A. J. (2018). Heat and mass transfer on magnetohydrodynamic chemically reacting flow of a micropolar fluid through a porous medium with Hall effects. *Special Topics & Reviews in Porous Media: An International Journal, 9*(4).

Mehmood, A., Afsar, K., Zameer, A., Awan, S. E., & Raja, M. A. Z. (2019). Integrated intelligent computing paradigm for the dynamics of micropolar fluid flow with heat transfer in a permeable walled channel. *Applied Soft Computing*, 79, 139–162. DOI: 10.1016/j.asoc.2019.03.026

Mirgolbabaee, H., Ledari, S. T., & Ganji, D. D. (2017). Semi-analytical investigation on micropolar fluid flow and heat transfer in a permeable channel using AGM. *Journal of the Association of Arab Universities for Basic and Applied Sciences*, 24(1), 213–222. DOI: 10.1016/j.jaubas.2017.01.002

Mirzaei, A., Jalili, P., Afifi, M. D., Jalili, B., & Ganji, D. D. (2023). Convection heat transfer of MHD fluid flow in the circular cavity with various obstacles: Finite element approach. *International Journal of Thermofluids*, 20, 100522. DOI: 10.1016/j.ijft.2023.100522

Mohamed, R. A., & Abo-Dahab, S. M. (2009). Influence of chemical reaction and thermal radiation on the heat and mass transfer in MHD micropolar flow over a vertical moving porous plate in a porous medium with heat generation. *International Journal of Thermal Sciences*, 48(9), 1800–1813. DOI: 10.1016/j.ijthermalsci.2009.01.019

Rashidi, M., Keimanesh, M., Bég, O. A., & Hung, T. K. (2011). Magnetohydrodynamic biorheological transport phenomena in a porous medium: A simulation of magnetic blood flow control and filtration. *International Journal for Numerical Methods in Biomedical Engineering*, 27(6), 805–821. DOI: 10.1002/cnm.1420

Reddy Gorla, R. S., Pender, R., & Eppich, J. (1983). Heat transfer in micropolar boundary layer flow over a flat plate. *International Journal of Engineering Science*, 21(7), 791–798. DOI: 10.1016/0020-7225(83)90062-9

Turkyilmazoglu, M. (2016). Flow of a micropolar fluid due to a porous stretching sheet and heat transfer. *International Journal of Non-linear Mechanics*, 83, 59–64. DOI: 10.1016/j.ijnonlinmec.2016.04.004

Section 4
Applications of MHD in Microchannels and Heat Transfer

This section discusses the application of MHD principles in microchannels and ways to improve heat transfer efficiency in MHD systems.

Chapter 7
MHD in Micro Channel

ABSTRACT

This chapter analyzes exponential heat source and induced magnetic field on second-class convection of Casson fluid in a microchannel. The flow is in a vertical microchannel organized by two vertical plates. The study of the natural convection flow of a fluid in the presence of an induced magnetic field has always been of considerable importance due to its many applications in various fields of science, technology, and industry, such as the operation of magnetohydrodynamic generators. Diagrams describing the conclusions of physical variables influencing velocity, temperature, current density, induced magnetic field, and skin friction coefficient distribution are shown.

7.1. CONVECTION WITH AN EXPONENTIAL HEAT SOURCE

Imagine two infinite vertical planes, as shown in Figure 1. A second kind of convection with an exponential heat source in the presence of CF is assumed in the vertical microchannel. Flows are considered unidirectional and non-transitory. The assumed coordinate axes are the x¢-axis opposite to gravity's direction and the y¢-axis in the direction perpendicular to the plate. The heating of the plates is asymmetric, and the left and right plate temperatures are maintained at T_1 and $T_{2,}$ respectively ($T_1 > T_2$). As a result, nonlinear convection occurs within the microchannel due to the large temperature difference between the two plates.

Based on these hypotheses, the governing equations of the magnetic field, energy, and motion are as follows (Nisar et al., 2022) (Deebani *et al.*, 2023) (Jalili *et al.*, 2023a) (Dawar *et al.*, 2021) (Shah *et al.*, 2020) (Jalili *et al.*, 2023b) (Uddin *et al.*, 20230 (Uddin *et al.*, 2019) (Jalili *et al.*, 2023c) (Zohra *et al.*, 2019) (Shateri *et al.*, 2023) (Jalili *et al.*, 2023d) (Patil *et al.*, 2012) (Pasha *et al.*, 2023) (Patil *et al.*, 2010) (Patil *et al.*, 2012) (Jalili *et al.*, 2023e) (Patil *et al.*, 2019) (Shamshuddin *et al.*, 2023) (Humane *et al.*, 2023) (Mabood *et al.*, 2022) (Shamshuddin & Ibrahim, 2022):

DOI: 10.4018/979-8-3693-5335-6.ch007

$$\nu\left(1+\frac{1}{\beta}\right)\left(\frac{d^2u'}{dy'^2}\right)+\left(\frac{\mu_e H_0'}{\rho}\right)\left(\frac{dH_x'}{dy'}\right)+g\left[\beta_0(T'-T_0)+\beta_1(T'-T_0)^2\right]=0,$$

(7.1.1)

$$\frac{1}{\beta\mu_e}\left(\frac{d^2H_x'}{dy'^2}\right)+H_0'\left(\frac{du'}{dy'}\right)=0,$$

(7.1.2)

$$k\frac{d^2T'}{dy'^2}+q_e(T_1-T_0)exp\left(\frac{-ny'}{b}\right)=0,$$

(7.1.3)

The corresponding boundary conditions are:

$$u'(y')\Big)=\frac{2-\sigma_v}{\sigma_v}\lambda\left(\frac{du'}{dy'}\right),H_x'(y')=0,$$

(7.1.4)

$$k\frac{d^2T'}{dy'^2}+q_e(T_1-T_0)exp\left(\frac{-ny'}{b}\right)=0,$$

(7.1.5)

$$T'(y')\Big)=T_2+\frac{2-\sigma_t}{\sigma_t}\left(\frac{2\upsilon}{\nu+1}\right)\left(\frac{\lambda}{Pr}\right)\left(\frac{dT'}{dy'}\right)\ at\ y'=0,$$

(7.1.6)

$$T'(y')\Big)=T_2+\frac{2-\sigma_t}{\sigma_t}\left(\frac{2\upsilon}{\nu+1}\right)\left(\frac{\lambda}{Pr}\right)\left(\frac{dT'}{dy'}\right)\ at\ y'=0,$$

(7.1.7)

$$T'(y')\Big)=T_1-\frac{2-\sigma_t}{\sigma_t}\left(\frac{2\upsilon}{\nu+1}\right)\left(\frac{\lambda}{Pr}\right)\left(\frac{dT'}{dy'}\right)\ at\ y'=b.$$

(7.1.8)

Dimensionless quantities are defined as:

$$y=\frac{y'}{b},u=\left(\frac{\nu u'}{g\beta_0 b^2(T_1-T_0)}\right),\theta=\frac{T'-T_0}{T_1-T_0},Pm=\nu\sigma\mu_e,$$

$$\alpha = \frac{\beta_1}{\beta_0}(T_1 - T_0), Q = \frac{q_e b^2}{k}, M = \frac{H'_0 b}{\nu}\sqrt{\frac{\mu_e}{\rho}},$$

(7.1.9)

According to the coupled scheme of nonlinear differential equations described above, the boundary conditions are due to the basic idea of AGM. (1) to (5) were rewritten in the following order.

$$\left(1 + \frac{1}{\beta}\right)\left(\frac{d^2u}{dy^2}\right) + M\left(\frac{dH}{dy}\right) + \theta + \alpha\theta^2 = 0,$$

(7.1.10)

$$\frac{d^2H}{dy^2} + MPm\left(\frac{du}{dy}\right) = 0,$$

(7.1.11)

$$\frac{d^2\theta}{dy^2} + Qexp(-ny) = 0,$$

(7.1.12)

$$\frac{d^2\theta}{dy^2} + Qexp(-ny) = 0,$$

(7.1.13)

$$u(y) = \beta_v Kn\left(\frac{du}{dy}\right), H(y) = 0, \theta(y) = \xi + \beta_v Kn\ ln\left(\frac{d\theta}{dy}\right)\ at\ y = 0,$$

(7.1.14)

$$u(y) = -\beta_v Kn\left(\frac{du}{dy}\right), H(y) = 0, \theta(y) = 1 - \beta_v Kn\ ln\left(\frac{d\theta}{dy}\right)\ at\ y = 1,$$

(7.1.15)

Following the basic idea of AGM, suitable test functions were investigated as solutions to hypothesized differential equations. It is considered a finite series of polynomials with constant coefficients, as follows:

$$u(y) = \sum_{i=0}^{6} d_i y^i = d_0 + d_1 y + d_2 y^2 + d_3 y^3 + d_4 y^4 + d_5 y^5 + d_6 y^6$$

(7.1.16)

$$H(y) = \sum_{i=0}^{8} e_i y^i = e_0 + e_1 y + e_2 y^2 + e_3 y^3 + e_4 y^4 + e_5 y^5 + e_6 y^6 + e_7 y^7 + e_8 y^8$$

(7.1.17)

$$\theta(y) = \sum_{i=0}^{6} f_i y^i = f_0 + f_1 y + f_2 y^2 + f_3 y^3 + f_4 y^4 + f_5 y^5 + f_6 y^6$$

(7.1.18)

The fundamental objective of this portion is to examine the effect of Q, Pm, M, α, β, $\beta_v k_n$, n, and ln on $u(y)$, $H(y)$, $\theta(y)$, $J(y)$, τ, Q_m, and Nu profiles under three case of ξ ($\xi = 1$ for symmetrical heating, $\xi = 0$ for one heating and one not heating, $\xi = -1$ for one cooling and one heating). The provided investigation has been carried out in the slips and continuum flow regimes ($K_n \leq 0.1$). via the numerical calculation the other parameters are chosen as $M = 5$, $Pm = \alpha = \beta = n = 0.5$, $\beta_v k_n = 0.05$, $ln = 1.667$ and $Q = 2$. Figure 2 (a & b) represents the impact of ξ and Q on (y) and (y) when $M = 5$, $Pm = \beta = \alpha = n = 0.5$, $\beta_v k_n = 0.05$ & $ln = 1.667$. Here Q and ξ raise with increasing (y) and (y), this is due to the portion of more heat into the System under review so that the fluid particles have attracted the heat and therefore have an orientation to move faster. It is also seen that there is a junction point inside the microchannel, which causes the (y) to be autonomous of Q. Figures 3 (a) and 4 (a) show the effectiveness of M and Pm on (y), respectively. Increases as decreases. A force that represents the opposing force on the velocity field.

Also seen when the heating is asymmetric ($\xi = -1$), the velocity field is independent of M and Pm, but with a connection point inside the microchannel. The effects of Pm and M on the microchannel slip velocity become more pronounced as ξ increases. Figures 3 (b) and 4 (b) represent the variation of M and P_m with H, respectively. Also, within the vertical microchannel, there is a connection point that makes H(y) independent of Pm and M. The effects of Pm and M on the microchannel slip velocity become more pronounced as it increases. Figures 5 (a & b) and 6 (a & b) show the effect of β and α on (y) and (y), respectively. As α increases, it is observed that the velocity profile increases due to strong lift forces. As the velocity increases with β, the yield point decreases. An increase in α and CF leads to an increase in H near the microchannel wall at y=0, and the opposite feature is seen at y=1. In addition, existing connection points within the vertical microchannels of β and α make H autonomous. Figures 7(a) and 8(a) show the effect of βvkn & ln on (y). An increase in βvkn, ln & ξ leads to an increase in fluid slip and thus fluid velocity, where the velocity of the liquid increases in increments of ln. Furthermore, the effect of βvkn and ln on velocity becomes more pronounced as ξ increases. Figures 7 (b) and 8 (b) show the effect of βvkn and ln on (y). Note that (y) increases as βvkn, ln & ξ increase. The numerical values of Qm (rkf-45) for various M, Pm, β, n, and ln for $Q = 2$ and $\alpha = 0.5$ are shown in Table 1 for $\xi = 1$ and $\xi = 0$, increase. A linear regression slope using the data points is fitted to know the value of decreasing or increasing Q_m. It is observed that Q_m is a decreasing function of Pm, M, and n, and Qm is an increasing function of β and ln. The effect of β on Qm is greater than ln. Tables 2 and 3 show the numerical values of t at $y = 0$ and $y = 1$ for different M, P_m, β, n, and ln for $Q = 2$ and $\alpha = 0.5$, respectively. It turns out that τ_0 is an increasing function of P_m, M, β, n, ln. The effect of n on τ0 is considerably more important than P_m, M, β, and ln. Table 4 shows that τ_1 is an increasing function of Pm, M, and n and a decreasing function of β and ln.

Figure 1. The physical adjustment of the microchannel

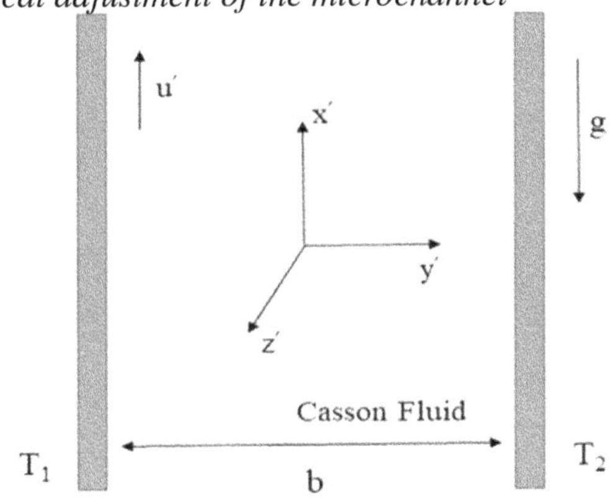

Figure 2. (y) and (y) for various Q

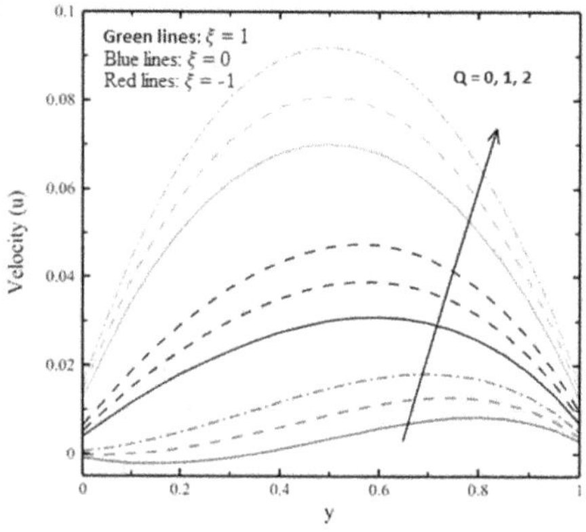

a

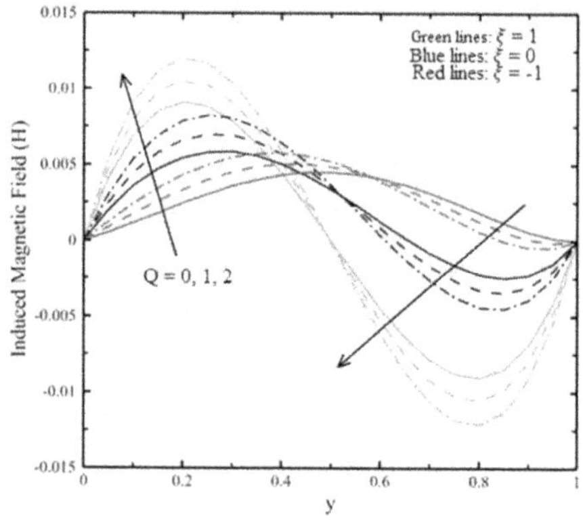

b

Figure 3. (y) and (y) for various M

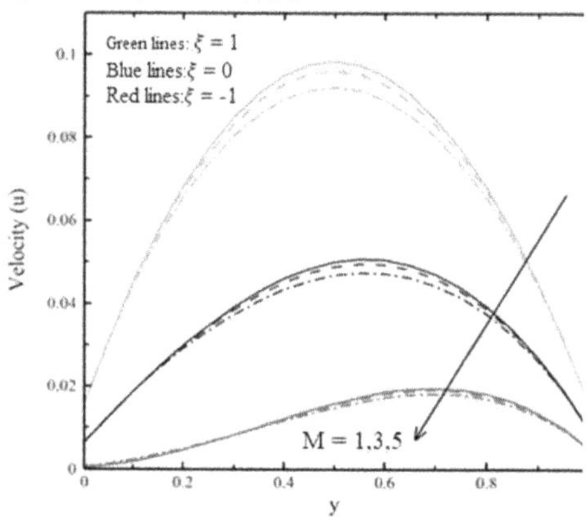

a

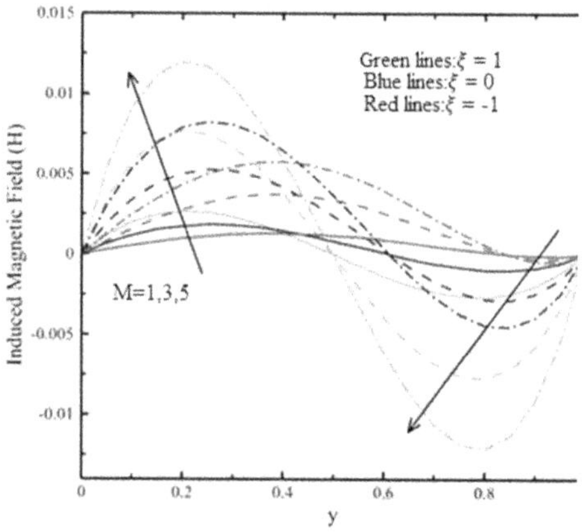

b

Figure 4. (y) and (y) for various Pm

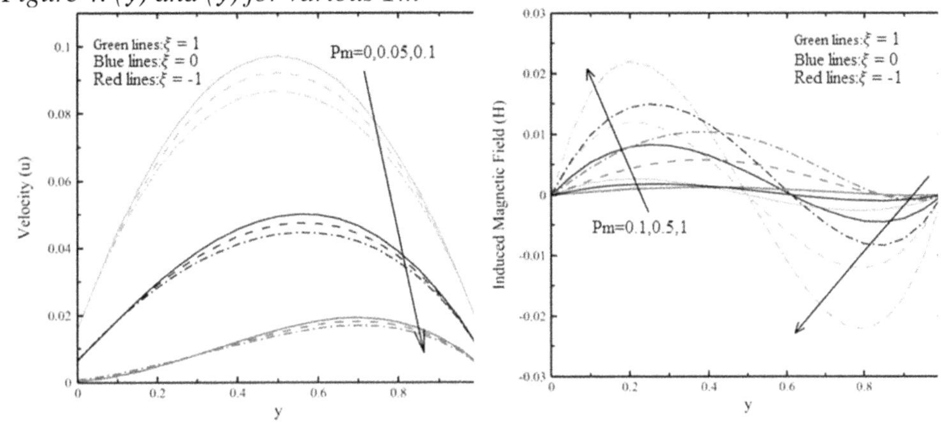

a **b**

Figure 5. (y) and (y) for various β

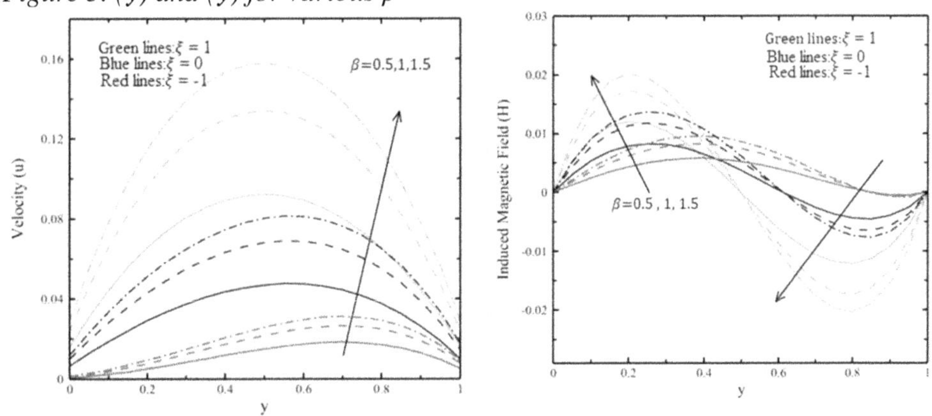

a **b**

Figure 6. (y) and (y) for various α

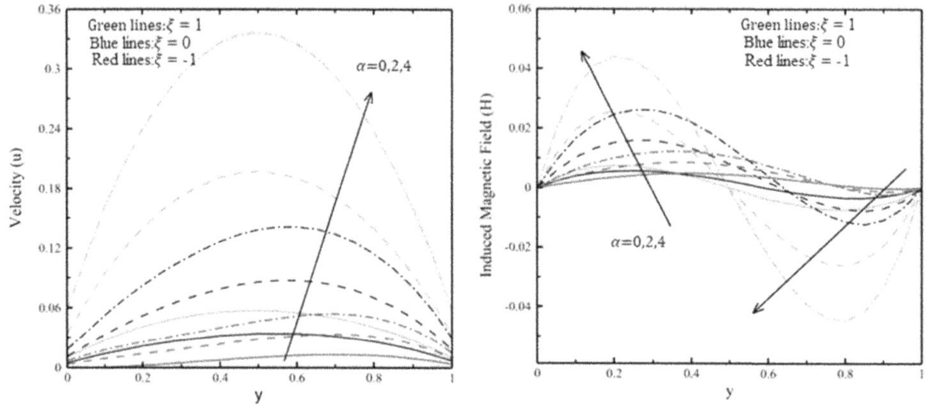

a b

Figure 7. (y) and (y) for various $\beta_v k_n$

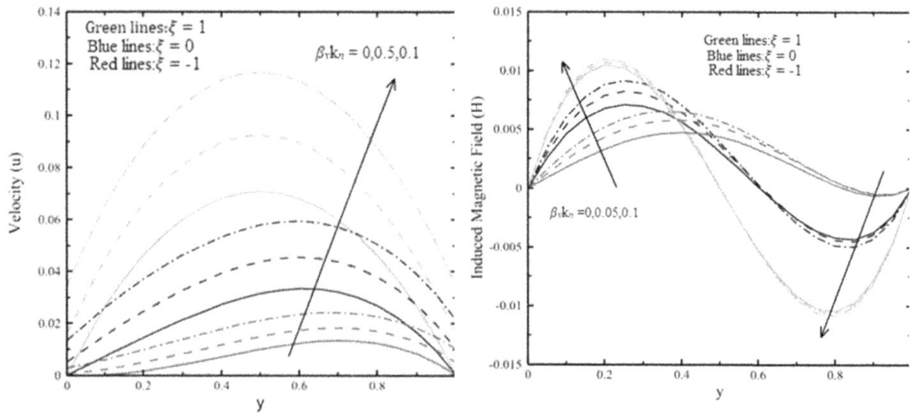

a b

Figure 8. (y) and (y) for various ln

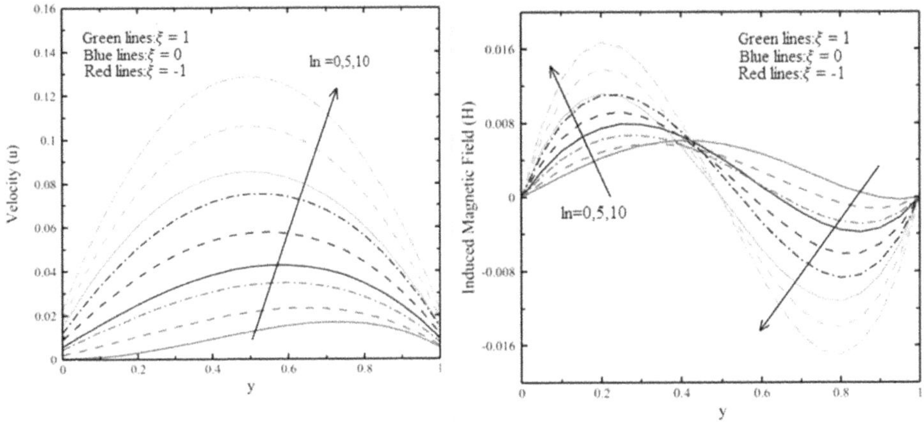

a b

Table 1. Comparison of AGM and rkf-45 method solution for Nu_0 and Nu_1 when Pm = β = n = α = 0.5, ln =1.667, M = 5 and ξ = 1

Q	Nu_0		Nu_1	
	AGM	RKFM	AGM	RKFM
0.1	0.0420859260	0.0421457082	-0.0364650960	-0.0365481598
0.2	0.0841718520	0.0842914165	-0.0729301920	-0.0730963197
0.3	0.1264482569	0.1264371247	-0.1095846862	-0.1096444795
0.4	0.1683437039	0.1685828329	-0.1458603841	-0.1461926393

Table 2. Numerical (rkf-45) values of Q_m for various M, Pm, β, n, ln when Q = 2, $\beta_v k_n$ =0.05 and α = 0.5

M	Pm	β	n	ln	Q_m for (ξ = 0)	Q_m for (ξ = 1)
1	0.5	0.5	2	1.667	0.02974	0.06374
1.5	0.5	0.5	2	1.667	0.02966	0.06357
2	0.5	0.5	2	1.667	0.02955	0.06334
5	0.5	1	2	1.667	0.04136	0.08825
5	0.5	1.5	2	1.667	0.04891	0.10409

continued on following page

Table 2. Continued

M	*Pm*	β	n	In	Q_m for (ξ = 0)	Q_m for (ξ = 1)
5	0.5	2	2	1.677	0.05382	0.11431
5	1	0.5	2	1.667	0.02691	0.05715
5	1.5	0.5	2	1.677	0.02568	0.05379
5	2	0.5	2	1.677	0.02456	0.05044
5	0.5	0.5	2	1	0.05958	0.02757
5	0.5	0.5	2	1.5	0.06028	0.02809
5	0.5	0.5	2	2	0.06098	0.02864

Table 3. Numerical (rkf-45) values of τ_0 for various M, Pm, β, n, In when Q = 2, α = 0.5

M	*Pm*	β	n	In	$_0$ for (ξ = 0)	$_0$ for (ξ = 1)
1	0.5	0.5	2	1.667	0.10408	0.28838
1.5	0.5	0.5	2	1.667	0.10424	0.28835
2	0.5	0.5	2	1.667	0.10445	0.28831
5	0.5	1	2	1.667	0.16058	0.43141
5	0.5	1.5	2	1.667	0.19323	0.51737
5	0.5	2	2	1.677	0.21500	0.57459
5	1	0.5	2	1.667	0.10750	0.28729
5	1.5	0.5	2	1.677	0.107728	0.28645
5	2	0.5	2	1.677	0.107169	0.28509
5	0.5	0.5	2	1	0.101305	0.28241
5	0.5	0.5	2	1.5	0.105093	0.28650
5	0.5	0.5	2	2	0.108857	0.29060

Table 4. Numerical (rkf-45) values of τ_1 for various M, Pm, β, n, In when Q = 2, α = 0.5

M	*Pm*	β	n	In	$_1$ for (ξ = 0)	$_1$ for (ξ = 1)
1	0.5	0.5	2	1.667	-0.16845	-0.28382
1.5	0.5	0.5	2	1.667	-0.16830	-0.28382
2	0.5	0.5	2	1.667	-0.16807	-0.28384

continued on following page

Table 4. Continued

M	*Pm*	β	n	In	$_1$ for (ξ = 0)	$_1$ for (ξ = 1)
5	0.5	1	2	1.667	-0.24667	-0.42666
5	0.5	1.5	2	1.667	-0.29487	-0.51254
5	0.5	2	2	1.677	-0.32686	-0.56998
5	1	0.5	2	1.667	-0.16343	-0.28499
5	1.5	0.5	2	1.677	-0.16194	-0.28677
5	2	0.5	2	1.677	-0.16109	-0.28948
5	0.5	0.5	2	1	-0.16428	-0.27931
5	0.5	0.5	2	1.5	-0.16527	-0.28289
5	0.5	0.5	2	2	-0.16646	-0.28651

7.2. MICROPOLAR FLUIDS

In the theory of micropolar fluids, two tensors exhibit asymmetry. The first tensor represents the asymmetric stress and is denoted as T={T_{ij} }. The second tensor corresponds to the couple stress and is denoted as C={C_{ij} }. These two tensors can be expressed mathematically as shown in equations (7.2.1, 7.2.2):

$$T_{ij} = \left(-P + \lambda V_{kk}\right)\delta_{ij} + \bar{\mu}\left(V_{ij} + V_{ji}\right) + \bar{\kappa}\left(V_{ji} - \varepsilon_{ijk}\omega_k\right),$$

(7.2.1)

$$C_{ij} = \alpha_1 \omega_{kk}\delta_{ij} + \alpha_2 \omega_{ij} + \alpha_3 \omega_{ji},$$

(7.2.2)

where pressure is denoted by *p*, the micro rotational field is denoted by ω= (ω_i,ω_j,ω_k), the velocity field is denoted by V= (V_i, V_j, V_k), the bulk viscosity coefficient is denoted by λ, the shear viscosity coefficient is denoted by $\bar{\mu}$, the vortex viscosity coefficient is denoted by $\bar{\kappa}$, the spin gradient viscosity coefficients are denoted by α_1, α_2, and α_3, the Levi-Civita tensor is denoted by ε_{ijk}, the Kronecker delta function is denoted by δ_{ij}.

The current research focuses on the methodology for investigating the steady and laminar flow of a two-dimensional micropolar fluid between two parallel permeable porous walls. The problem involves uniform injection or suction of the fluid at a velocity of v_0, and is illustrated in Figure (7.2.1).

The governing equations for the problem, as stated by Mirgolbabaee et al., Mirzaaghaian and Ganji, Mehmood et al., and Sibanda and Awad, can be expressed in the Cartesian coordinate system as follows:

$$\frac{\partial u}{\partial x} + \frac{\partial v}{\partial y} = 0 , \tag{7.2.3}$$

$$\rho\left(u\frac{\partial u}{\partial x} + v\frac{\partial u}{\partial y}\right) = -\frac{\partial P}{\partial x} + (\mu + \kappa)\left(\frac{\partial^2 u}{\partial x^2} + \frac{\partial^2 u}{\partial y^2}\right) + \kappa\left(\frac{\partial N}{\partial y}\right), \tag{7.2.4}$$

$$\rho\left(u\frac{\partial v}{\partial x} + v\frac{\partial v}{\partial y}\right) = -\frac{\partial P}{\partial y} + (\mu + \kappa)\left(\frac{\partial^2 v}{\partial x^2} + \frac{\partial^2 v}{\partial y^2}\right) - \kappa\left(\frac{\partial N}{\partial x}\right), \tag{7.2.5}$$

$$\rho\left(u\frac{\partial N}{\partial x} + v\frac{\partial N}{\partial y}\right) = -\frac{\kappa}{j}\left(2N + \frac{\partial u}{\partial y} - \frac{\partial v}{\partial x}\right) + \frac{\nu_s}{j}\left(\frac{\partial^2 N}{\partial x^2} + \frac{\partial^2 N}{\partial y^2}\right), \tag{7.2.6}$$

$$\rho\left(u\frac{\partial T}{\partial x} + v\frac{\partial T}{\partial y}\right) = \frac{k_1}{c_p}\frac{\partial^2 T}{\partial y^2}, \tag{7.2.7}$$

$$\rho\left(u\frac{\partial C}{\partial x} + v\frac{\partial C}{\partial y}\right) = D^*\frac{\partial^2 C}{\partial y^2}, \tag{7.2.8}$$

Equations (7.2.3) involve several variables, where u and v represent the velocity components in the Cartesian system, and x and y denote the Cartesian coordinate components. The fluid temperature is represented by T, and C denotes the fluid concentration. Other variables include the fluid density, which is denoted by ρ, the dynamic viscosity, denoted as μ, and the specific heat at constant pressure, denoted by c_p. The angular or micro-rotation velocity is denoted by N, while the thermal conductivity is represented as k_1. There are also additional parameters such as molecular diffusivity, denoted by D*; micro-inertia density, denoted by j; micro rotation viscosity, represented by ν_s=(μ+k/2)j; a coupling coefficient, denoted by κ, and the fluid pressure, denoted by P.

And also, the boundary conditions related to the physics of the problem for Eqs. (7.2.8) are as follows:

$$v = -v_0, u = 0, N = -s\frac{\partial u}{\partial y}\bigg|_{y=-h}, T = T_1, C = C_1 \text{ when } y = -h,$$

(7.2.9)

$$v = v_0, u = 0, N = -s\frac{\partial u}{\partial y}\bigg|_{y=h}, T = T_2, C = C_2 \text{ when } y = h,$$

In Equation (7.2.9), the parameter $v_0>0$ corresponds to suction, while $v_0<0$ corresponds to injection. The boundary parameter, denoted as s, determines the extent to which microelements are allowed to rotate near the walls of the channel. When s=0, the microelements cannot rotate near the walls. However, when s=0.5, it indicates a weak concentration and the absence of the asymmetric part of the stress tensor. Finally, when s=1, it represents turbulent flows.

To make the governing equations of Eqs. (7.2.8) dimensionless, similarity variables and their corresponding boundary conditions from Eq. (7.2.9) were employed, as presented in.

$$\xi = \frac{y}{h}, \psi = -v_0 x f(\xi), N = \frac{v_0 x}{h^2} g(\xi)$$

(7.2.10)

$$\theta(\xi) = \frac{T - T_2}{T_1 - T_2}, \phi(\xi) = \frac{C - C_2}{C_1 - C_2},$$

Equation (7.2.10) introduces various dimensionless variables: f(ξ) represents the dimensionless stream function, g(ξ) represents the dimensionless microrotation function, θ(ξ) represents the dimensionless temperature, and ϕ(ξ) represents the dimensionless concentration. N corresponds to the microrotation or angular velocity, while h denotes the width of the channel. Additionally, ψ represents the stream function, ξ stands for the dimensionless independent variable, T represents the temperature function, C represents the concentration function, T_1 and C_1 refer to the temperature and concentration of the lower plate, respectively, and T_2 and C_2 represent the temperature and concentration of the upper plate, respectively.

Using the similarity variables mentioned above, the governing equations of Eqs. (7.2.11) along with their corresponding boundary conditions from Eq. (7.2.9), were transformed into a dimensionless form as follows:

$$(1+N_1)f^{(4)} - N_1 g - Re(ff''' - ff') = 0,$$

(7.2.11)

$$N_2 g'' + N_1(f'' - 2g) - N_3 Re(fg' - gf) = 0,$$

(7.2.12)

$$\theta'' + Pe_h(f'\theta - f\theta') = 0,$$

(7.2.13)

$$\phi'' + Pe_m(f'\phi - f\phi') = 0,$$

(7.2.14)

In Eqs. (7.2.11-7.2.14), various parameters are defined: $N_1=\kappa/\mu$ represents the coupling parameter, $N_2=v_s/\mu h^2$ denotes the spin-gradient viscosity parameter, and $N_3=j/h^2$ represents the ratio of micro-inertia density over the width of the channel. Additionally, Pe_h=Pr Re is the Peclet number for heat diffusion, where Pr is the Prandtl number, and Pe_m=Sc Re is the Peclet number for mass diffusion, where Sc is the Schmidt number. The Reynolds number, Re, is also present, and it was shown by Mirgolbabaee et al. and Sibanda and Awad to be negative in the case of injection and positive for suction. Furthermore, the transformation variables introduced in Eq. (7.2.10) result in a modified form of Eq. (7.2.9) given by Eq. (7.2.15).

$$f = f' = g = 0, \theta = \phi = 1, \text{ when } \xi = -1$$

(7.2.15)

$$f = \theta = \phi = 0, f' = -1, ,g = 1 \text{ when } \xi = 1$$

The HAN (Homotopy Analysis Method) was developed by Jalili et al. to approximate analytical solutions for differential equations. Here, we provide an explanation of the HAN method:

Let's consider a differential equation of mth order given by:

$$\Gamma(u(\xi), u'(\xi), u''(\xi), \dots, u^{(m)}(\xi)) = 0.$$

(7.2.16)

Eq. (7.2.16) is nonlinear ODE, and Γ is the function of u and its derivatives with respect to ξ. The parameter u is the function of the independent variable ξ. The derivatives of the function $u(\xi)$ with respect to ξ at ξ=0 and ξ=L are denoted in Eq. (7.2.17) as follows:

$$\begin{cases} u(\xi) = u_0\,, u'(\xi) = u_1\,,\ldots,u^{(m-1)}(\xi) = u_{m-1} \; when\; \xi = 0, \\ u(\xi) = u_{L_0}\,, u'(\xi) = u_{L_1}\,,\ldots,u^{(m-1)}(\xi) = u_{L_{m-1}} \; when\; \xi = L\,. \end{cases} \tag{7.2.17}$$

The solution of Eq. (7.2.17) is considered as follows:

$$u(\xi) = \sum_{i=0}^{n} a_i \xi^i = a_0 + a_1 \xi^1 + a_2 \xi^2 + \ldots + a_n \xi^n, \tag{7.2.18}$$

In equation (7.2.18), a_0, a_1, ..., a_n are constant coefficients where n>m. By solving a system of n+1 equations with n+1 unknowns, fixed coefficients are determined. The boundary conditions of the problem can be used to construct some of these equations, which can be expressed as follows:

$$\begin{cases} u(0) = a_0 = u_0, \\ u'(0) = a_1 = u_1\,, \\ u''(0) = a_2 = u_2, \end{cases} \tag{7.2.19}$$

$$\begin{cases} u(L) = a_0 + a_1 L + a_2 L^2 + \ldots + a_n L^n = u_{L_0}, \\ u'(L) = a_1 + 2a_2 L + 3a_3 L^2 + \ldots + n a_n L^{n-1} = u_{L_1}, \\ u''(L) = 2a_2 + 6a_3 L + 12a_4 L^2 + \ldots + n(n-1) a_n L^{n-2} = u_{L_2}, \end{cases} \tag{7.2.20}$$

The equations obtained from the boundary conditions of the problem, as shown in Eqs. (7.2.19, 7.2.20), Since it is initially assumed that the value of n in this methodology is greater than m, they have limitations. However, additional boundary equations are required, and these can be approximated using a numerical method (regardless of which specific method). Thus, the newly approximated boundary conditions are given as follows:

$$
\begin{cases}
u(\xi) = \alpha_0 \, , \, u'(\xi) = \alpha_1 \, , \ldots , u^{(m-1)}(\xi) = \alpha_{m-1} \; at \; \xi = L_0 \\
u(\xi) = \beta_0 \, , \, u'(\xi) = \beta_1 \, , \ldots , u^{(m-1)}(\xi) = \beta_{m-1} \; at \; \xi = L_1 \\
u(\xi) = \gamma_0 \, , \, u'(\xi) = \gamma_1 \, , \ldots , u^{(m-1)}(\xi) = \gamma_{m-1} \; at \; \xi = L_2 \\
\cdots \\
u(\xi) = \varepsilon_0 \, , \, u'(\xi) = \varepsilon_1 \, , \ldots , u^{(m-1)}(\xi) = \varepsilon_{m-1} \; at \; \xi = L_z
\end{cases}
\tag{7.2.21}
$$

For instance, the following equations are constructed from approximated boundary conditions of Eq. (7.2.21):

$$
\begin{cases}
u(L_0) = a_0 + a_1(L_0) + a_2(L_0)^2 + \ldots + a_n(L_0)^n = \alpha_0 \\
u'(L_0) = a_1 + 2a_2(L_0) + 3a_3(L_0)^2 + \ldots + na_n(L_0)^{n-1} = \alpha_1 \\
u''(L_0) = 2a_2 + 6a_3(L_0) + 12a_4(L_0)^2 + \ldots + n(n-1)a_n(L_0)^{n-2} = \alpha_2 , \\
\cdots \\
\left(u^{(m-1)}(\xi)\right)_{\xi=L_0} = \alpha_{m-1}
\end{cases}
\tag{7.2.22}
$$

$$
\begin{cases}
u(L_1) = a_0 + a_1(L_1) + a_2(L_1)^2 + \ldots + a_n(L_1)^n = \beta_0 \\
u'(L_1) = a_1 + 2a_2(L_1) + 3a_3(L_1)^2 + \ldots + na_n(L_1)^{n-1} = \beta_1 \\
u''(L_1) = 2a_2 + 6a_3(L_1) + 12a_4(L_1)^2 + \ldots + n(n-1)a_n(L_1)^{n-2} = \beta_2 \\
\cdots \\
\left(u^{(m-1)}(\xi)\right)_{\xi=L_1} = \beta_{m-1}
\end{cases}
\tag{7.2.23}
$$

$$
\begin{cases}
u(L_2) = a_0 + a_1(L_2) + a_2(L_2)^2 + \ldots + a_n(L_2)^n = \gamma_0 \\
u'(L_2) = a_1 + 2a_2(L_2) + 3a_3(L_2)^2 + \ldots + na_n(L_2)^{n-1} = \gamma_1 \\
u''(L_2) = 2a_2 + 6a_3(L_2) + 12a_4(L_2)^2 + \ldots + n(n-1)a_n(L_2)^{n-2} = \gamma_2 \\
\cdots \\
\left(u^{(m-1)}(\xi)\right)_{\xi=L_2} = \gamma_{m-1}
\end{cases}
\tag{7.2.24}
$$

$$
\begin{cases}
u(L_z) = a_0 + a_1(L_z) + a_2(L_z)^2 + \ldots + a_n(L_z)^n = \varepsilon_0 \\
u'(L_z) = a_1 + 2a_2(L_z) + 3a_3(L_z)^2 + \ldots + na_n(L_z)^{n-1} = \varepsilon_1 \\
u''(L_z) = 2a_2 + 6a_3(L_z) + 12a_4(L_z)^2 + \ldots + n(n-1)a_n(L_z)^{n-2} = \varepsilon_2 \\
\cdots \\
\left(u^{(m-1)}(\xi)\right)_{\xi=L_z} = \varepsilon_{m-1}
\end{cases}
\tag{7.2.25}
$$

The Akbari-Ganji Method was previously developed for the analytical solution of nonlinear differential equations and has been successfully applied to solve many problems that cannot be solved using exact analytical methods. However, this paper proposes a modification to the method to obtain more accurate solutions. To explain the main idea behind the modified AGM, let's assume the general form of an mth-order differential equation as given in Eq. (7.2.26):

$$\Theta : f(u, u', u'', \ldots, u^{(m)}) = 0 \; ; u = u(\xi)$$

(7.2.26)

with boundary conditions:

$$\begin{cases} u(\xi) = u_0, u'(\xi) = u_1, \ldots, u^{(m-1)}(\xi) = u_{m-1}, aaat\ \xi = 0 \\ u(\xi) = u_{L_0}, u'(\xi) = u_{L_1}, \ldots, u^{(m-1)}(\xi) = u_{L_{m-1}}, aaat\ \xi = L \end{cases}$$

(7.2.27)

In Eq. (7.2.28) below, express the solution of Eq. (7.2.29) as a polynomial of degree n with unknown constant coefficients:

$$u(\xi) = \sum_{i=0}^{n} a_i \xi^i = a_0 + a_1 \xi^1 + a_2 \xi^2 + \ldots + a_n \xi^n,$$

(7.2.28)

Here, $a_0, a_1, \ldots, a_n$ are $n+1$ constant coefficients which $n>m$. By solving a system of $n+1$ unknowns and $n+1$ equations, constant coefficients will be determined. According to Eq. (7.2.27), the boundary conditions of the problem can construct some of these equations as follows:

$$\begin{cases} u(0) = a_0 = u_0, \\ u'(0) = a_1 = u_1, \\ u''(0) = a_2 = u_2, \end{cases}$$

(7.2.29)

$$\begin{cases} u(L) = a_0 + a_1 L + a_2 L^2 + \ldots + a_n L^n = u_{L_0}, \\ u'(L) = a_1 + 2a_2 L + 3a_3 L^2 + \ldots + n a_n L^{n-1} = u_{L_1}, \\ u''(L) = 2a_2 + 6a_3 L + 12 a_4 L^2 + \ldots + n(n-1) a_n L^{n-2} = u_{L_2}, \end{cases}$$

(7.2.30)

This section of the paper presents the results and discussions of the problem involving steady, laminar, two-dimensional micropolar fluid flow between two parallel permeable porous walls. The main physical parameters considered in this study are N_1, N_2, N_3, Re, Pe_h, and Pe_m. These parameters were chosen based on Eqs. (7.2.11-7.2.15) as they are the only ones that can impact the physical quantities, such as the dimensionless stream function, micro rotation, temperature, and concentration. The results and discussions presented in this section were obtained using the solutions of the modified AGM and HAN methods. The impact of different coupling parameters on the dimensionless stream function, microrotation, temperature, and concentration profiles is examined in Figures 10 to 13 with Re=5, N_2=1, N_3=1, Pe_h=2.5, and Pe_m=7.5. Tables 5 and 6 provide the average values of each profile as the coupling parameter varies. Additionally, Tables 7 to 9 illustrate Nusselt number, Sherwood number, and skin friction coefficient variations as the coupling parameter ranges from 0 to 6.

Tables 5 and 6 display average results that exhibit changes as the coupling parameter transitions from case 0 to case 6. In some cases, these values increase, while in others, they decrease. The calculation of the decrease or increase of these values can be determined using the following relationship:

$$\Omega = \frac{Z_2 - Z_1}{|Z_1|} \times 100, \tag{7.2.31}$$

The equation (7.2.31) describes the relationship where Ω represents the percentage change in values, Z_2 is the second value, and Z_1 is the first value. Based on the results obtained from the modified AGM solutions presented in Table 5, when the coupling parameter changes from 0 to 6, the average value of the dimensionless stream function increases by 18.45225856%. In contrast, the average value of the dimensionless microrotation decreases by 93.54091441%. Additionally, the average value of the dimensionless temperature increases by 5.078799162%, while the average value of the dimensionless concentration increases by 24.21210464%.

According to the results obtained from the HAN method solutions in Table 6, when the coupling parameter changes from 0 to 6, the average value of the dimensionless stream function increases by 18.39008647%. However, the average value of the dimensionless microrotation decreases by 93.75079947%. Furthermore, the average value of the dimensionless temperature increases by 5.087382342%, while the average value of the dimensionless concentration increases by 24.28531617%.

Figures 14 to 19 examine the impact of different spin-gradient viscosity parameters on the dimensionless stream function, microrotation, temperature, and concentration profiles when Re=5, N_1=0.1, N_3=1, Pe_h=2.5, and Pe_m=7.5. Tables 10 and 11 provide the average value of each profile as the spin-gradient viscosity parameter varies. Additionally, Tables 10 and 11 illustrate the variations of the Nusselt number, Sherwood number, and skin friction coefficient as the spin-gradient viscosity parameter ranges from 0.1 to 4.

Table 10 presents the results obtained from the modified AGM solutions, indicating that as the spin-gradient viscosity parameter varies from 0.1 to 4, the average value of the dimensionless stream function increases by 0.8061658368%, while the average value of the dimensionless microrotation significantly increases by 1080.256194%. Moreover, the average value of the dimensionless temperature increases by 0.2268230056%, and the average value of the dimensionless concentration increases by 0.9311482285%.

Based on the findings from Table 11 regarding the HAN method solutions, it can be observed that as the spin-gradient viscosity parameter ranges from 0.1 to 4, the average value of the dimensionless stream function increases by 0.9205635298%. Additionally, there is a substantial increase in the average value of the dimensionless microrotation by 704.5221970%. Moreover, the average value of the dimensionless temperature shows a rise of 0.2438746051%, while the average value of the dimensionless concentration increases by 0.9977525134%. Figures 20 and 21 explore the impact of different ratios of micro-inertia density over the width of the channel on the dimensionless stream function, microrotation, temperature, and concentration profiles when Re=5, N_1=0.1, N_2=1, Pe_h=2.5, and Pe_m=7.5. Tables 13 and 14 present the average value of each profile as the ratio of micro-inertia density changes. Additionally, Tables 15 to 17 illustrate the variations of the Nusselt number, Sherwood number, and skin friction coefficient as the ratio of micro-inertia density over the width of the channel ranges from 0 to 4. Figures 22 to 25 examine the impact of different Peclet numbers on the dimensionless stream function, microrotation, temperature, and concentration profiles when Re=5, N_1=1, N_2=1, N_3=1, and Pe_m =7.5. These figures aim to investigate the diffusion of heat in the system and how it affects the mentioned variables. Figures 26 to 29 examine the impact of different Peclet numbers on the dimensionless stream function, microrotation, temperature, and concentration profiles when Re=5, N_1=1, N_2=1, N_3=1, and Pe_h =2.5. These figures aim to investigate the diffusion of mass in the system and how it affects the mentioned variables. Additionally, Tables 18 and 19 present the average value of each profile as the Peclet number for the diffusion of mass undergoes changes.

This section aims to qualitatively summarize the quantitative results and discussions presented in the previous section, in order to draw a clear and general conclusion. Based on the findings discussed it can be concluded that if the problem's physics are related to steady, laminar, two-dimensional micropolar fluid flow between two parallel permeable porous walls, certain variables such as the dimensionless stream function, microrotation, temperature, and concentration profiles are affected by factors like spin-gradient viscosity parameter, the ratio of micro-inertia density over the width of the channel, and Peclet numbers for mass and heat diffusion. These findings contribute to a better understanding of micropolar fluid dynamics and have implications for various engineering applications that involve porous materials and micropolar fluids.

- With an increase in the coupling parameter, the average values of the dimensionless parameters (dimensionless stream function, temperature, and concentration) generally increase. However, the dimensionless microrotation profile does not show a significant change with the increase in the coupling parameter.
- The average values of the dimensionless parameters (dimensionless stream function, microrotation, temperature, and concentration) tend to increase with an increase in the spin-gradient viscosity parameter.

Figure 9. (a) Geometry of the problem in 3D. (b) The geometry of the problem in 2D

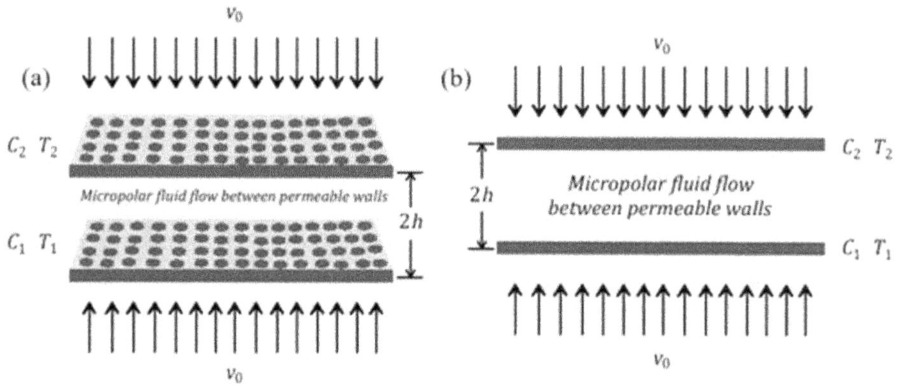

Table 5. The results from the adjusted AGM solutions' average values as the coupling parameter change from 0 to 6

	N_1=0	N_1=1	N_1=2	N_1=3	N_1=6
$\bar{f}_{avg}$	0.12063042	0.134453653	0.138574068	0.140554005	0.142889457
$\bar{g}_{avg}$	0.417677048	0.192191466	0.122505709	0.084784902	0.026978118
$\bar{\theta}_{avg}$	0.659779545	0.680670189	0.686936143	0.689894739	0.693288423
$\bar{\phi}_{avg}$	1.490148388	1.710292683	1.781892398	1.814652416	1.850944675

Table 6. The results from the average values of the HAN method solutions as the coupling parameter change from 0 to 6

	N_1=0	N_1=1	N_1=2	N_1=3	N_1=6
$\bar{f}_{avg}$	0.120646616	0.134455754	0.138586272	0.140535741	0.142833633
$\bar{g}_{avg}$	0.416751149	0.19122096	0.121318593	0.083493848	0.026043615
$\bar{\theta}_{avg}$	0.659682362	0.680638695	0.686957543	0.689894584	0.693242926
$\bar{\phi}_{avg}$	1.491130918	1.712051537	1.783704098	1.816922507	1.853256776

Table 7. The variation of Nusselt number when the coupling parameter changes from 0 to 6

N_1	The HAN method solutions		The modified AGM solutions	
	$Nu_{x(+1)}$	$Nu_{x(-1)}$	$Nu_{x(+1)}$	$Nu_{x(-1)}$
0	0.892935247597681	0.0272835490250930	0.8984060000	0.030234
1	0.934504714709690	-0.0466555983635985	0.9386960000	-0.043836
2	0.946937439753559	-0.0693065520902910	0.9506980000	-0.066418
3	0.952648832927337	-0.0800699299486687	0.9562480000	-0.077188
6	0.959030643646631	-0.0927957738973633	0.9626120000	-0.089952

Table 8. The variation of Sherwood number when the coupling parameter changes from 0 to 6

N_1	The HAN method solutions		The modified AGM solutions	
	$Sh_{x(+1)}$	$Sh_{x(-1)}$	$Sh_{x(+1)}$	$Sh_{x(-1)}$
0	3.21240952959046	-2.09309271584898	3.290310000	-2.05821
1	3.75175365228856	-2.71427155705224	3.829060000	-2.67266
2	3.92544073077383	-2.91717806111481	4.001990000	-2.87281
3	4.00516138438528	-3.01272149299149	4.081350000	-2.96745
6	4.09077332542568	-3.12037066249956	4.169130000	-3.07687

Table 9. The variation of skin friction coefficient when the coupling parameter changes from 0 to 6

N_1	The HAN method solutions		The modified AGM solutions	
	$C_{fx(+1)}$	$C_{fx(-1)}$	$C_{fx(+1)}$	$C_{fx(-1)}$
0	-2.516030859	0.582662113	-2.518347662	0.586009076
1	-4.286806843	1.441918942	-4.286362711	1.443473826
2	-6.108122798	2.299071502	-6.106387451	2.299636062
3	-7.949880516	3.158715421	-7.944636467	3.158552275
6	-13.53730874	5.744222553	-13.52105817	5.743525745

Table 10. The results from the adjusted AGM solutions' average values as the spin-gradient viscosity parameter change from 0.1 to 4

	N_2=0.1	N_2=1	N_2=2	N_2=3	N_2=4
$\bar{f}_{avg}$	0.122524294	0.123317212	0.123480319	0.123527738	0.123512043
$\bar{g}_{avg}$	- 0.047329958	0.371381657	0.430458971	0.452523955	0.463954845
$\bar{\theta}_{avg}$	0.662476452	0.663690497	0.66393716	0.663998999	0.663979101
$\bar{\phi}_{avg}$	1.519512422	1.531532275	1.533214155	1.533861155	1.533661335

Table 11. The results from the average values of the HAN method solutions as the spin-gradient viscosity parameter changes from 0.1 to 4

	N_2=0.1	N_2=1	N_2=2	N_2=3	N_2=4
$\bar{f}_{avg}$	0.122419362	0.123351821	0.123475512	0.123522058	0.12354631
$\bar{g}_{avg}$	- 0.076748226	0.37136963	0.43047525	0.452512314	0.463960062
$\bar{\theta}_{avg}$	0.662387951	0.663722115	0.663901075	0.663968323	0.664003347
$\bar{\phi}_{avg}$	1.518689334	1.531162191	1.532868627	1.533508801	1.533842095

Table 12. The variation of Nusselt number when the spin-gradient viscosity parameter changes from 0.1 to 4

N_2	The HAN method solutions		The modified AGM solutions	
	$Nu_{x(+1)}$	$Nu_{x(-1)}$	$Nu_{x(+1)}$	$Nu_{x(-1)}$
0.1	0.898437458408870	0.0180034344605650	0.8986035000	0.01762870
1	0.900941017862617	0.0130184414091202	0.9008456000	0.01307420
2	0.901278677090900	0.0123561682437580	0.9012135000	0.01233410
3	0.901405362111217	0.0121068384659770	0.9013287000	0.01217750
4	0.901471311339760	0.0119769111323805	0.9013897000	0.01197670

Table 13. The variation of Sherwood number when the spin-gradient viscosity parameter changes from 0.1 to 4.

N_2	The HAN method solutions		The modified AGM solutions	
	$Sh_{x(+1)}$	$Sh_{x(-1)}$	$Sh_{x(+1)}$	$Sh_{x(-1)}$
0.1	3.28103229567621	-2.16904089739150	3.280720000	-2.171660
1	3.31010322074959	-2.20609600208883	3.307630000	-2.206110
2	3.31410181920878	-2.21111989257703	3.312020000	-2.211640
3	3.31559978120628	-2.21300780541482	3.312150000	-2.212410
4	3.31637933459360	-2.21399119669594	3.313370000	-2.214030

Table 14. The variation of skin friction coefficient when the spin-gradient viscosity parameter changes from 0.1 to 4

N_2	The HAN method solutions		The modified AGM solutions	
	$C_{fx(+1)}$	$C_{fx(-1)}$	$C_{fx(+1)}$	$C_{fx(-1)}$
0.1	-2.706334586	0.6589169989	-2.704898958	0.6593778859
1	-2.687343953	0.6700982464	-2.687378556	0.6700011618
2	-2.684794583	0.6715441269	-2.685009614	0.6717187822
3	-2.683843559	0.6720907485	-2.683861969	0.6718245090
4	-2.683349392	0.6723759200	-2.683664979	0.6726049343

Table 15. The variation of Nusselt number when the ratio of micro-inertia density over the width of the channel changes from 0 to 4

N_2	The HAN method solutions		The modified AGM solutions	
	$Nu_{x(+1)}$	$Nu_{x(-1)}$	$Nu_{x(+1)}$	$Nu_{x(-1)}$
0	0.936700247639827	-0.0507631679243867	0.9367224200	-0.05087758
1	0.934504714709690	-0.0466555983635985	0.9386960000	-0.043836
2	0.933424342253513	-0.0447819624256969	0.9376280000	-0.042148
3	0.933571786006221	-0.0454129080513491	0.9382200000	-0.043720
4	0.936103872887036	-0.0510720727748737	0.9361721596	-0.0510427180

Table 16. The variation of the Sherwood number is when the ratio of micro-inertia density over the width of the channel changes from 0 to 4

N_3	The HAN method solutions		The modified AGM solutions	
	$Sh_{x(+1)}$	$Sh_{x(-1)}$	$Sh_{x(+1)}$	$Sh_{x(-1)}$
0	3.78249018377029	-2.75104380985881	3.780160000	-2.751980
1	3.75175365228856	-2.71427155705224	3.829060000	-2.67266
2	3.73597519094272	-2.69644311145973	3.814180000	-2.65614
3	3.73615165968189	-2.69918238626337	3.818080000	-2.66480
4	3.76644440505942	-2.74202173743628	3.771252009	-2.7413137

Table 17. The variation of skin friction coefficient when the ratio of micro-inertia density over the width of the channel changes from 0 to 4

N_3	The HAN method solutions		The modified AGM solutions	
	$C_{fx(+1)}$	$C_{fx(-1)}$	$C_{fx(+1)}$	$C_{fx(-1)}$
0	-4.257115604	1.456853797	-4.256913402	1.456433976
1	-4.286806843	1.441918942	-4.286362711	1.443473826
2	-4.302706794	1.435965225	-4.300990531	1.438369867
3	-4.303457473	1.440449814	-4.297405668	1.446169272
4	-4.275540124	1.466952097	-4.275519230	1.466951778

Table 18. The results from the average values of the modified AGM solutions as the Peclet number for the diffusion of mass change from 0 to 4

	Pe_m=0	Pe_m=1	Pe_m=2	Pe_m=3	Pe_m=4
$\bar{f}_{avg}$	0.134455754	0.134455754	0.134455754	0.134455754	0.134455754
$\bar{g}_{avg}$	0.19122096	0.19122096	0.19122096	0.19122096	0.19122096
$\bar{\theta}_{avg}$	0.680638693	0.680638693	0.680638693	0.680638693	0.680638693
$\bar{\phi}_{avg}$	0.500000005	0.56050238	0.635907226	0.731056233	0.852954772

Table 19. The results from the average values of the HAN method solutions as the Peclet number for the diffusion of mass change from 0 to 4

	Pe_m=0	Pe_m=1	Pe_m=2	Pe_m=3	Pe_m=4
$\bar{f}_{avg}$	0.134455754	0.134455754	0.134455754	0.134455754	0.134455754
$\bar{g}_{avg}$	0.19122096	0.19122096	0.19122096	0.19122096	0.19122096
$\bar{\theta}_{avg}$	0.680638693	0.680638693	0.680638693	0.680638693	0.680638693
$\bar{\phi}_{avg}$	0.500000005	0.56050238	0.635907226	0.731056233	0.852954773

Figure 10. The impact of the coupling parameter when it changes from 0 to 6 for the dimensionless stream function

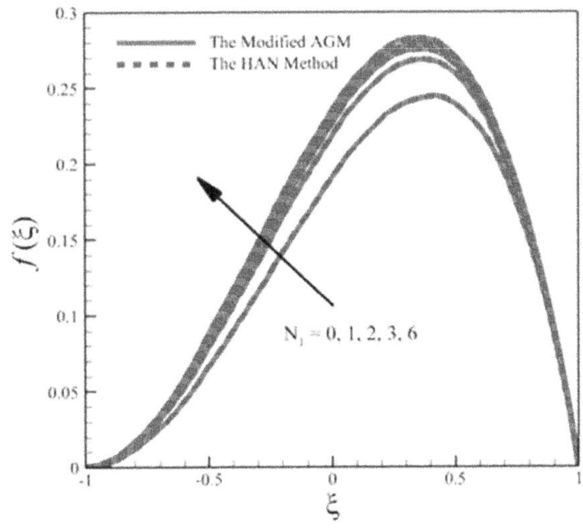

Figure 11. The impact of the coupling parameter when it changes from 0 to 6 for the dimensionless microrotation function

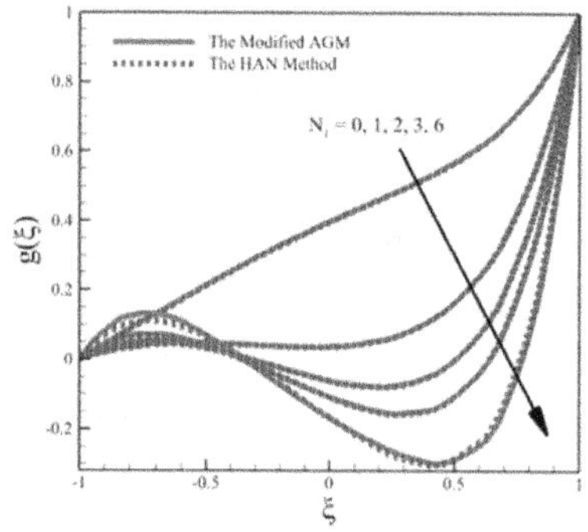

Figure 12. The impact of the coupling parameter when it changes from 0 to 6 for the dimensionless temperature

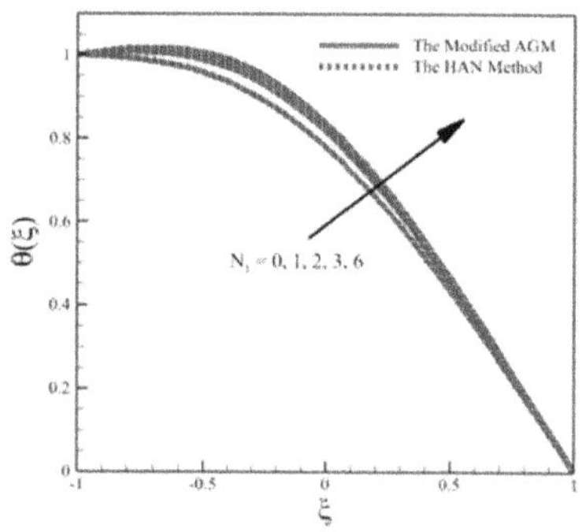

Figure 13. The impact of the coupling parameter when it changes from 0 to 6 for the dimensionless concentration

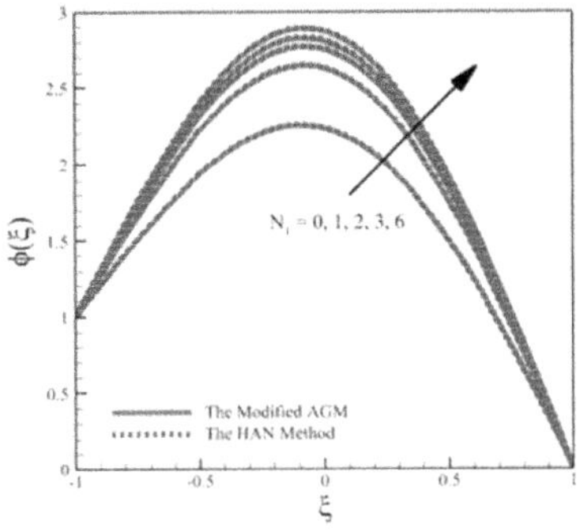

Figure 14. The impact of the spin-gradient viscosity parameter when it changes from 0 to 4 for the dimensionless stream function

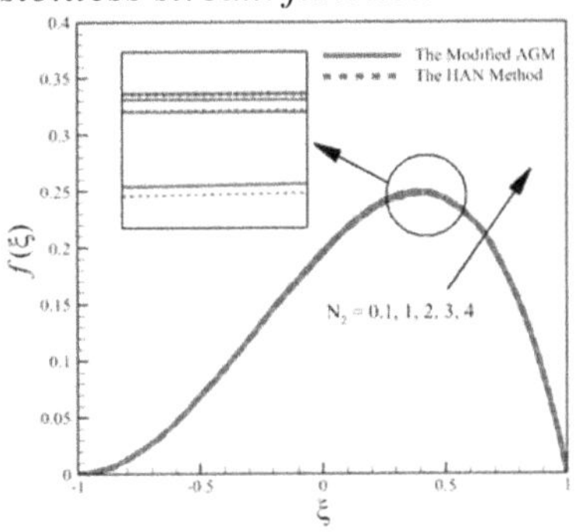

Figure 15. The impact of the spin-gradient viscosity parameter when it changes from 0 to 4 for the dimensionless microrotation function

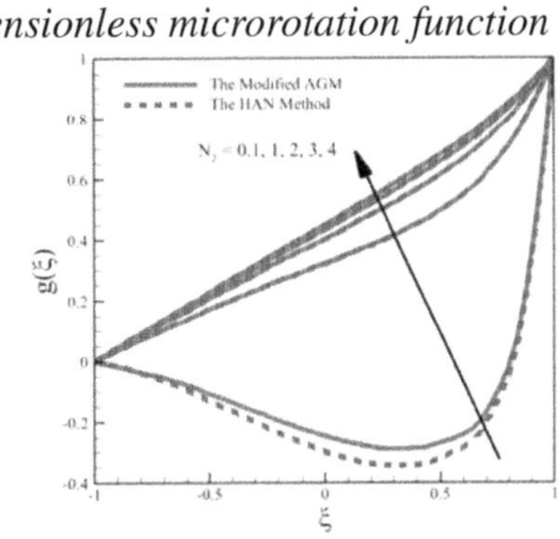

Figure 16. The impact of the spin-gradient viscosity parameter when it changes from 0 to 4 for the dimensionless temperature

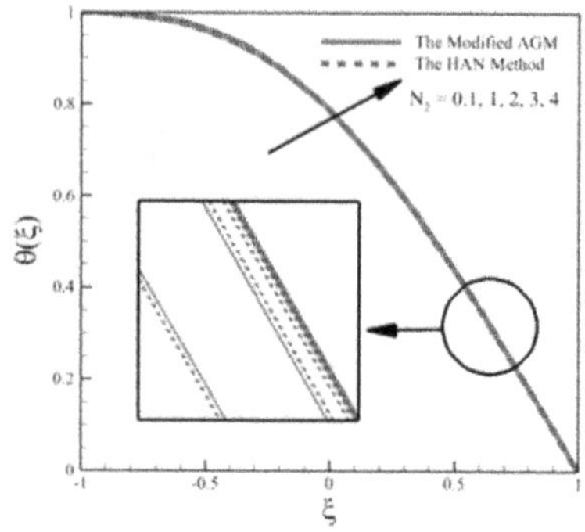

Figure 17. The impact of the spin-gradient viscosity parameter when it changes from 0 to 4 for the dimensionless concentration

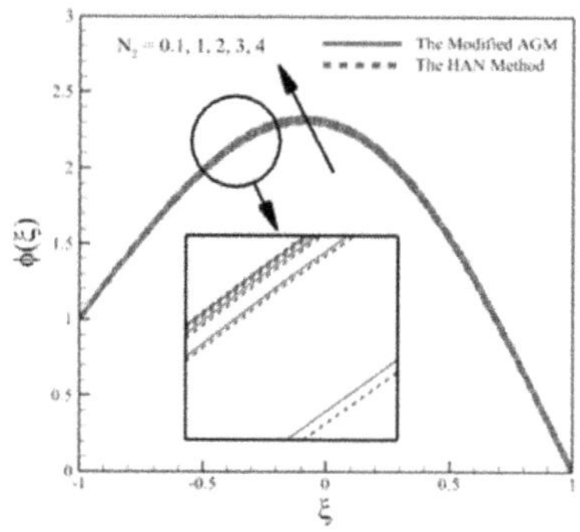

Figure 18. The impact of the ratio of micro-inertia density over the width of the channel when it changes from 0 to 4 for the dimensionless stream function

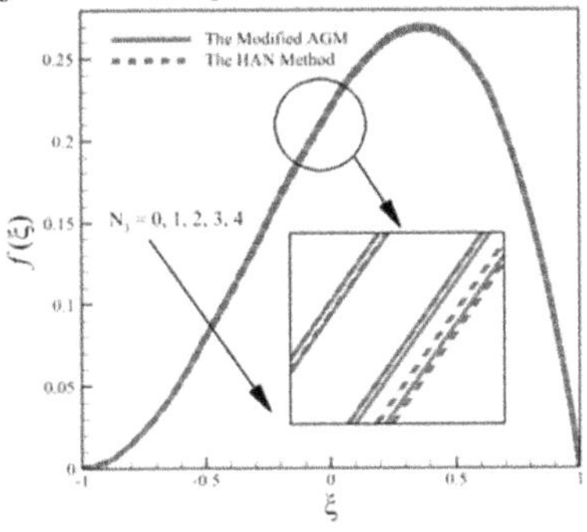

Figure 19. The impact of the ratio of micro-inertia density over the width of the channel when it changes from 0 to 4 for the dimensionless microrotation function

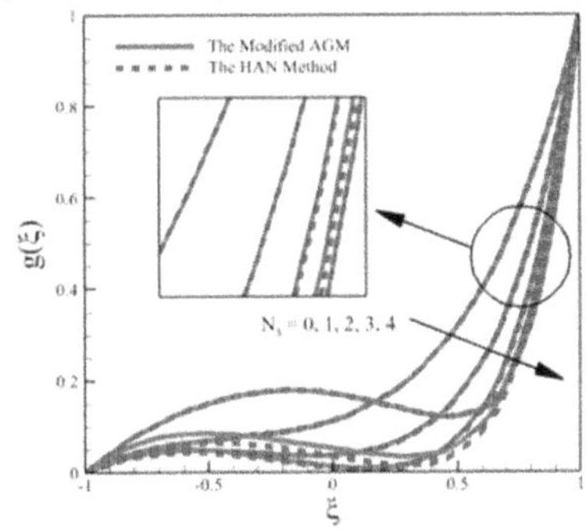

Figure 20. The impact of the ratio of micro-inertia density over the width of the channel when it changes from 0 to 4 for the dimensionless temperature

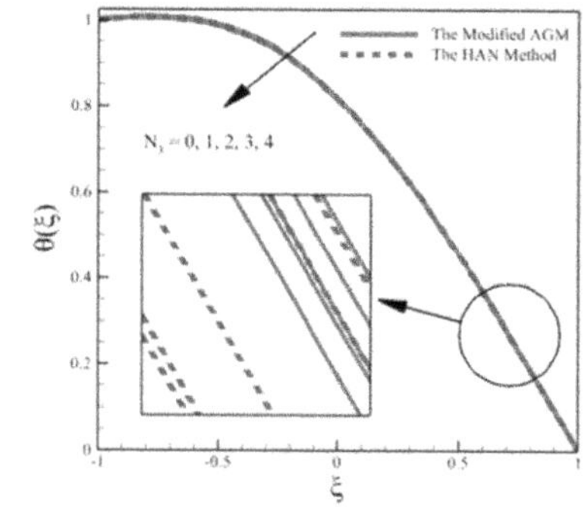

Figure 21. The impact of the ratio of micro-inertia density over the width of the channel when it changes from 0 to 4 for the dimensionless concentration

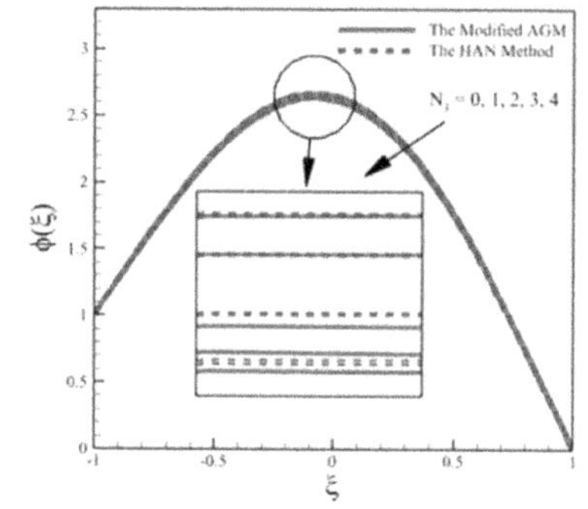

Figure 22. The impact of the Peclet number for the diffusion of heat when it changes from 0 to 4 for the dimensionless stream function

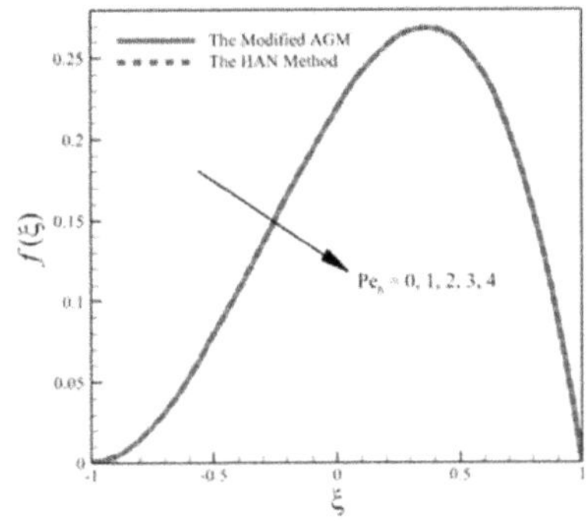

Figure 23. The impact of the Peclet number for the diffusion of heat when it changes from 0 to 4 for the dimensionless microrotation function

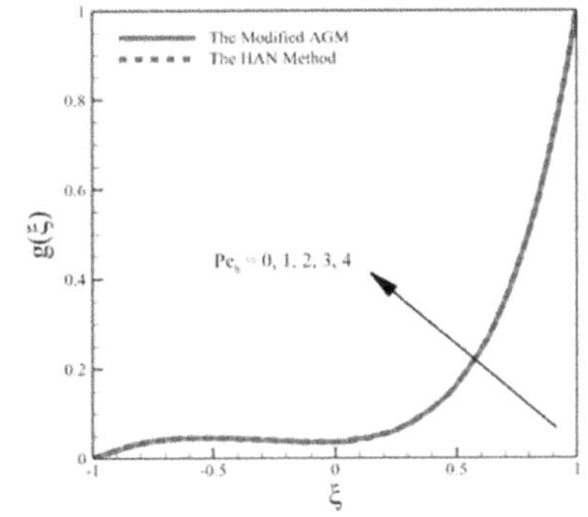

Figure 24. The impact of the Peclet number for the diffusion of heat when it changes from 0 to 4 for the dimensionless temperature

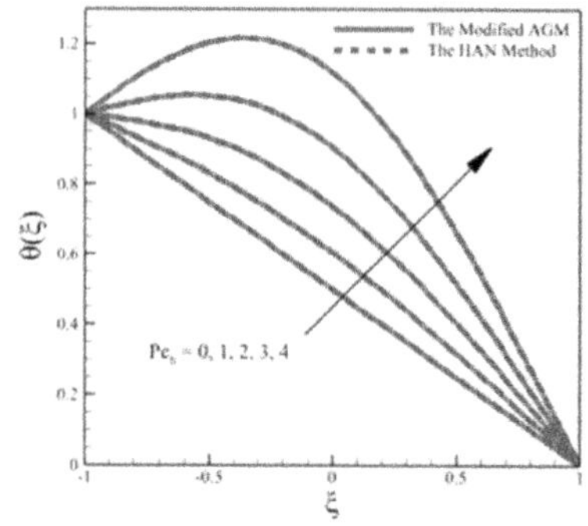

Figure 25. The impact of the Peclet number for the diffusion of heat when it changes from 0 to 4 for the dimensionless concentration

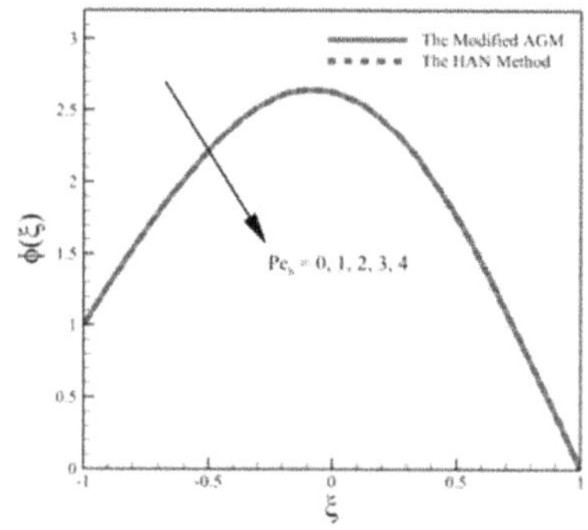

Figure 26. The impact of the Peclet number for the diffusion of mass when it changes from 0 to 4 for the dimensionless stream function

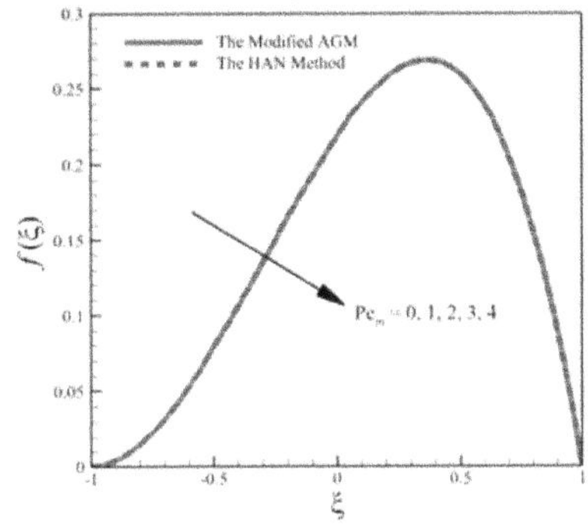

Figure 27. The impact of the Peclet number for the diffusion of mass when it changes from 0 to 4 for the dimensionless microrotation

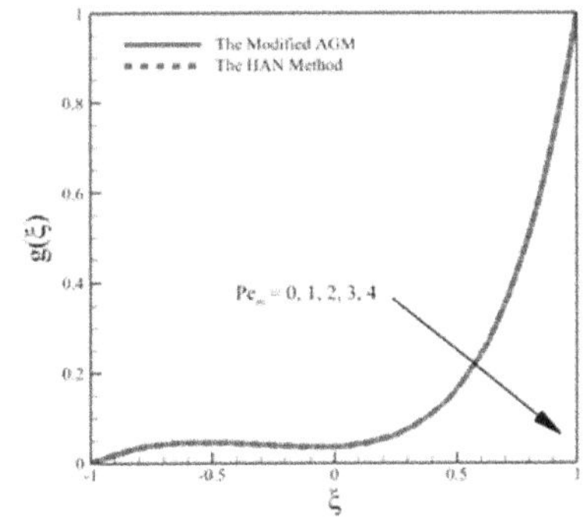

Figure 28. The impact of the Peclet number for the diffusion of mass when it changes from 0 to 4 for the dimensionless temperature

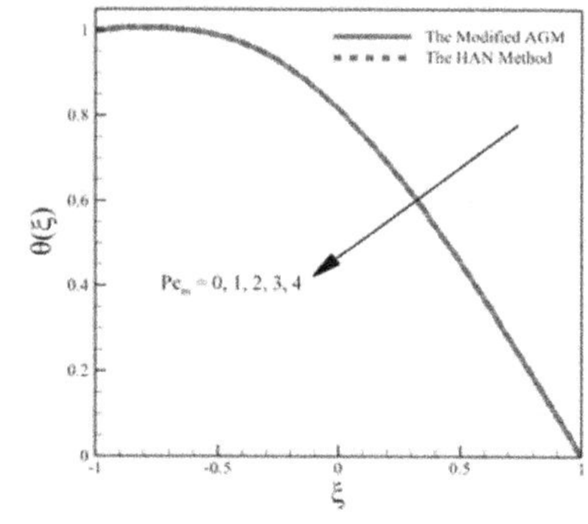

Figure 29. The impact of the Peclet number for the diffusion of mass when it changes from 0 to 4 for the dimensionless concentration

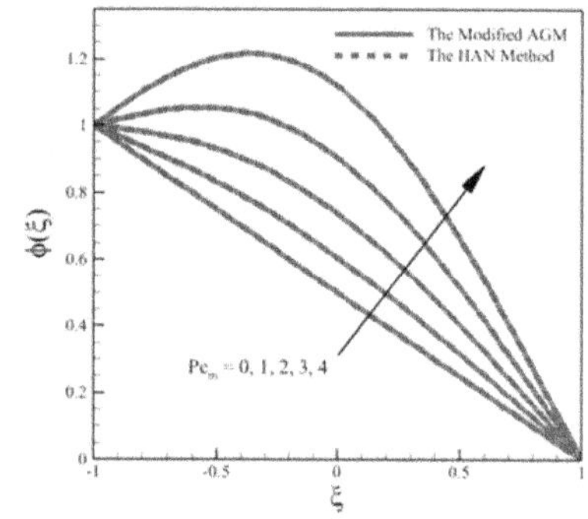

REFERENCES

Dawar, A., Shah, Z., Tassaddiq, A., Islam, S., & Kumam, P. (2021). Joule heating in magnetohydrodynamic micropolar boundary layer flow past a stretching sheet with chemical reaction and microstructural slip. *Case Studies in Thermal Engineering*, 25, 100870. DOI: 10.1016/j.csite.2021.100870

Deebani, W. (2023). Convective micropolar fluid over inclined surface with thermal radiation and velocity slip condition effects: Duality and stability. *International Journal of Modern Physics B*, 2450114.

Humane, P. P., Patil, V. S., Patil, A. B., & Shamshuddin, M. D. (2023). Buongiorno modelled nanoliquid consequence of thermal and solutal convection on the magneto-micropolar fluid inside an inclined porous stretching device. *Journal of Nanofluids*, 12(1), 211–222. DOI: 10.1166/jon.2023.1949

Jalili, B., Emad, M., Jalili, P., Ganji, D. D., Saleem, S., & Tag-eldin, E. M. (2023). Thermal analysis of boundary layer nanofluid flow over the movable plate with internal heat generation, radiation, and viscous dissipation. *Case Studies in Thermal Engineering*, 49, 103203. DOI: 10.1016/j.csite.2023.103203

Jalili, B., Rezaeian, A., Jalili, P., Ganji, D. D., & Khan, Y. (2023). Squeezing flow of Casson fluid between two circular plates under the impact of solar radiation. *ZAMM-Journal of Applied Mathematics and Mechanics. Zeitschrift für Angewandte Mathematik und Mechanik*, 103(9), e202200455. DOI: 10.1002/zamm.202200455

Jalili, P., Ahmadi Azar, A., Jalili, B., & Domiri Ganji, D. (2023). A novel technique for solving unsteady three-dimensional brownian motion of a thin film nanofluid flow over a rotating surface. *Scientific Reports*, 13(1), 13241. DOI: 10.1038/s41598-023-40410-3 PMID: 37582817

Jalili, P., Narimisa, H., Jalili, B., & Ganji, D. D. (2023). Micro-polar nanofluid in the presence of thermophoresis, hall currents, and Brownian motion in a rotating system. *Modern Physics Letters B*, 37(01), 2250197. DOI: 10.1142/S0217984922501974

Jalili, P., Narimisa, H., Jalili, B., Shateri, A., & Ganji, D. D. (2023). A novel analytical approach to micro-polar nanofluid thermal analysis in the presence of thermophoresis, Brownian motion and Hall currents. *Soft Computing*, 27(2), 677–689. DOI: 10.1007/s00500-022-07643-2

Mabood, F., Shamshuddin, M. D., & Mishra, S. R. (2022). Characteristics of thermophoresis and Brownian motion on radiative reactive micropolar fluid flow towards continuously moving flat plate: HAM solution. *Mathematics and Computers in Simulation*, 191, 187–202. DOI: 10.1016/j.matcom.2021.08.004

Nisar, K. S., Faridi, A. A., Ahmad, S., Khan, N., Ali, K., Jamshed, W., Abdel-Aty, A.-H., & Yahia, I. S. (2022). Cumulative impact of micropolar fluid and porosity on MHD channel flow: A numerical study. *Coatings*, 12(1), 93. DOI: 10.3390/coatings12010093

Pasha, P., Taghinia, F., Nadalinia Chari, F., Jalili, B., Jalili, P., & Domiri Ganji, D. (2023). Chemical reaction-diffusion model around a vessel for studying temperature and concentration of three chemical species by finite element method. *International Journal of Engineering*, 36(1), 171–181. DOI: 10.5829/IJE.2023.36.01A.19

Patil, P. M., Doddagoudar, S. H., & Hiremath, P. S. (2019). Impacts of surface roughness on mixed convection nanofluid flow with liquid hydrogen/nitrogen diffusion. *International Journal of Numerical Methods for Heat & Fluid Flow*, 29(6), 2146–2174. DOI: 10.1108/HFF-11-2018-0703

Patil, P. M., Pop, I., & Roy, S. (2012). Unsteady heat and mass transfer over a vertical stretching sheet in a parallel free stream with variable wall temperature and concentration. *Numerical Methods for Partial Differential Equations*, 28(3), 926–941. DOI: 10.1002/num.20665

Patil, P. M., Roy, S., & Pop, I. (2010). Unsteady mixed convection flow over a vertical stretching sheet in a parallel free stream with variable wall temperature. *International Journal of Heat and Mass Transfer*, 53(21), 4741–4748. DOI: 10.1016/j.ijheatmasstransfer.2010.06.018

Patil, P. M., Roy, S., & Pop, I. (2012). Flow and heat transfer over a moving vertical plate in a parallel free stream: Role of internal heat generation or absorption. *Chemical Engineering Communications*, 199(5), 658–672. DOI: 10.1080/00986445.2011.614978

Shah, Z., Shutaywi, M., Dawar, A., Kumam, P., Thounthong, P., & Islam, S. (2020). Impact of Cattaneo-Christov heat flux on non-isothermal convective micropolar fluid flow in a hall MHD generator system. *Journal of Materials Research and Technology*, 9(3), 5452–5462. DOI: 10.1016/j.jmrt.2020.03.071

Shamshuddin, M., & Ibrahim, W. (2022). Finite element numerical technique for magneto-micropolar nanofluid flow filled with chemically reactive Casson fluid between parallel plates subjected to rotatory system with electrical and Hall currents. *International Journal of Modelling and Simulation*, 42(6), 985–1004. DOI: 10.1080/02286203.2021.2012634

Shamshuddin, M., Mabood, F., Khan, W. A., & Rajput, G. R. (2023). Exploration of thermal Péclet number, vortex viscosity, and Reynolds number on two-dimensional flow of micropolar fluid through a channel due to mixed convection. *Heat Transfer*, 52(1), 854–873. DOI: 10.1002/htj.22719

Shateri, A., Moghaddam, M. M., Jalili, B., Khan, Y., Jalili, P., & Ganji, D. D. (2023). Heat transfer analysis of unsteady nanofluid flow between moving parallel plates with magnetic field: Analytical approach. *Journal of Central South University*, 30(7), 2313–2323. DOI: 10.1007/s11771-023-5388-3

Uddin, M., Rana, P., Gupta, S., & Uddin, M. N. (2023). Bio-nanoconvective micropolar fluid flow in a Darcy porous medium past a cone with second-order slips and Stefan blowing: FEM solution. *Iranian Journal of Science and Technology. Transaction of Mechanical Engineering*, 47(4), 1–15. DOI: 10.1007/s40997-023-00626-0

Uddin, M. J., Kabir, M. N., Alginahi, Y., & Bég, O. A. (2019). Numerical solution of bio-nano-convection transport from a horizontal plate with blowing and multiple slip effects. *Proceedings of the Institution of Mechanical Engineers. Part C, Journal of Mechanical Engineering Science*, 233(19-20), 6910–6927. DOI: 10.1177/0954406219867985

Zohra, F. T., Uddin, M. J., & Ismail, A. I. M. (2019). Magnetohydrodynamic bio-nanoconvective Naiver slip flow of micropolar fluid in a stretchable horizontal channel. *Heat Transfer, Asian Research*, 48(8), 3636–3656. DOI: 10.1002/htj.21560

Chapter 8
Heat Transfer Improvement in MHD

ABSTRACT

This chapter examines a steady two-dimensional ferrofluid flow on a stretched film. The impact of utilizing blades with different shapes on nanofluid heat transfer and nanofluid flow rate is discussed. In industry, heat transfer can be optimized and improved by utilizing fins with the same distance from each other in various positions. The shape of the surface contains triangles, rectangles, and chamfered ribs. In this chapter, the finite element method is used for solving. Also, a coherent nanofluidic flux is described by passing a Helmholtz coil over the thread tension surface. The impacts of magnetic parameters are evaluated through surfaces and coils. This chapter investigates the influence of magnetic pressure on the metal sheet. This chapter analyzes the influence of magnetic forces on sheet metal. In this chapter, two software (Design-Expert and FlexPDE) are used to check physical simulation and differential equations of flow parameters. The results demonstrate that FlexPDE can be used as suitable software to solve various heat transfer issues.

8.1. STEADY TWO-DIMENSIONAL MICROPOLAR/ CLASSICAL FERROFLUID FLOW

In Figure 2, a steady two-dimensional micropolar/classical ferrofluid flow is assumed on a stretch film with ten different ribs in the direction y>0. In the geometry of this paper, the effect of using fins of different shapes on nanofluid heat transfer and the nanofluid flow rate has been discussed. In industry, heat transfer can be optimized and increased by using fans that are the same distance from each other in different positions. The surface uses triangles, rectangles, and chamfered ribs. The Nanofluidic flow is input at 25 degrees, and the angular velocity is input from one direction, passing through the ribs and surface. The surface temperature is 30

DOI: 10.4018/979-8-3693-5335-6.ch008

degrees, and the surface is stretched by the velocity in the X direction ($U_W(x) = ax$, $a>0$). The velocity in the x direction is two m/s, and the velocity in the y direction is 0.2 m/s. This work investigates the effect of surface magnetic forces (B) (Table 1). The finite element method is used for the solution in this article. The equations related to Ferrofluids are simulated below.

The finite element method is used for the solution in this article. The governing equation for magnetic fluid is (Harwin, 2007) (Berman, 1953) (Ziabakhsh & Domairry, 2008) (Abdul Latiff *et al.,* 2016) (Jalili *et al.,* 2022a) (Jalili *et al.,* 2023a) (Jalili *et al.,* 2023b) (Jalili *et al.,* 2022b) (Berkan, Hoseini, & Ganji, 2017) (Mirgolbabaee, Ledari, & Ganji, 2017) (Jalili *et al.,* 2019) (Jalili *et al.,* 2022c) (Jalili *et al.,* 2023d) (Abdollahi *et al.,* 2023) (Jalili *et al.,* 2022e):

$$\frac{d}{dt}(\rho_{nf}) = \nabla.(\rho_{nf}v) \tag{8.1.1}$$

$$\rho_{nf}\left(\frac{dv}{dt}\right) = -\nabla_p + \left(2\mu_{nf} + \aleph\right) \nabla\,(\nabla.v) - \left(\mu_{nf} + \aleph\right) \nabla\,.(\nabla.v) + \aleph(\nabla.N) + j.B + \rho_{nf}g \tag{8.1.2}$$

$$\rho_{nf}j\left(\frac{dN}{dt}\right) = \left(\varphi + \lambda + Y_{nf}\right) \nabla\,.\left(\rho_{nf}.N\right) - Y_{nf}\,\nabla\,.(\nabla.N) + \aleph(\nabla.v) - 2\aleph N + \rho_{nf}I \tag{8.1.3}$$

Effective density, effective dynamic viscosity, and spin gradient viscosity are presented by Brinkman, Bourantas and Loukopoulos.

$$\rho_{nf} = (1-\varphi)\rho_f + \phi\rho_s \tag{8.1.4}$$

$$\mu_{nf} = \frac{\mu_f}{(1-\varphi)^{2.5}} \tag{8.1.5}$$

$$Y_{nf} = \left(\mu_{nf} + {}^{\aleph}\!/_{2}\right)j \tag{8.1.6}$$

Furthermore, the flow is incompressible without external forces. Therefore, it becomes difficult to solve the equation.

$$\rho_{nf}\left(\frac{dv}{dt} + (v.\nabla)v\right) = \left(\mu_{nf} + \aleph\right)\nabla^2 v + \aleph(\nabla.N) + j.B$$

(8.1.7)

$$\rho_{nf} j\left(\frac{dN}{dt} + (N.\nabla)N\right) = Y_{nf}.\nabla^2 N + \aleph(\nabla.v) - 2\aleph N$$

(8.1.8)

The body force proposed by hussanan et al. is:

$$j.B = -\sigma_{nf} B_0^2 v$$

(8.1.9)

The Ferrofluid electric conductivity represents:

$$\sigma_{nf} = \left[1 + \frac{3(\sigma - 1)\varnothing}{(\sigma + 2) - (\sigma - 1)\varnothing}\right]\sigma_f$$

(8.1.10)

The governing equations go to:

$$\rho_{nf}\left(u\frac{du}{dx} + v\frac{du}{dy}\right) = \left(\mu_{nf} + \aleph\right)\frac{d^2 u}{dy^2} + \aleph\frac{dN}{dy} - \sigma_{nf} B_0^2 u$$

(8.1.11)

$$\rho_{nf} j\left(u\frac{dN}{dx} + v\frac{dN}{dy}\right) = Y_{nf}\frac{d^2 N}{dy^2} - \aleph\left(2N + \frac{du}{dy}\right)$$

(8.1.12)

The specific heat capacity and the effective thermal conductivity researched by Din and Khan et al.

$$C_{p,nf} = \frac{(C_p)s + (1 - \varnothing)(\rho C_p)f}{\rho_{nf}}$$

(8.1.13)

The energy equation is represented as follows:

$$\frac{k_{nf}}{k_f} = \frac{k_s + 2k_f - 2\varnothing\left(k_f - k_s\right)}{k_s + 2k_f + \varnothing\left(k_f - k_s\right)}$$

(8.1.14)

$$u\frac{dT}{dx} + v\frac{dT}{dy} = \frac{k_{nf}}{C_p)nf}\frac{d^2T}{dy^2} - \frac{dq_r}{dy}$$

(8.1.15)

It is stated as according to Roseland's approximation as:

$$u\frac{dT}{dx} + v\frac{dT}{dy} = \frac{1}{(C_p)nf}\left(k_{nf} + \frac{16\sigma^* T_\infty^3}{3k^*}\right)\frac{d^2T}{dy^2}$$

(8.1.16)

The boundary conditions are:

$u= au_w(x)$, $v=v_w$ at $y=0$, $u\leftrightarrow 0$, $y\rightarrow\infty$ (8.1.17)

$$N = -\delta\frac{du}{dy}aty = 0, N \leftrightarrow 0, y \rightarrow \infty$$

(8.1.18)

$T=T_w$ at $y=0$, $T\leftrightarrow T_w$, $y\rightarrow\infty$ (8.1.19)

For convenience in these calculations, the velocity components in the x and y directions are referred to as u and v, respectively. Positive and negative values of the surface mass transfer velocity v_w represent aspiration and injection. N. μ represents angular velocity, which is the kinematic viscosity, and δ is a constant between 0 and 1. A setting of 0 specifies microelements. It is shown that when δ=1/2, the concentration of trace elements is weak. Furthermore, if turbulent boundary layer flow is desired, δ = 1. j represents the inertia of the small element per unit mass, where is the spin gradient viscosity. To easily solve these nonlinear differential equations from similar variables such as:

$$\eta = y\sqrt{\frac{a}{v_f}}, u = axF'(\eta), v = -\sqrt{av_f}F(\eta), N = ax\sqrt{\frac{a}{v_f}}G(\eta, \theta(\eta) = \frac{T - T_\infty}{T_w - T_\infty},$$

(8.1.20)

By substituting the variables (20) into the Eq. (11), (12), (16):

$$\left(\frac{1}{(1-\varnothing)^{2.5}}+k\right)F'''(\eta)+\left(1-\varnothing+\varnothing\frac{\rho_s}{\rho_f}\right)F(\eta)F''(\mu)$$
$$-\left(1-\varnothing+\varnothing\frac{\rho_s}{\rho_f}\right)F'^2(\eta)-M\left(1+\frac{3(\sigma-1)\varnothing}{\sigma+2-(\sigma-1)\varnothing}\right)F'(\eta)+kG'(\eta)=0 \tag{8.1.21}$$

$$\left(\frac{1}{(1-\varnothing)^{2.5}}+\frac{k}{2}\right)G''(\eta)+\left(1-\varnothing+\varnothing\frac{\rho_s}{\rho_f}\right)F(\eta)G'(\eta)$$
$$-\left(1-\varnothing+\varnothing\frac{\rho_s}{\rho_f}\right)F'(\eta)G(\eta)-k(2G(\eta)+F''(\eta))=0 \tag{8.1.22}$$

$$\frac{1}{\Pr}\left(\frac{k_s+2k_f-2\varnothing\,(k_f-k_s)}{k_s+2k_f+\varnothing\,(k_f-k_s)}+R\right)\theta''(\eta)+\left(1-\varnothing+\frac{C_p)s}{C_p)f}\right)F(\eta)\theta'(\eta)=0 \tag{8.1.23}$$

And here the new boundary conditions:

$$F(\eta)=S,F'(\eta)=\alpha,G(\eta)=-\delta F''(\eta),\theta(\eta)=1,at\eta=0 \tag{8.1.24}$$

$$F'(\eta)\leftrightarrow 0,G(\eta)\leftrightarrow 0,\theta(\eta)\rightarrow 0at\eta\rightarrow\infty \tag{8.1.25}$$

The finite element method is an important numerical method and one of the practical applications of this method in Flexpde software for solving nonlinear partial and ordinary differential equations. Output of Results FlexPDE is also a "problem-solving environment" that performs all functions for solving systems of partial differential equations. The finite element method is a numerical technique for solving partial differential equations defined in terms of one or two spatial variables. This method divides a large system into smaller, simpler pieces called finite elements to solve the problem. This spatial discretization requires defining the object or environment in question as a network or mesh.

According to Figure 2, the largest meshes are at the edges. This is due to the changing temperature and velocity characteristics of the Nanofluidic flow around the vane. The computational error in our work is very small compared to other works. The highest number of errors occurred at η=1.5 and the lowest number of errors occurred at η=1.In addition to Figure 3, the nanofluid velocities are lower in the surface and rib regions. The maximum velocity of the nanofluid is attached to the walls of the ribs. The maximum velocity value is 2 (m/s). Increasing the distance of the nanofluid from the surface decreases its velocity, while decreasing its distance from the surface increases its velocity. Comparing the a-(3) and b-(3) modes, the use of triangular blades instead of chamber blades increases the velocity gradient. Figure 4 shows the angular velocity of the Nanofluidic stream passing through the elongated surface. There are many ribs on the surface, such as angular ribs, triangular ribs, and diagonal ribs. Below the surface, the angular velocity of the nanofluid is very small, but near the surface and ribs, the angular velocity increases. In general, the angular velocities near rectangular and chamfered ribs (b-(4)) have additional values because the lateral surface area is larger than that of triangular ribs than triangular ribs. The angular velocity of the nanofluid has a maximum value when the diversity of surface fins (d-(4)) increases. Maximum value is 340. Figure 5 shows the temperature change of a nanofluid passing through a strained surface. The temperature of the nanofluid inside the fin space is heat and the same as the surface temperature. The temperature value is 30. The temperature maximum of the nanofluid resides on the last fin of the surface. By passing the nanofluid flow from the first rib of the surface, the temperature of the flow goes from 25 to 31, with higher temperatures at the edge of the surface. It can be seen that using the chamfer blade at the top of the page produces a larger temperature gradient than the other modes due to the different position of the ribs at the top of the page. The angular velocity vectors of the nanofluid flow are shown in Figure 6. The maximum spread of velocities (N) is shown in Figure d-8. Like the middle of the surface near the 3rd and 8th fins, the angular velocity values are low, while the first and last fins have high velocity values. Figure 7 shows the variation of Nanofluidic flow around the rectangular and triangular and chamfered ribs in the figure. The minimum velocity value in the x direction of the nanofluid flow is from $x = 0$ to $x = 0.1$, and the maximum velocity in the $u = 0.3$ m/s.x direction is 2 m/s. The velocity variation between different fins is very high. For the triangular-chamfered rib condition (b-(8)), the maximum angular velocity is at $x=0.2$, close to the first rib. The velocity value at $x=0.2$ is 400. At the locations of different rib geometries (triangular, rectangular, and chamfered) adjacent to each other, most angular velocity variations are produced in the range $x=0$ to $x=1$. Even in the same case, the velocity varies depending on the mode. The angular velocity value at $x=0$ is 30 and the velocity value at $x=0.4$ is $N=150$. Figure 9 shows the temperature change of a nanofluid passing through a strained surface. Temperature readings start at 28 to 31 degrees across the various slats. The temperature of the nanofluid from $x=0.2$ to $x=0.8$ is a constant value ($T=30$) and does not change around the rib. According to Table 2, the maximum ethylene glycol temperature occurs at $x=0.1$ for three different fins and the minimum temperature occurs at $x=0$ for three different fins. The maximum temperature of the nanofluid is

T=30.36°C, and the minimum temperature is T=28.13°C. The maximum temperature change is near three different (rectangular, triangular, and distorted) fins, where the thermal boundary layer is larger than elsewhere. The Nanofluidic temperature of the three different fins is 4.5% higher than the angled and square fins, and 5.34% higher than the angled and triangular fins. It can be seen that using the chamfered blade at the top of the page produces a larger temperature gradient than the other modes due to the different rib positions at the top of the page. According to Table 3, the ethylene glycol velocity maximum occurs at x=0.3 at T=399.98 for tilt and triangle modes, and the temperature minimum occurs at x=0.8 at T=33.34 degrees for the three different fins. The angular velocity of the nanofluid in triangular and tilted modes is 34.5% higher than three different vanes and 58.54% higher than tilted and rectangular vanes. Most of the angular velocity boundary conditions are shown around the first fin of the stretch foil.

Figure 1. Schematic of different fins on the stretching sheet

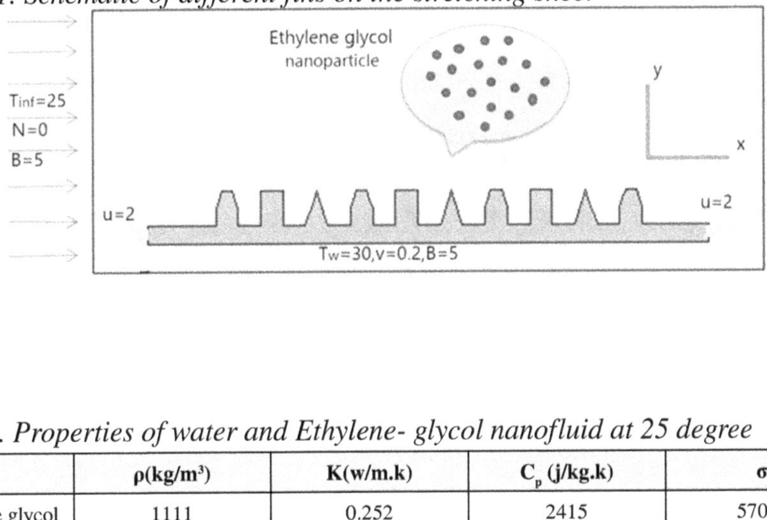

Table 1. Properties of water and Ethylene- glycol nanofluid at 25 degree

	ρ(kg/m³)	**K(w/m.k)**	**C_p (j/kg.k)**	**σ**
Ethylene glycol	1111	0.252	2415	57000
H_2O	997	0.613	4179	0.05

Figure 2. Comparison of mesh geometry for stretching surface

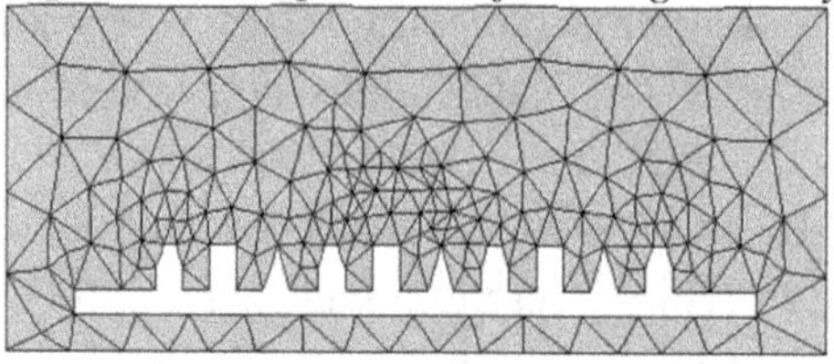

a) rectangular, triangular and chamfer fins

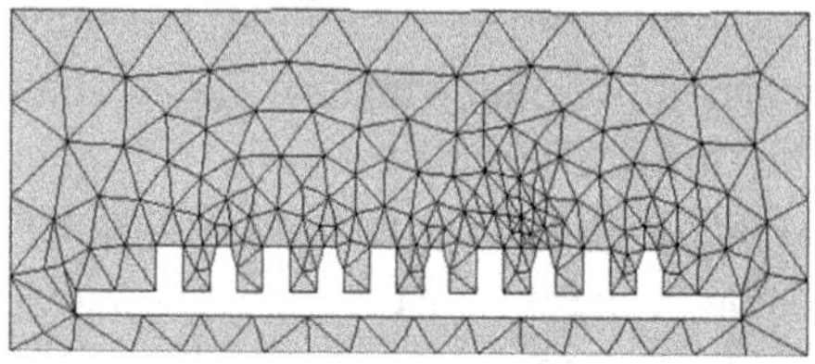

b) rectangular and chamfer fins

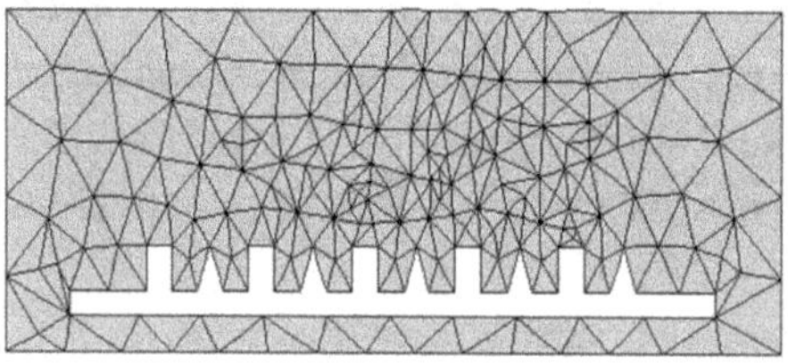

c) rectangular and triangular fins

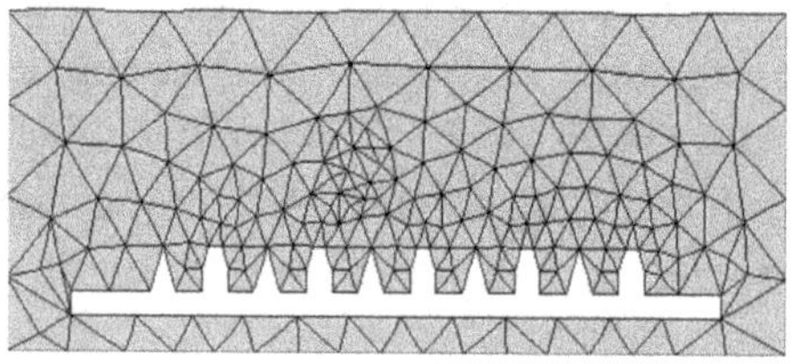

d) triangular and chamfer fins

Figure 3. Comparison of velocity changes in the X direction in different fins for stretching surface

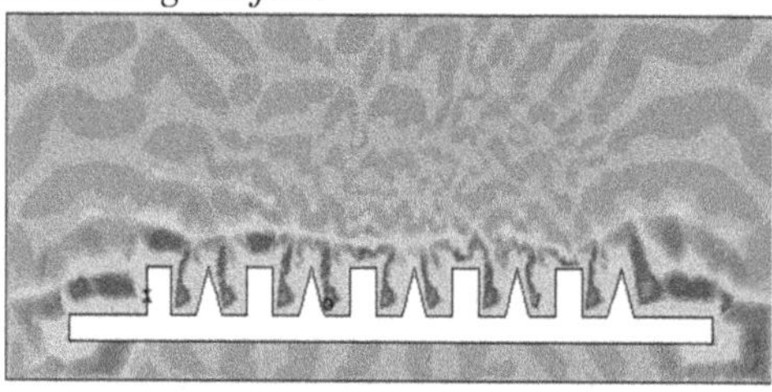

a) Triangular and rectangular fins

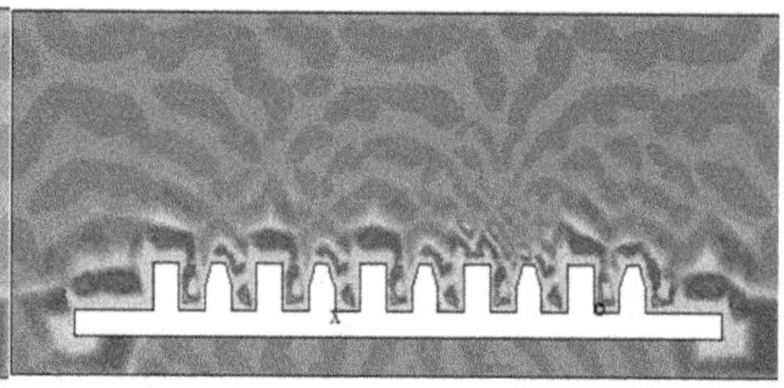

b) rectangular and chamfer fins

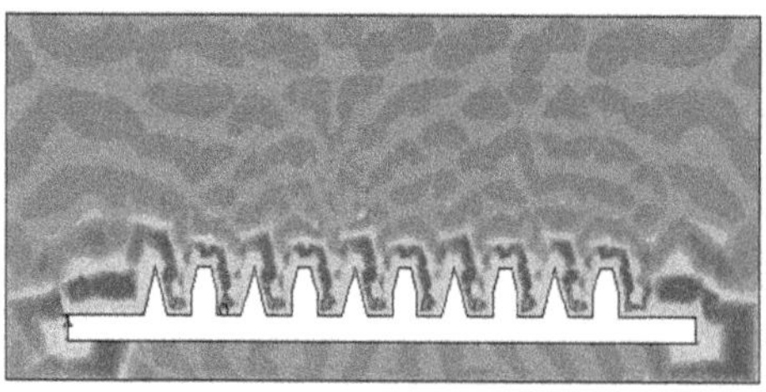

c) triangular and chamfer fins

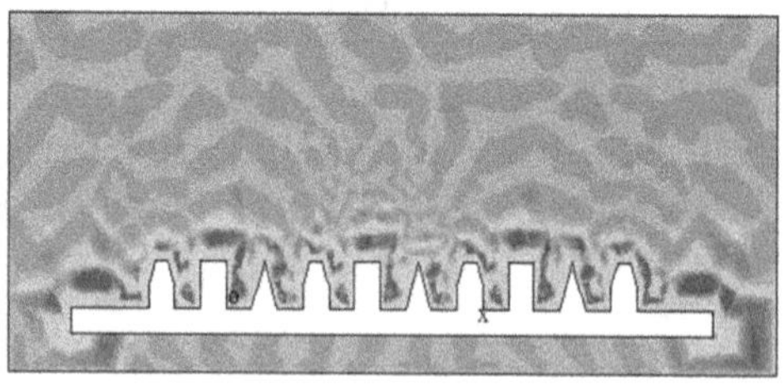

d) rectangular, triangular and chamfer fins

Figure 4. Comparison of angular velocity changes in different fins for stretching surface

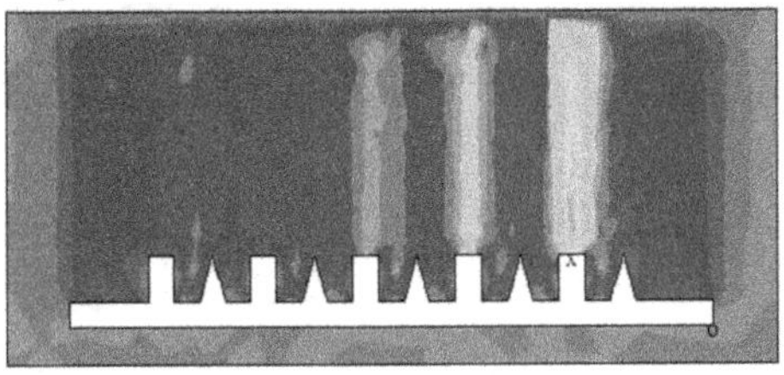

a) Triangular and rectangular fins

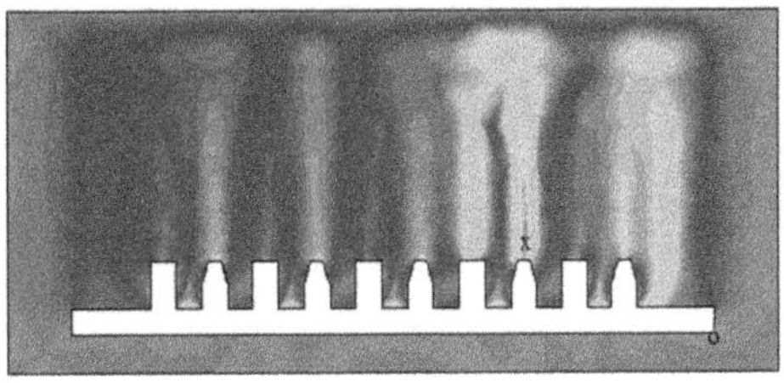

b) rectangular and chamfer fins

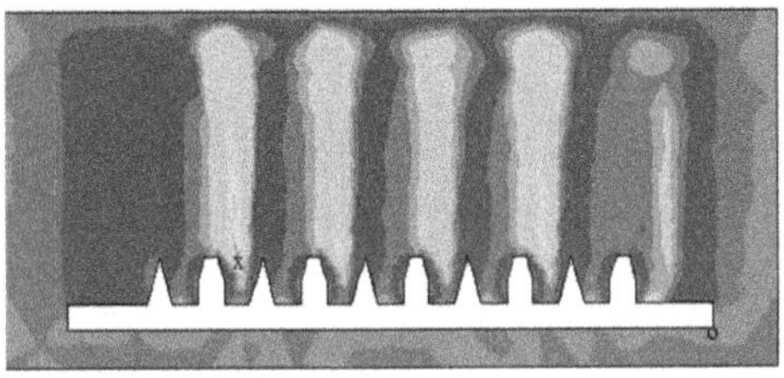

c) triangular and chamfer fins

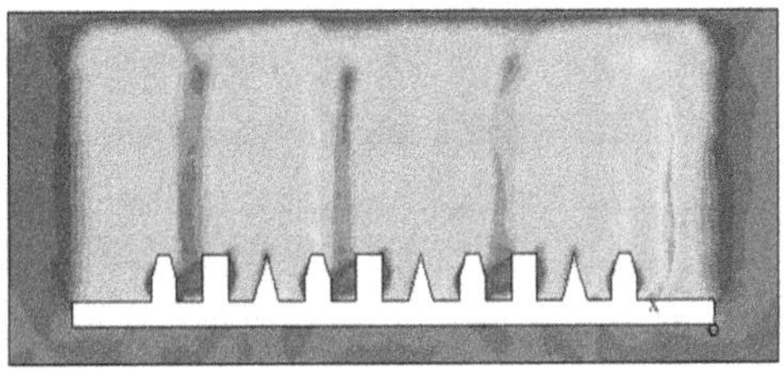

d) rectangular, triangular and chamfer fins

Figure 5. Comparison of temperature profiles for threated stretching surface with different fins

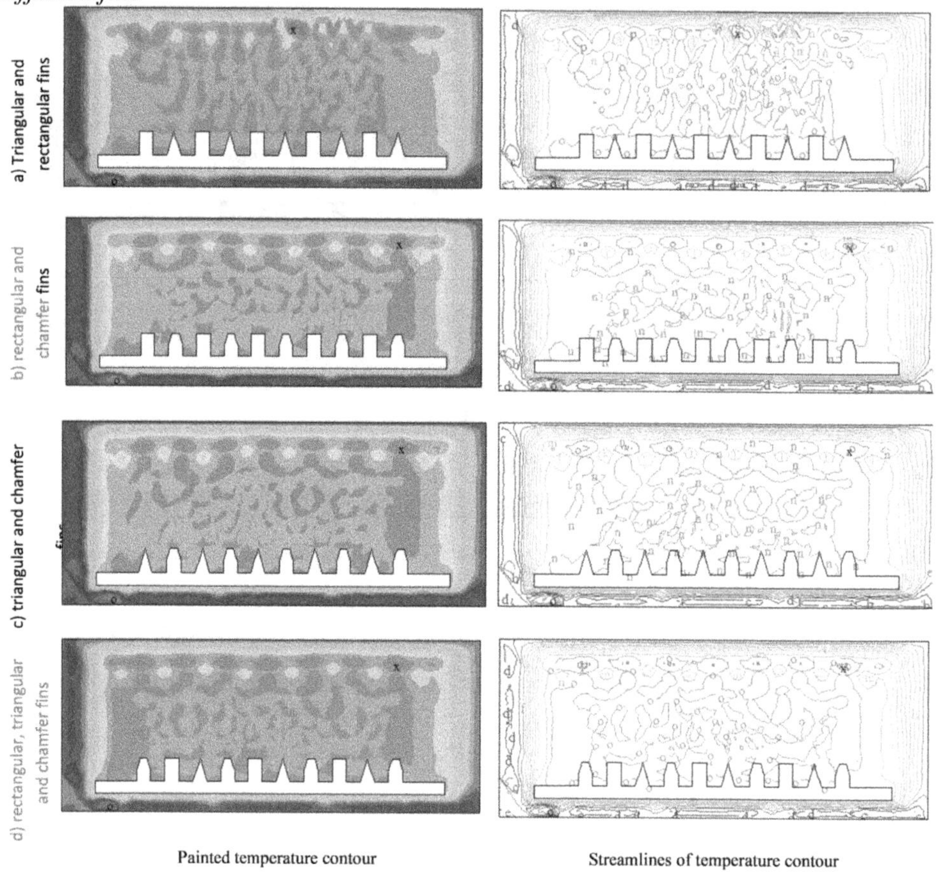

Figure 6. Comparison of vector grid temperature changes in different fins for stretching surface

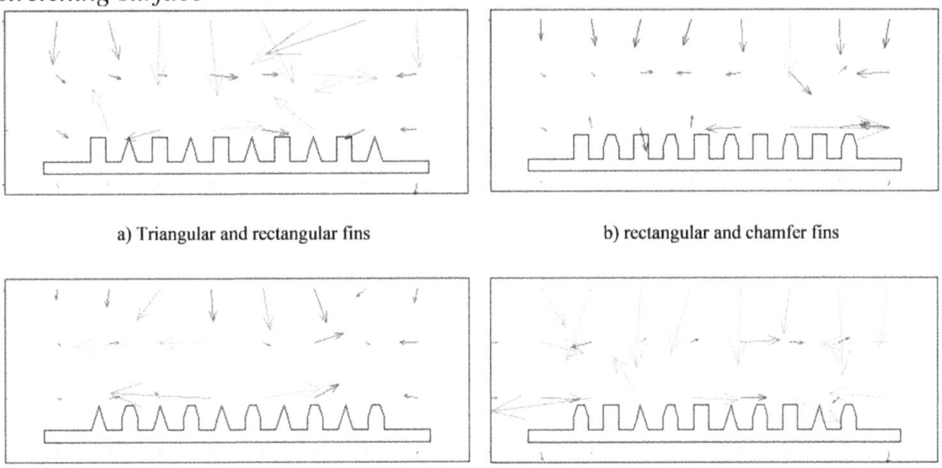

a) Triangular and rectangular fins

b) rectangular and chamfer fins

c) triangular and chamfer fins

d) rectangular, triangular and chamfer fins

Figure 7. Comparison of velocity (u) changes in in rectangular and triangular fins for threated stretching surface

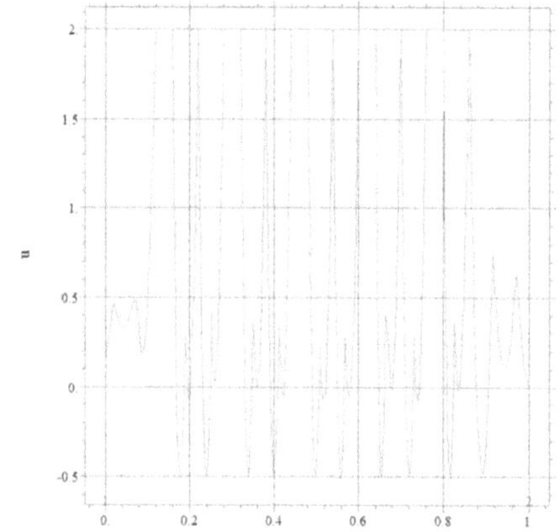

a) Triangular and rectangular fins

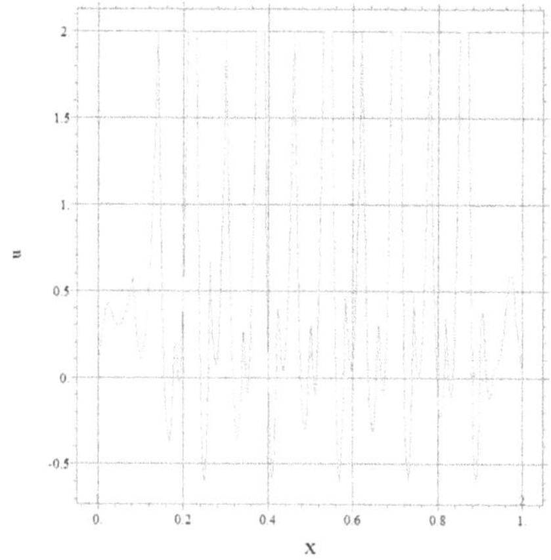

b) triangular and chamfer fins

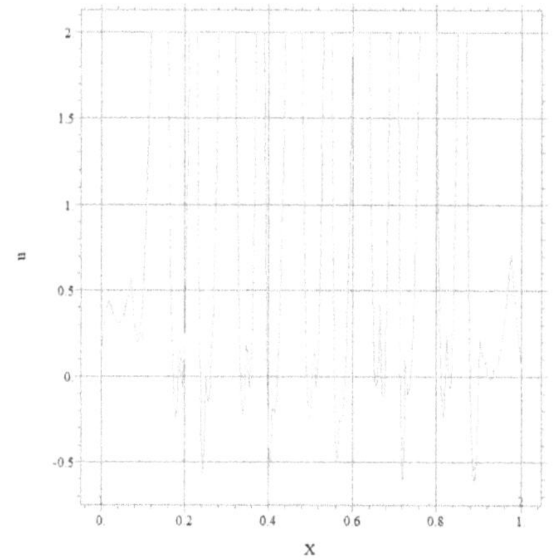

c) Rectangular and chamfer fins

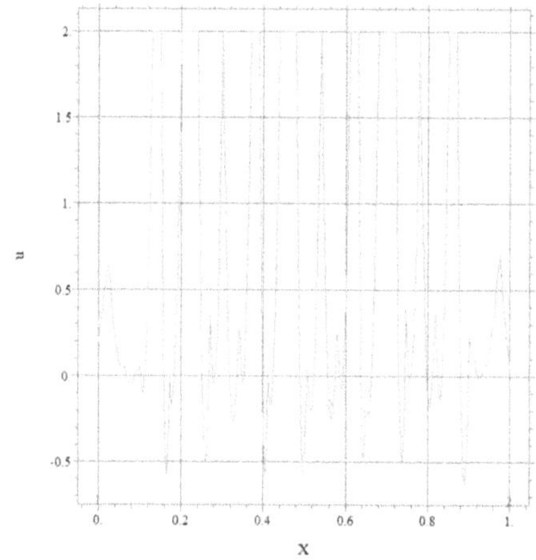

d) rectangular, triangular and chamfer fins

Figure 8. Comparison of angular velocity changes in in rectangular and triangular fins for threated stretching surface

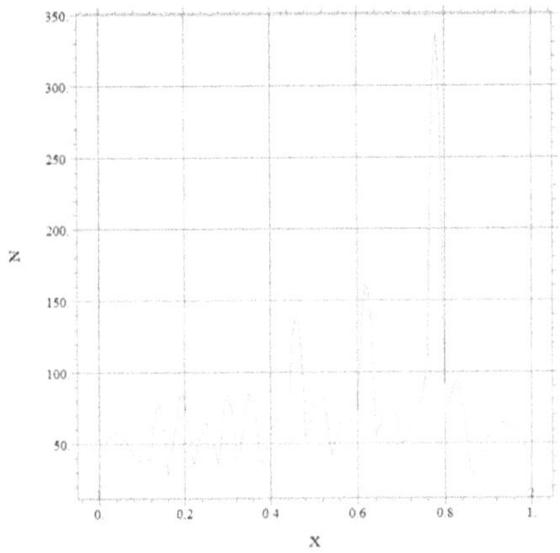

a) Triangular and rectangular fins

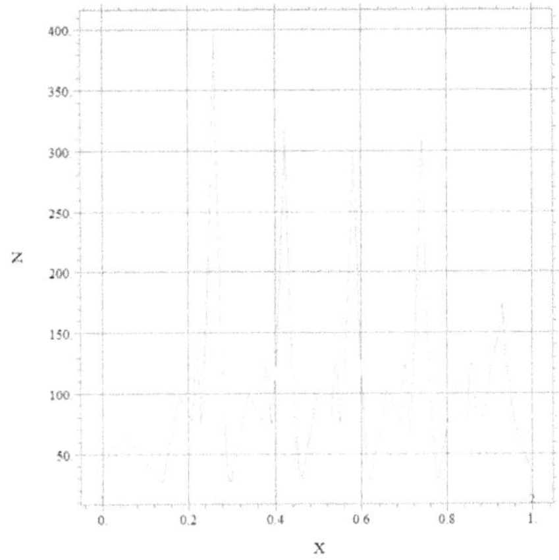

b) triangular and chamfer fins

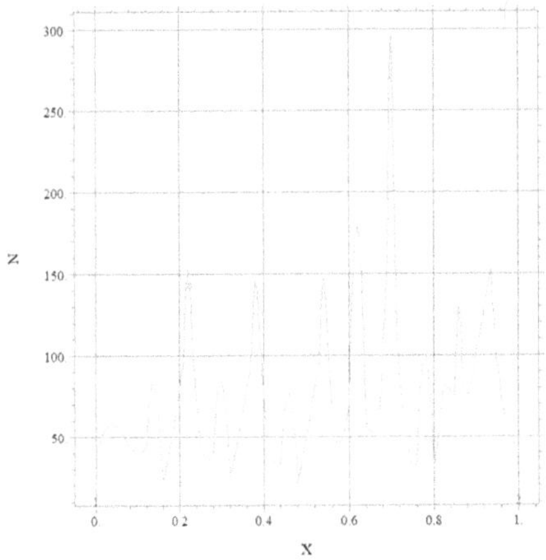

c) Rectangular and chamfer fins

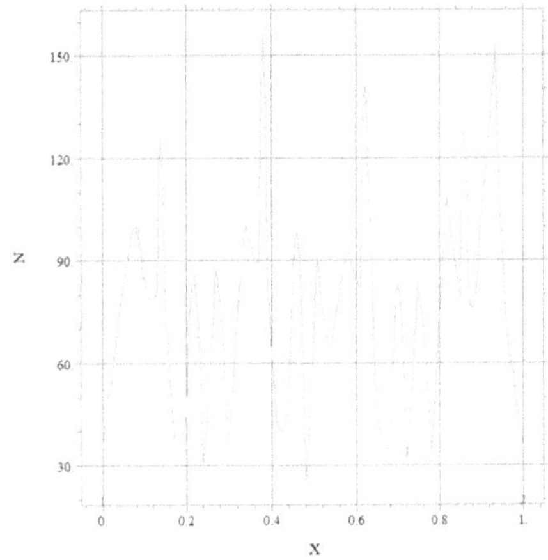

d) rectangular, triangular and chamfer fins

Figure 9. Comparison of temperature changes in in rectangular and triangular fins for treated stretching surface

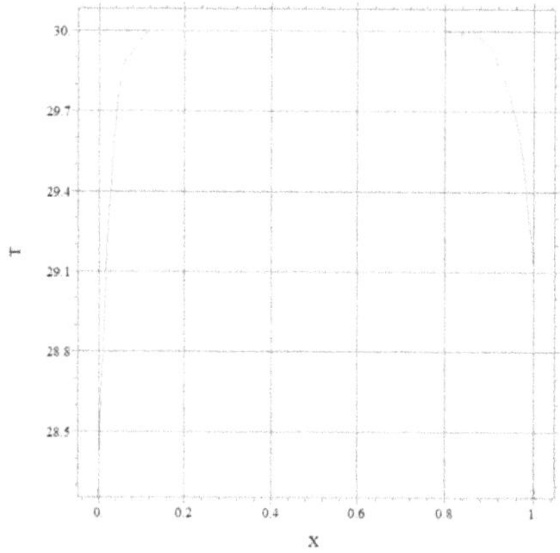

a) Triangular and rectangular fins

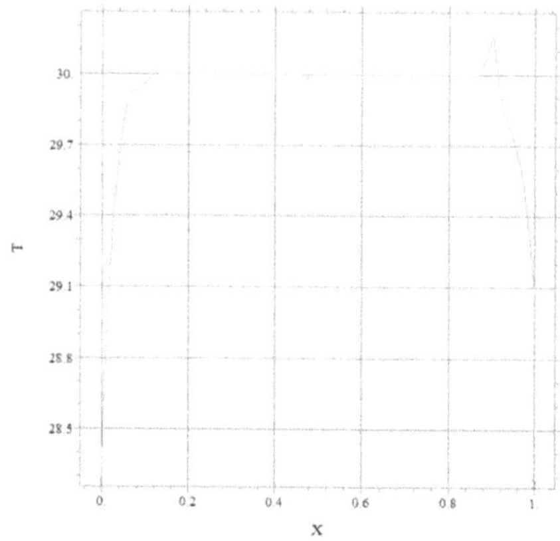

b) triangular and chamfer fins

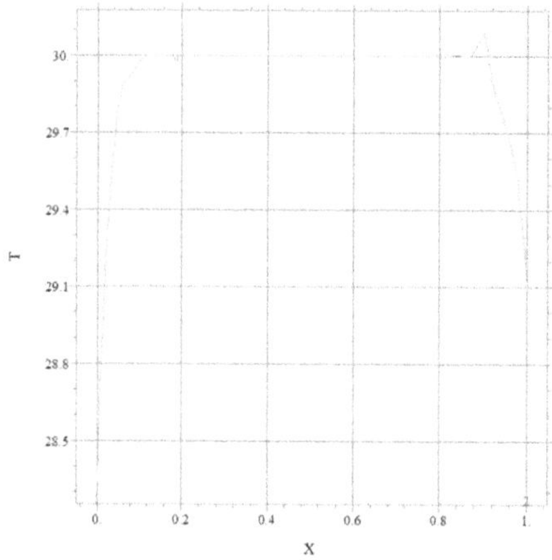

c) Rectangular and chamfer fins

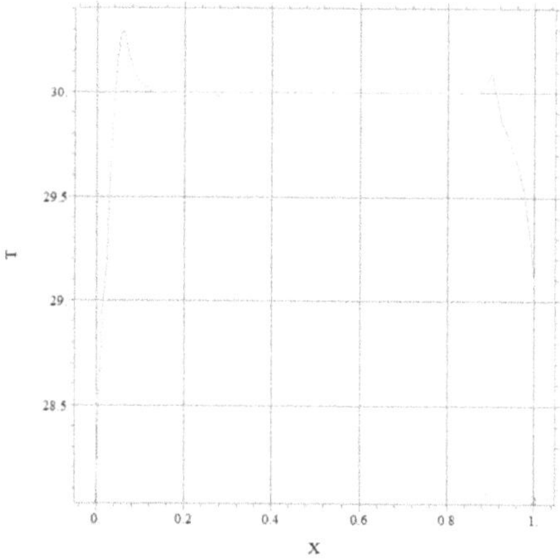

d) rectangular, triangular, and chamfer fins

Table 2. Numerical comparison of temperature between different modes of fins

x	$T_{3different\ fins}$	$T_{chamfer\ and\ rectangular}$	$T_{chamfer\ and\ triangular}$	$T_{rectangular\ and\ triangular}$
0.0	**28.13**	**28.33**	**28.34**	**28.33**
0.1	**30.36**	**29.94**	**29.96**	**29.89**
0.2	**30.00**	**30.00**	**29.99**	**29.97**

continued on following page

Table 2. Continued

x	$T_{3different\ fins}$	$T_{chamfer\ and\ rectangular}$	$T_{chamfer\ and\ triangular}$	$T_{rectangular\ and\ triangular}$
0.5	**30.00**	**30.00**	**29.99**	**29.97**
0.8	**29.99**	**30.00**	**30.00**	**29.99**
0.9	**30.09**	**30.08**	**30.34**	**29.83**
1.0	**29.09**	**29.09**	**29.09**	**29.11**

Table 3. Numerical comparison of angular velocity between different modes of fins

x	$N_{3different\ fins}$	$N_{chamfer\ and\ rectangular}$	$N_{chamfer\ and\ triangular}$	$N_{rectangular\ and\ triangular}$
0.0	**37.85**	**49.97**	**49.95**	**49.95**
0.1	**100.99**	**48.87**	**48.01**	**47.76**
0.3	**43.33**	**45.25**	**399.98**	**46.06**
0.5	**57.87**	**47.89**	**69.33**	**76.55**
0.8	**33.34**	**46.76**	**70.90**	**77.89**
0.9	**89.08**	**120.43**	**89.05**	**55.48**
1.0	**34.04**	**47.56**	**48.02**	**48.76**

8.2. EFFECT OF MAGNETIC FORCES ON SHEET METAL

A coherent Nanofluidic flux (MWCNT) is described by passing a Helmholtz coil over the filament stretch surface. Magnetic parameter effects are evaluated via surfaces and coils. After entering at 30 degrees with a unidirectional y-direction angular velocity, the Nanofluidic flow continues through the sheets and coils. The surface temperature was 10 degrees, and the surface was stretched in the x direction by the velocity ($U_{W(x)}$ =ax, a>0). The x-course step size is 1m/s, and the y-course step size is 0.3m/s. This study analyzes the magneto pressure effect in sheet metal (Figure 10).

This work investigates the effect of magnetic forces on sheet metal. A finite element approach was used to modify the current work. This article used two software programs (Design-Expert and FlexPDE) to investigate flow parameters' physical simulation and differential equations. FlexPDE can be presented as a stand-alone authoring framework that analyzes the presentation of a topic and approaches it by creating a joint framework.

The finite element method is used to solve this work. In the following text, the applicable equations for Ferrofluid are:

$$\frac{D\rho_{nf}}{dt} = \left(u\frac{d\rho_{nf}}{dx} + v\frac{d\rho_{nf}}{dy} + w\frac{d\rho_{nf}}{dz}\right).\left(\rho_{nf}\frac{du}{dx} + \rho_{nf}\frac{dv}{dy} + \rho_{nf}\frac{dw}{dz}\right)$$

(8.2.1)

$$\rho_{nf}\left(\frac{DV}{dt}\right) = -\nabla_p + (2\mu_{nf} + \kappa)\ \nabla\ \left(\rho_{nf}\frac{du}{dx} + \rho_{nf}\frac{dv}{dy} + \rho_{nf}\frac{dw}{dz}\right)$$
$$-(\mu_{nf} + \kappa)\ \nabla\ .\left(\rho_{nf}\frac{du}{dx} + \rho_{nf}\frac{dv}{dy} + \rho_{nf}\frac{dw}{dz}\right) + \kappa(\nabla .N)j.B + \rho_{nf}g$$

(8.2.2)

$$j.B_x = -\sigma_{nf}B_0^2 u$$

(8.2.3)

$$j.B_y = -\sigma_{nf}B_0^2 v$$

(8.2.4)

$$j.B_z = -\sigma_{nf}B_0^2 w$$

(8.2.5)

$$N = -\delta\left(\frac{du}{dy} + \frac{dv}{dz} + \frac{dw}{dx}\right)$$

(8.2.6)

div(1/mu. grad(Az))+ Jz=0 (8.2.7)

Equation (8.2.1) is a continuous equation. Equation (8.2.2) is the momentum equation. Boolantas et al. In equations (8.2.3) to (8.2.5), N is the angular velocity (6). By placing equations (8.2.3) through (8.2.6) into the momentum equation (8.2.2), the following momentum equations are evaluated in the x, y, and z directions. Where B is the magnetic flux density, and Jz is the current density. AZ is the magnetic potential:

$$\rho_{nf}\left(\frac{Du}{dt}\right) = -\frac{dp}{dx} + (2\mu_{nf} + \kappa)\ \nabla\ \left(\rho_{nf}\frac{du}{dx} + \rho_{nf}\frac{dv}{dy} + \rho_{nf}\frac{dw}{dz}\right) - (\mu_{nf} + \kappa)\ \nabla\ .$$
$$\left(\rho_{nf}\frac{du}{dx} + \rho_{nf}\frac{dv}{dy} + \rho_{nf}\frac{dw}{dz}\right) + \kappa\left(\nabla .-\delta\left(\frac{du}{dy} + \frac{dv}{dz} + \frac{dw}{dx}\right)\right) - \sigma_{nf}B_0^2 u + \rho_{nf}g$$

(8.2.8)

$$\rho_{nf}\left(\frac{Dv}{dt}\right) = -\frac{dp}{dy} + (2\mu_{nf} + \kappa)\ \nabla\ \left(\rho_{nf}\frac{du}{dx} + \rho_{nf}\frac{dv}{dy} + \rho_{nf}\frac{dw}{dz}\right) - (\mu_{nf} + \kappa)\ \nabla\ .$$
$$\left(\rho_{nf}\frac{du}{dx} + \rho_{nf}\frac{dv}{dy} + \rho_{nf}\frac{dw}{dz}\right) + \kappa\left(\nabla.-\delta\left(\frac{du}{dy} + \frac{dv}{dz} + \frac{dw}{dx}\right)\right) - \sigma_{nf}B_0^2 v + \rho_{nf}g$$

(8.2.9)

$$\rho_{nf}\left(\frac{Dw}{dt}\right) = -\frac{dp}{dz} + (2\mu_{nf} + \kappa)\ \nabla\ \left(\rho_{nf}\frac{du}{dx} + \rho_{nf}\frac{dv}{dy} + \rho_{nf}\frac{dw}{dz}\right) - (\mu_{nf} + \kappa)\ \nabla\ .$$
$$\left(\rho_{nf}\frac{du}{dx} + \rho_{nf}\frac{dv}{dy} + \rho_{nf}\frac{dw}{dz}\right) + \kappa\left(\nabla.-\delta\left(\frac{du}{dy} + \frac{dv}{dz} + \frac{dw}{dx}\right)\right) - \sigma_{nf}B_0^2 w + \rho_{nf}g$$

(8.2.10)

The energy equation is considered below:

$$\rho_{nf}j\left(\frac{dN}{dt}\right) = (\varphi + \lambda + Y_{nf})\ \nabla\ .\left(\rho_{nf}.-\delta\left(\frac{du}{dy} + \frac{dv}{dz} + \frac{dw}{dx}\right)\right)$$
$$-Y_{nf}\ \nabla\ .\left(\nabla.-\delta\left(\frac{du}{dy} + \frac{dv}{dz} + \frac{dw}{dx}\right)\right) + \kappa\left(\rho_{nf}\frac{du}{dx} + \rho_{nf}\frac{dV}{dy} + \rho_{nf}\frac{dw}{dz}\right)$$
$$-2\kappa\left(-\delta\left(\frac{du}{dy} + \frac{dv}{dz} + \frac{dw}{dx}\right)\right) + \rho_{nf}I \qquad (8.2.11)$$

where the effective density, effective dynamic viscosity, and spin gradient viscosity are presented by Brinkman, Bourantas and Loukopoulos and, respectively.

$$\rho_{nf} = \rho_f - \varphi\rho_f + \varphi\rho_s \qquad (8.2.12)$$

$$\mu_{nf} = \mu_f(1+39.11\phi+533.9\phi^2)$$
(8.2.13)

$$Y_{nf} = \left(\mu_f(1 + 39.11\varnothing + 533.9\varnothing^2) + {}^{\kappa}/_{2}\right)j$$

(8.2.14)

In fact, the flow is incompressible in the absence of external forces. Thus, it becomes difficult to elucidate the condition:

$$\rho_{nf}\left(\frac{dv}{dt} + \left(\rho\frac{du}{dx} + \rho\frac{dV}{dy} + \rho\frac{dw}{dz}\right)v\right) = (\mu_{nf} + \kappa)\nabla^2 v + \kappa(\nabla.N) + -\sigma_{nf}B_0^2 v$$

(8.2.15)

$$\rho_{nf} j\left(\frac{dN}{dt} + (N.\nabla)N\right) = Y_{nf}.\nabla^2 N + \kappa\left(\rho\frac{du}{dx} + \rho\frac{dV}{dy} + \rho\frac{dw}{dz}\right)2\kappa N$$

(8.2.16)

The Ferrofluid stream electric conductivity represents:

$$\sigma_{nf} = \left[\sigma_f + \sigma_f \frac{3\sigma\varnothing - 3\varnothing}{(\sigma+2) - (\sigma\varnothing - \varnothing)}\right]$$

(8.2.17)

The governing equations go to:

$$\rho_{nf}\left(u\frac{du}{dx} + v\frac{du}{dy}\right) = (\mu_{nf} + \kappa)\frac{d^2u}{dy^2} + \kappa\frac{dN}{dy} - \sigma_{nf} B_0^2 u$$

(8.2.18)

$$\rho_{nf} j\left(u\frac{dN}{dx} + v\frac{dN}{dy}\right) = Y_{nf}\frac{d^2N}{dy^2} - \kappa\left(2N + \frac{du}{dy}\right)$$

(8.2.19)

The specific heat capacity and the effective thermal conductivity were examined by Vajravelu and Khan et al.

$$C_{p,nf} = \frac{(C_p)s + (1-\varnothing)\rho_f Cp_f}{\rho_f - \varphi\rho_f + \varphi\rho_s}$$

(8.2.20)

$$\frac{k_{nf}}{k_f} = \frac{2k_1 + k_p - 2\varnothing\,(k_f - k_p)(1+\eta)^3}{2k_1 + k_p + \varnothing\,(k_f - k_p)(1+\eta)^3}$$

(8.2.21)

It is expressed as concurring to Rosseland's guess as:

$$u\frac{dT}{dx} + v\frac{dT}{dy} = \frac{1}{\frac{(C_p)s + (1-\varnothing)\rho_f Cp_f}{\rho_f - \varphi\rho_f + \varphi\rho_s}}\left(\frac{(C_p)s + (1-\varnothing)\rho_f Cp_f}{\rho_f - \varphi\rho_f + \varphi\rho_s} + \frac{16\sigma^* T_\infty^3}{3k^*}\right)\frac{d^2T}{dy^2}$$

(8.2.22)

The boundary conditions are:

$u{=}au_w(x)$, $v{=}v_w$ at $y{=}0{\to}u{\to}0$, $Jz{=}1$ at $y{\to}\infty$ (8.2.23)

$$N = -\delta\frac{du}{dy}aty = 0 \to N \leftrightarrow 0aty \to \infty$$

(8.2.24)

$T{=}T_w$. at $y{=}0{\to}T{\leftrightarrow}T_w$ at $y{\to}\infty$ (8.2.25)

To fathom comfortably these nonlinear differential conditions, utilized closeness factors such as:

$$\eta = y\sqrt{\frac{a}{v_f}}, u = axF'(\mu), v = -\sqrt{av_f}F(\eta), N = ax\sqrt{\frac{a}{v_f}}G(\eta, \theta(\eta) = \frac{T - T_\infty}{T_w - T_\infty},$$

(8.2.26)

By substituting the factors (27) into the Eq. (8.2.15), (8.2.16), (8.2.22), have:

$$\left(\frac{1}{(1-\varnothing)^{2.5}} + k\right)F'''(\eta) + \left(1 - \varnothing + \varnothing\frac{\rho_s}{\rho_f}\right)F(\eta)F''(\eta)$$
$$-\left(1 - \varnothing + \varnothing\frac{\rho_s}{\rho_f}\right)F'^2(\eta) - M\left(1 + \frac{3(\sigma - 1)\varnothing}{\sigma + 2 - (\sigma - 1)\varnothing}\right)F'(\eta) + kG'(\eta) = 0$$

(8.2.27)

$$\left(\frac{1}{(1-\varnothing)^{2.5}} + \frac{k}{2}\right)G''(\eta) + \left(1 - \varnothing + \varnothing\frac{\rho_s}{\rho_f}\right)F(\eta)G'(\eta)$$
$$-\left(1 - \varnothing + \varnothing\frac{\rho_s}{\rho_f}\right)F'(\eta)G(\eta) - k(2G(\eta) + F''(\eta)) = 0$$

(8.2.28)

And here the unused boundary conditions:

$$F(\eta) = S, F'(\eta) = \alpha, G(\eta) = -\delta F''(\eta), \theta(\eta) = 1, at\eta = 0$$

(8.2.29)

$$F'(\eta) \leftrightarrow 0, G(\eta) \leftrightarrow 0, \theta(\eta) \to 0at\eta \to \infty$$

(8.2.30)

The finite element method (abbreviated FEM) is a numerical method for obtaining putative solutions to a set of problems governed by elliptic partial differential equations. Such problems are called boundary value problems because they involve partial differential equations. (FEM) is a precious design tool for numerically approximating complex physical structures, even in standard explanatory configurations. In statistics, RSM analyzes the relationship between some explanatory variable and one or more response variables. K. B. Wilson and George E.P. Box used this approach in 1951. This approach uses a series of careful experiments to obtain the most effective response. For this purpose, Wilson and Box advocate using a second-order polynomial model. In your opinion, such models are only approximations. However, these authors used this model because it is easy to evaluate and apply even when little is known about the preparation.

The computational error in our work is very small compared to other works. The largest type of error occurred at $\eta=1.5$, and the smallest extent of error occurred at $\eta=4$.

Figure 11 shows the temperature variation of the MWCNT nanofluid flowing through the coil with three different current modes. The highest heat transfer and nanofluid temperature occurs around the coils with magnetic currents on the front surface of the plate. Hence, the heat transfer lines in these regions are closer to each other, and the nanoparticle molecules collide more. In the image above, there is more heat around the square coils at the top of the plate than around the bottom because the maximum heat flow between the nanofluid and the plate temperature is at the top. Indicates that it occurs in the plate. In general, the heat transfer between the fluid and the plate is low at the beginning of the plate with the onset of the thermal boundary layer, and the further away from the beginning, the thicker the boundary layer and the higher the heat and temperature transition. Here, the heat flow is directly related to the magnetic flow.

Figure 12 shows the velocity variation of the MWCNT nanofluid passing through the current coil in three different modes. According to the Nanofluidic flow line in Fig. b-(12), the lowest fluid velocity around the Helmholtz coil is observed at the bottom of the plate where the fluid pressure is low and negative, producing a highly viscous wake. The highest velocities of the MWCNT nanofluids are observed around the electrical coils at the edge of the traction plate c-(12). In this region, the thickness of the velocity boundary layer is larger than that of the thermal boundary layer, and the thickness increases with increasing liquid distance from the edge of the plate. In Figure f-(12), the most prominent nanofluid velocity contours occur far from the magnetic flux, where the velocity values decrease as they approach the plane. The velocity at which an object's angular position or orientation changes over time is represented in physics by the terms "angular velocity" or "rotational velocity" (or), also known as "angular frequency vector and angular frequency vector". (i.e., how to spin or rotate an object about a point or axis). The direction of the pseudo-vector is perpendicular to the current plane of rotation or angular displacement, and its magnitude corresponds to the angular velocity of the object, or the velocity at which it rotates or orbits. The right-hand rule is commonly used to indicate the direction of angular motion. Figure 13 shows the variation of the angular velocity of the MWCNT nanofluid passing through the current coil in three different modes. According to the available geometries, the change in angular velocity of the nanofluid along the pull surface is insignificant and smooth. Maximum angular velocity occurs in c-(13) mode. If a coil of current is used at the bottom of the traction seat, the lowest angular velocity will occur when the coil is in the middle of the seat. Magnetic capacity is a scalar magnitude function of degrees in the magnetic field, the negative of which is commensurate with the strength and symbolizes the effort required to claim a unit pole from an infinitely distant angle to the intended application. The hardware uses the magnetic potential to calculate the electric field. The magnetic potential parameters are higher near the Helmholtz coils placed on the surface (Figure 14), and around the coils under the foil when MWCNT nanoparticles from various positions of the coils are directed to the stretched surface. be a low value. The potential magnetic parameter of the bottom coil reaches its lowest value when the coil is placed in the center of the disc. In contrast, the top coil has the highest magnetic force. The potential magnet strength increases with the passage of the Nanofluidic stream moving from the beginning of the plate to the end of the plate, reaching a maximum near the end coils of the plate in the on-plate state. Table 5 shows the calculated MWCNT nanofluid velocities at different positions of the two Helmholtz coils. The maximum velocity of the MWCNT nanofluid is the emerging center of the surface at x = 0.7. The velocity was U = 1.82 m/s. A minimum MWCNT velocity was observed at x=0.1. The average MWCNT nanofluid velocity around the two Helmholtz coils in the center of the surface was on average 12.03 percent higher than at the beginning of the surface and 14.06 percent higher than at the end of the surface. Table 6 shows that the MWCNT stream started from the MWCNT nanoparticle stream passing from the stretched surface and exhibited a high temperature at x = 0.8 around the current Helmholtz coil of the surface. In this mode the temperature is T=34.05. The edge turns of the filament stretch surface are shown to have the lowest MWCNT nanofluid temperature at x=0.00. The MWCNT nanofluid temperature around the Helmholtz coil at the beginning of the surface is on average 37.16 percent higher than around the coil at the center of the coil and 27.24 percent higher than around the coil at the end of the coil. Table 7 examines the angular velocities at various locations around the two Helmholtz coils. The flow velocity of MWCNTs is higher near the current coil at the sheet edge at x = 0.6 because the MWCNT nanoparticles flow out from the stretching surface. The angular velocity of the MWCNT nanofluid is on average 25.46% faster around the current coil at the edge of the layer than the middle two Helmholtz coils and 41% faster than around the first Helmholtz coil of the layer. Table 8 examines the magnetic potential effect at various locations around the two Helmholtz coils. The angular velocity of the MWCNT nanofluid is on average 25.46% faster around the current coil at the edge of the layer than the middle two Helmholtz coils and 41% faster than around the first Helmholtz coil of the layer. Table 6 examines the magnetic potential effect at various locations around the two Helmholtz coils. The magnetic potential of the MWCNT flux has a huge value at x = 0.8 near the two coils in the middle of the surface, allowing the MWCNT nanoparticle flux to pass through the stretched surface. The average magnetic potential of the MWCNT nanofluid around the current coil at the center of the surface is 33.10 percent higher than the magnetic potential around the two Helmholtz coils at the beginning of the surface and higher than the magnetic potential around the surrounding wire. 79.90 percent higher plate edge. In this article, the RSM method was used to optimize nanoparticle velocity and heat exchange, as well as magnetic field passage in the extended plane. Twenty experiments were performed in the Design Expert software according to the RSM strategy to obtain the optimal algorithmic results. According to Figure 15 the magnetic effect of the MWCNT nanofluids was considered between a range of surface length variations (0.2 to 0.6), simulating the distance between the maximum and minimum potentials of the magnetic MWCNT nanofluids. The maximum magnetic potential of the MWCNT nanofluid occurred when the sheet length was 0.2 and the spacing between the coils was 0.3. This figure shows that shortening the sheet length and increasing the distance between the coils increases the magnetic effect. Figure 16 shows his quadratic effect of the magnetic counter changing as the height of the Helmholtz coil changes. According to this cantor, the magnetic parameters of MWCNT nanofluids increased from AZ=-1 to AZ=-0.2 by increasing the distance between current coils. This interpretation states that the magnetism of nanoparticles can be enhanced by increasing the distance between coils on the stretched surface. The coil spacing of the stretched slab increases with potential magnetic parameters, and additional magnetic generating force is provided by increasing the height of the coils on the top and bottom of the slab. The temperature change rate and the height change rate between the coils of the 3D MWCNT nanoparticles with respect to the distance between the coils were calculated in Figure 17. This Cantor claims that reducing the height of the coil increased the temperature from T=22 to T=29. This interpretation indicates that the temperature of the nanofluid increased as the distance of the coil from the strain surface increased. This model has good heat convection and high hydraulic pressure. In Figure 18, the maximum and minimum angular velocities of the MWCNT nanofluids are simulated and the MWCNT Nanofluidic velocities are investigated over a range of turn height variation (0.05–0.15) and inter-turn distance (0.1–0.3). When the maximum velocity of the MWCNT nanofluid was observed, the layer length size was 0.54 times N=4 and the gap size between the Helmholtz coils was 0.25. The magnitude of the angular velocity increases and reverses as the distance between the coils increases. Figure 19 shows the maximum and minimum MWCNT Nanofluidic velocities simulated for different coil height changes (0.05 to 0.15), inter-coil distances (0.1 to 0.3), and vertical MWCNT Nanofluidic velocities. Indicates that The MWCNT velocity parameter was increased from u=0 to u=2 by increasing the coil height from 0.07 to 0.15. MWCNT nanofluid Velocity Diagrams, Centers, and Magnetic Potentials in Hot Transfer Design – Expert Software.

Figure 10. The geometry of the problem for Helmholtz coil and stretching sheet

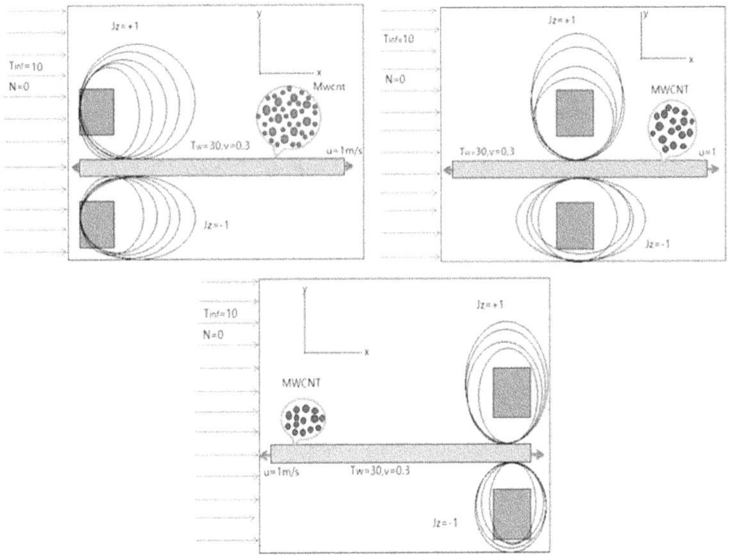

a) The passage of the nanofluid in different positions of the Helmholtz coil and stretching sheets.

b) Helmholtz coil.

Figure 11. Temperature alterations around the two circular Helmholtz coils on the thread-stretching surface

Invalid image: \\IGI-vFS01\Production_Image\Books\2025\9798369353356\979-8-3693-5335-6.ch008.f11.tif (Import failed.).

Figure 12. Velocity alterations around the two Helmholtz colis on the thread stretching sheet

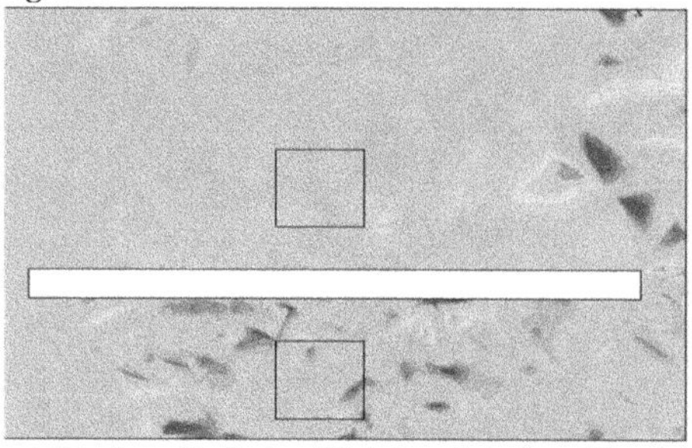

a) Velocity profile for two Helmholtz coils in the middle of the stretching surface.

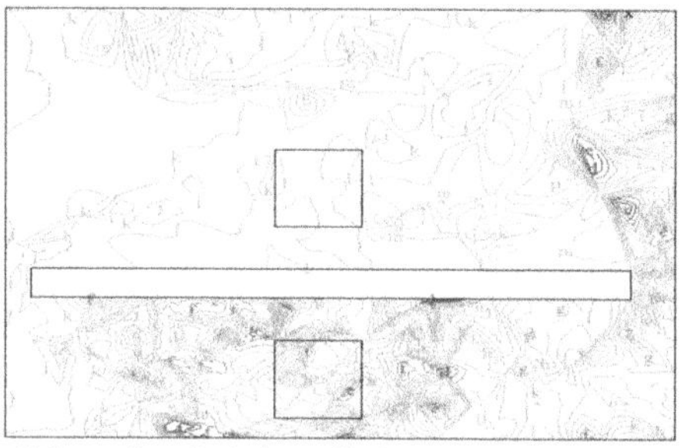

b) Stream line of velocity profile for two Helmholtz coils in the middle of the stretching surface.

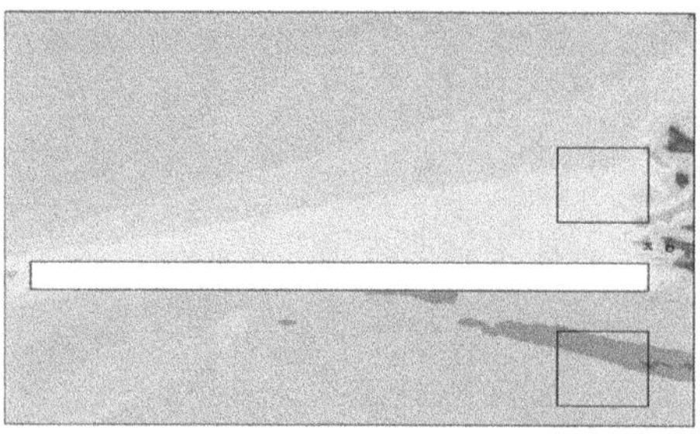

c) Velocity profile for two Helmholtz coils in the end of stretching surface.

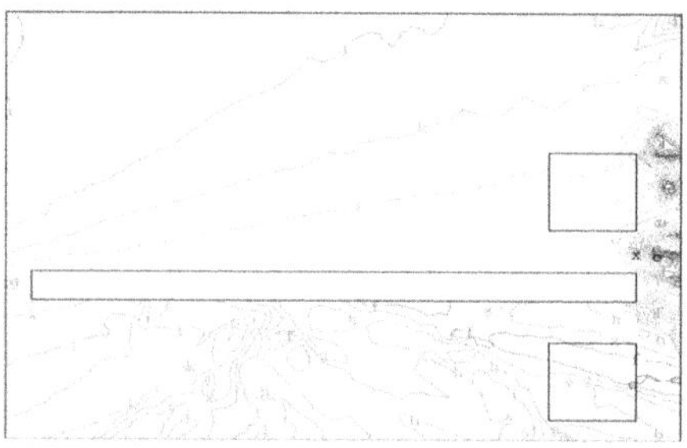

d) Stream line of velocity profile for two Helmholtz coils in the end of the stretching surface.

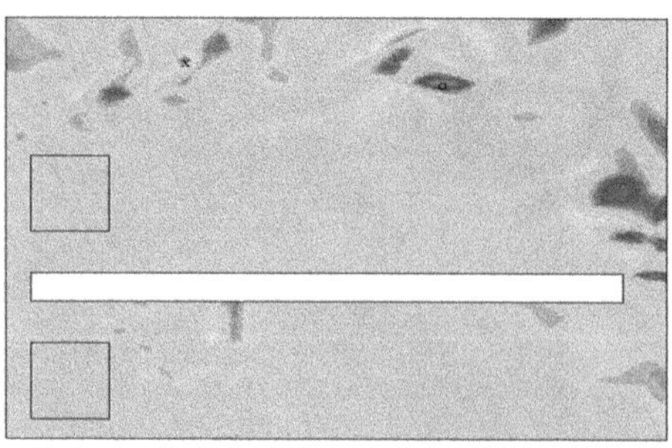

e) Velocity profile for two Helmholtz coils in the starting of stretching surface.

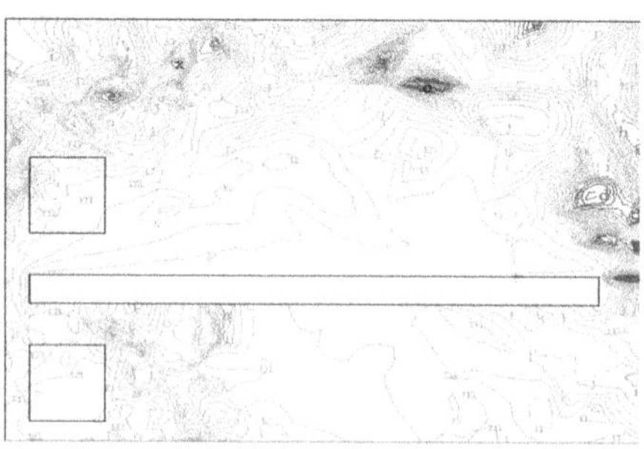

f) Stream line of Velocity profile for two Helmholtz coils in the starting of stretching surface.

Figure 13. Angular velocity alterations around the two Helmholtz coils on the thread stretching sheet

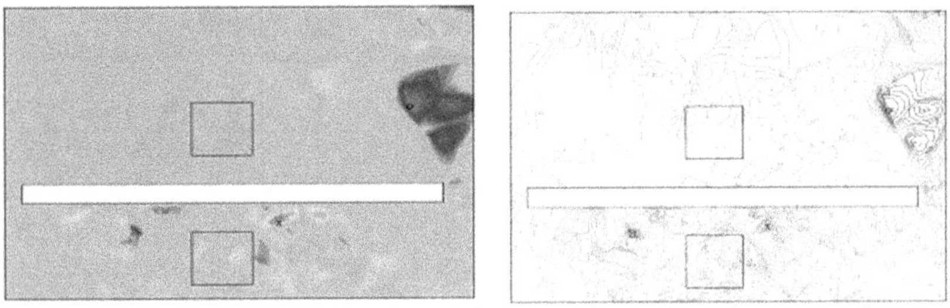

a) Angular velocity profile for two Helmholtz coils in the middle of stretching surface.

b) Stream line of angular velocity profile for two Helmholtz coils in the middle of the stretching surface.

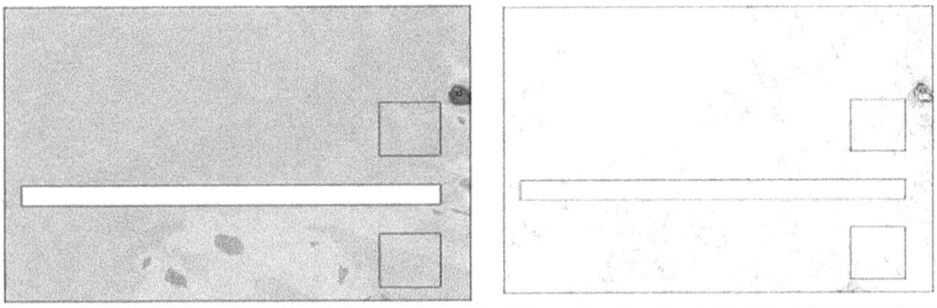

c) Angular velocity profile for two Helmholtz coils in the end of stretching surface.

d) Stream line of angular velocity for two Helmholtz coils in the end of stretching surface.

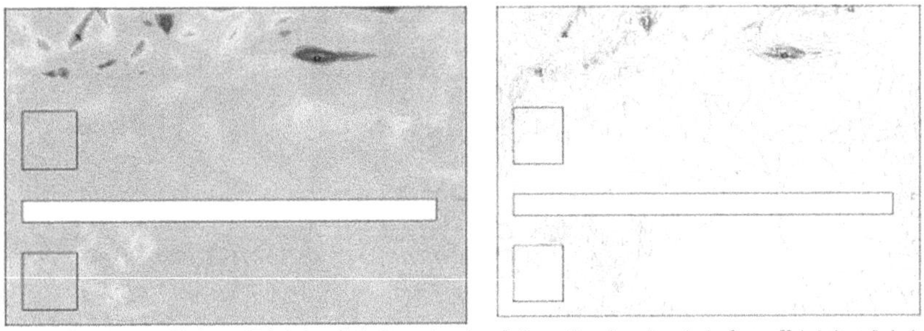

e) Angular velocity profile for two Helmholtz coils in the starting of stretching surface.

f) Stream line of angular velocity for two Helmholtz coils in the starting of stretching surface.

Figure 14. Three-dimensional images potential magnetic alterations around the two Helmholtz coils on the thread stretching surface

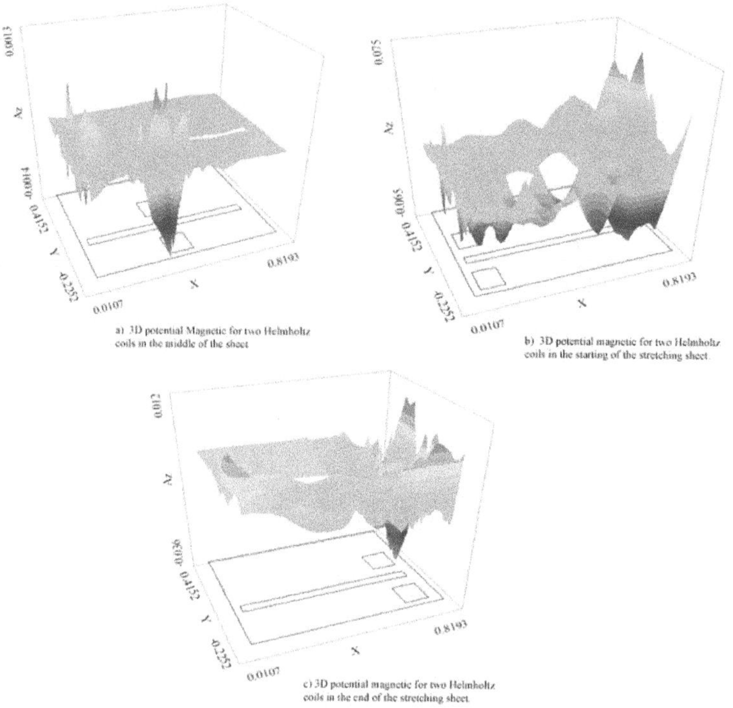

Figure 15. 3D graph RSM method in the Magnetism potential parameter for range of distance between coils

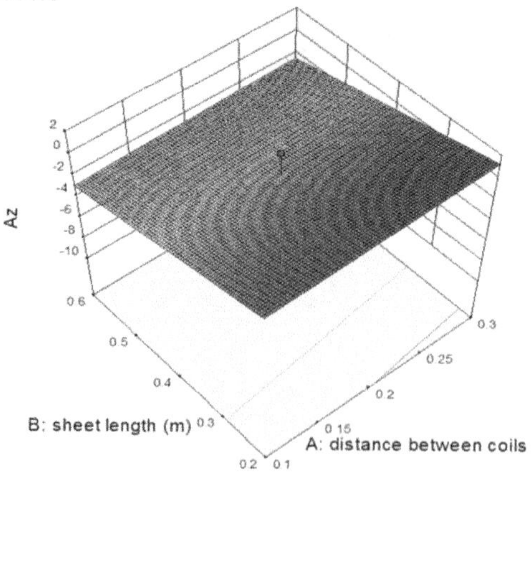

Figure 16. Two-dimensional graph RSM method in the magnetic parameter for range of coil height

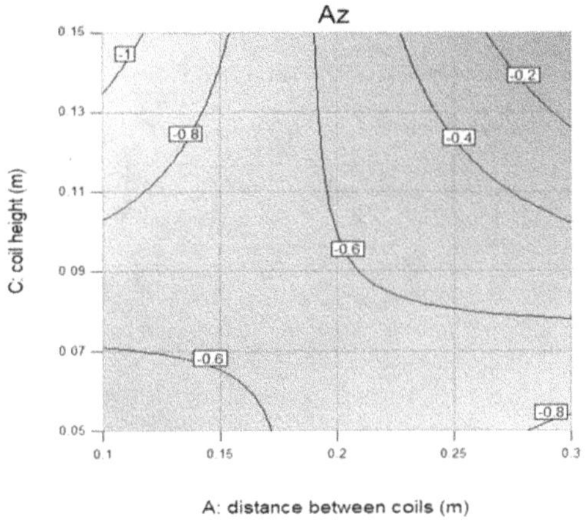

Figure 17. Three dimensional graph RSM method in the temperature parameter for range of distance between coils

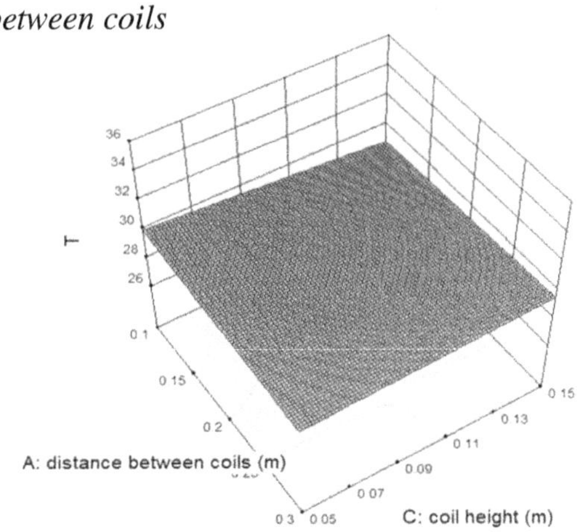

Figure 18. Three-dimensional graph RSM method in the angular velocity for range of distance between coils

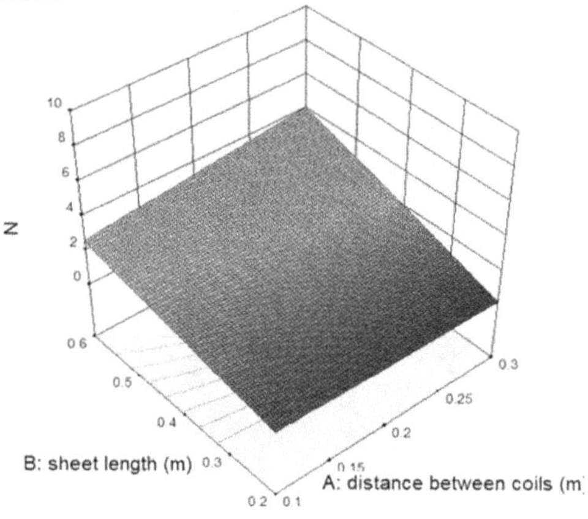

Figure 19. Three-dimensional graph RSM method in the velocity parameter for range of distance between coils

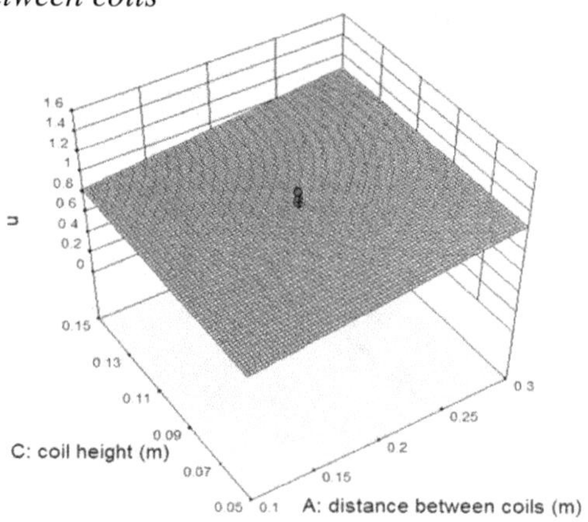

Table 4. Properties of MWCNT and H_2O at 30

	ρ(kg/m³)	K(w/m.k)	C_p (j/kg.k)	σ
H_2O	997	0.613	4179	0.05
MWCNT	1600	3000	796	0.00002

Table 5. Numerical comparison MWCNT nanofluid velocity between different modes of current Helmholtz coils

	$U_{starting\ of\ sheet}$	$U_{middle\ of\ sheet}$	$U_{end\ of\ sheet}$
0	**0.19**	**0.2**	**0.70**
0.1	**0.65**	**0.00**	**0.87**
0.2	**0.9**	**1.25**	**0.96**
0.5	**0.99**	**1.10**	**0.98**
0.6	**1.10**	**1.79**	**1.02**
0.7	**0.85**	**1.82**	**1.07**
0.8	**1.50**	**0.21**	**1.11**

Table 6. Numerical comparison MWCNT nanofluid temperature between different modes of current Helmholtz coils

	$T_{starting\ of\ sheet}$	$T_{middle\ of\ sheet}$	$T_{end\ of\ sheet}$
0.0	**27.45**	**5.03**	**5.01**
0.1	**28.25**	**26.06**	**24.90**
0.2	**32.52**	**24.70**	**25.61**
0.5	**27.17**	**29.39**	**28.37**
0.6	**28.04**	**30.00**	**30.07**
0.7	**27.23**	**28.80**	**28.10**
0.8	**34.05**	**33.33**	**27.54**

Table 7. Numerical comparison MWCNT nanofluid angular velocity between different modes of current Helmholtz coils

x	$N_{starting\ of\ sheet}$	$N_{middle\ of\ sheet}$	$N_{end\ of\ sheet}$
0	**2.69**	**-0.48**	**2.70**
0.1	**-6.02**	**-0.22**	**-6.11**
0.2	**0.02**	**0.26**	**0.00**
0.5	**3.74**	**0.56**	**4.12**
0.6	**6.74**	**0.49**	**6.88**
0.7	**0.0**	**0.50**	**3.15**
0.8	**2.03**	**0.53**	**1.11**

Table 8. Numerical comparison MWCNT nanofluid Magnetic potential between different modes of current Helmholtz coils

x	$AZ_{starting\ of\ sheet}$	$AZ_{middle\ of\ sheet}$	$AZ_{end\ of\ sheet}$
0	**0.93**	**-2.28**	**-1.83**
0.1	**0.00**	**-17.17**	**0.00**
0.2	**-0.06**	**-4.19**	**-0.10**
0.5	**0.00**	**2.18**	**-6.00**
0.6	**3.13**	**2.50**	**-7.08**
0.7	**1.04**	**2.12**	**-12**
0.8	**0.09**	**3.70**	**-13**

REFERENCES

Abdollahi, S. A., Jalili, P., Jalili, B., Nourozpour, H., Safari, Y., Pasha, P., & Ganji, D. D. (2023). Computer simulation of Cu: AlOOH/water in a microchannel heat sink using a porous media technique and solved by numerical analysis AGM and FEM. *Theoretical & Applied Mechanics Letters*, 13(3), 100432. DOI: 10.1016/j.taml.2023.100432

Abdul Latiff, N. A., Uddin, M. J., Bég, O. A., & Ismail, A. I. (2016). Unsteady forced bioconvection slip flow of a micropolar nanofluid from a stretching/shrinking sheet. *Proceedings of the Institution of Mechanical Engineers. Part N, Journal of Nanomaterials, Nanoengineering and Nanosystems*, 230(4), 177–187. DOI: 10.1177/1740349915613817

Berkan, S., Hoseini, S., & Ganji, D. (2017). Analytical investigation of steady three-dimensional problem of condensation film on inclined rotating disk by Akbari-Ganji's method. *Propulsion and Power Research*, 6(4), 277–284. DOI: 10.1016/j.jppr.2016.08.001

Berman, A. S. (1953). Laminar flow in channels with porous walls. *Journal of Applied Physics*, 24(9), 1232–1235. DOI: 10.1063/1.1721476

Harwin, D. A. (2007). *Flows in porous channels* (Doctoral dissertation, University of Bath, United Kingdom).

Jalili, B., Ganji, A. D., Jalili, P., Salman Nourazar, S., & Ganji, D. D. (2022). Thermal analysis of Williamson fluid flow with Lorentz force on the stretching plate. *Case Studies in Thermal Engineering*, 39, 102374. DOI: 10.1016/j.csite.2022.102374

Jalili, B., Jalili, P., Shateri, A., & Ganji, D. D. (2022). Rigid plate submerged in a Newtonian fluid and fractional differential equation problems via Caputo fractional derivative. *Partial Differential Equations in Applied Mathematics : a Spin-Off of Applied Mathematics Letters*, 6, 100452. DOI: 10.1016/j.padiff.2022.100452

Jalili, B., Rezaeian, A., Jalili, P., Ommi, F., & Ganji, D. D. (2023). Numerical modeling of magnetic field impact on the thermal behavior of a microchannel heat sink. *Case Studies in Thermal Engineering*, 45, 102944. DOI: 10.1016/j.csite.2023.102944

Jalili, B., Sadighi, S., Jalili, P., & Ganji, D. D. (2019). Characteristics of ferrofluid flow over a stretching sheet with suction and injection. *Case Studies in Thermal Engineering*, 14, 100470. DOI: 10.1016/j.csite.2019.100470

Jalili, J., Sadighi, S., Jalili, P., & Ganji, D. D. (2022). Numerical analysis of MHD nanofluid flow and heat transfer in a circular porous medium containing a Cassini oval under the influence of the Lorentz and buoyancy forces. *Heat Transfer*, 51(7), 6122–6138. DOI: 10.1002/htj.22582

Jalili, P., Ahmadi Azar, A., Jalili, B., Asadi, Z., & Domiri Ganji, D. (2022). Heat transfer analysis in cylindrical polar system with magnetic field: A novel hybrid analytical and numerical technique. *Case Studies in Thermal Engineering*, 40, 102524. DOI: 10.1016/j.csite.2022.102524

Jalili, P., Azar, A. A., Jalili, B., & Ganji, D. D. (2023). Study of nonlinear radiative heat transfer with magnetic field for non-Newtonian Casson fluid flow in a porous medium. *Results in Physics*, 48, 106371. DOI: 10.1016/j.rinp.2023.106371

Jalili, P., Azar, A. A., Jalili, B., & Ganji, D. D. (2023). The HAN method for a thermal analysis of forced non-Newtonian MHD Reiner-Rivlin viscoelastic fluid motion between two disks. *Heliyon*, 9(6), e17535. DOI: 10.1016/j.heliyon.2023.e17535 PMID: 37408890

Jalili, P., Sadeghi Ghahare, A., Jalili, B., & Domiri Ganji, D. (2023). Analytical and numerical investigation of thermal distribution for hybrid nanofluid through an oblique artery with mild stenosis. *SN Applied Sciences*, 5(4), 95. DOI: 10.1007/s42452-023-05312-z

Mirgolbabaee, H., Ledari, S., & Ganji, D. (2017). Semi-analytical investigation on micropolar fluid flow and heat transfer in a permeable channel using AGM. *Journal of the Association of Arab Universities for Basic and Applied Sciences*, 24(1), 213–222. DOI: 10.1016/j.jaubas.2017.01.002

Ziabakhsh, Z., & Domairry, G. (2008). Homotopy analysis solution of micro-polar flow in a porous channel with high mass transfer. *Advances in Theoretical and Applied Mechanics*, 1(2), 79–94.

Compilation of References

Abd-Alla, A., Abbas, I. A., Abo-Dhab, S. M., Elmhedy, Y., Sapoor, H., & Abdel-hafez, M. A. (2022). Effect of magnetic field and heat transfer on peristaltic flow of a micropolar fluid through a porous medium. *Waves in Random and Complex Media*, 1–12. DOI: 10.1080/17455030.2022.2058111

Abdel-Aziz, M. H. (2023). One-dimensional ternary conducting polymers blend with 9.26% power conversion efficiency for photovoltaic devices applications. *Alexandria Engineering Journal*, 66, 475–488. DOI: 10.1016/j.aej.2022.11.013

Abdollahi, S. A., Alizadeh, A., Esfahani, C., Zarinfar, M., & Pasha, P. (2023). Investigating heat transfer and fluid flow betwixt parallel surfaces under the influence of hybrid nanofluid suction and injection with numerical analytical technique. *Alexandria Engineering Journal*, 70, 423–439. DOI: 10.1016/j.aej.2023.02.040

Abdollahi, S. A., Jalili, P., Jalili, B., Nourozpour, H., Safari, Y., Pasha, P., & Ganji, D. D. (2023). Computer simulation of Cu: AlOOH/water in a microchannel heat sink using a porous media technique and solved by numerical analysis AGM and FEM. *Theoretical & Applied Mechanics Letters*, 13(3), 100432. DOI: 10.1016/j.taml.2023.100432

Abdul Latiff, N. A., Uddin, M. J., Bég, O. A., & Ismail, A. I. (2016). Unsteady forced bioconvection slip flow of a micropolar nanofluid from a stretching/shrinking sheet. *Proceedings of the Institution of Mechanical Engineers. Part N, Journal of Nanomaterials, Nanoengineering and Nanosystems*, 230(4), 177–187. DOI: 10.1177/1740349915613817

Abu-Nada, E., & Oztop, H. F. (2009). Effect of inclination angle on natural convection in enclosures filled with nanofluids. *International Journal of Heat and Fluid Flow*, 30(4), 669–678. DOI: 10.1016/j.ijheatfluidflow.2009.02.001

Adamo, B. C., Junger, A. S., Bressan, L. P., da Silva, J. A. F., Poppi, R. J., & de Jesus, D. P. (2020). Fast and straightforward in-situ synthesis of gold nanoparticles on a thread-based microfluidic device for application in surface-enhanced Raman scattering detection. *Microchemical Journal*, 156, 104985. DOI: 10.1016/j.microc.2020.104985

Agarwal, V., Singh, B., Kumari, A., Jamshed, W., Nisar, K. S., Almaliki, A. H., & Zahran, H. Y. (2021). Steady magnetohydrodynamic micropolar fluid flow and heat and mass transfer in permeable channel with thermal radiation. *Coatings*, 12(1), 11. DOI: 10.3390/coatings12010011

Ahmad, S., Ashraf, M., & Ali, K. (2020). Micropolar fluid flow with heat generation through a porous medium. *Punjab University Journal of Mathematics, 52*(4).

Ahmadi Azar, A., Jalili, B., Jalili, P., & Domiri Ganji, D. (2023). Investigating the effect of structural changes of two stretching disks on the dynamics of the MHD model. *Scientific Reports*, 13(1), 21833. DOI: 10.1038/s41598-023-48988-4 PMID: 38071256

Ahmad, S., Ashraf, M., & Ali, K. (2019). Numerical simulation of viscous dissipation in a micropolar fluid flow through a porous medium. *Journal of Applied Mechanics and Technical Physics*, 60(6), 996–1004. DOI: 10.1134/S0021894419060038

Ajmal, T. S., Arya, S. B., Maurya, P., & Shariff, S. M. (2022). Effect of hydrodynamics and laser surface melting on erosion-corrosion of X70 steel pipe elbow in oilfield slurry. *International Journal of Pressure Vessels and Piping*, 199, 104687. DOI: 10.1016/j.ijpvp.2022.104687

Alizadeh, A., Zekri, H. J., & Jafarmadar, S. (2020). Numerical simulation of the formation of vortices around rigid cylinders as a issue of fluid-structure interaction using immersed interface method. *Mechanics*, 26(1), 18–24. DOI: 10.5755/j01.mech.26.1.23186

Alizadeh, M., Dogonchi, A. S., & Ganji, D. D. (2018). Micropolar nanofluid flow and heat transfer between penetrable walls in the presence of thermal radiation and magnetic field. *Case Studies in Thermal Engineering*, 12, 319–332. DOI: 10.1016/j.csite.2018.05.002

Alizadeh, M., Nabizadeh, A., Fazlollahtabar, A., & Ganji, D. D. (2021). An optimization study of solidification procedure in a wavy-wall storage unit considering the impacts of NEPCM and curved fin. *International Communications in Heat and Mass Transfer*, 124, 105249. DOI: 10.1016/j.icheatmasstransfer.2021.105249

Alizadeh, M., Seyyedi, S. M., Rahni, M. T., & Ganji, D. D. (2017). Three-dimensional numerical simulation of rising bubbles in the presence of cylindrical obstacles, using lattice Boltzmann method. *Journal of Molecular Liquids*, 236, 151–161. DOI: 10.1016/j.molliq.2017.04.009

Alsabery, A. I., Hashim, I., Chamkha, A. J., Saleh, H., & Chanane, B. (2017). Effect of spatial side-wall temperature variation on transient natural convection of a nanofluid in a trapezoidal cavity. *International Journal of Numerical Methods for Heat & Fluid Flow*, 27(6), 1365–1384. DOI: 10.1108/HFF-11-2015-0488

Alsabery, A. I., Sheremet, M. A., Chamkha, A. J., & Hashim, I. (2018). MHD convective heat transfer in a discretely heated square cavity with conductive inner block using a two-phase nanofluid model. *Scientific Reports*, 8(1), 7410. DOI: 10.1038/s41598-018-25749-2 PMID: 29743641

Alsabery, A. I., Sheremet, M. A., Chamkha, A. J., & Hashim, I. (2019). Impact of nonhomogeneous nanofluid model on transient mixed convection in a double lid-driven wavy cavity involving a solid circular cylinder. *International Journal of Mechanical Sciences*, 150, 637–655. DOI: 10.1016/j.ijmecsci.2018.10.069

Al-Turki, Y. A. (2021). Numerical investigation of nanofluid flow and heat transfer in a pillow plate heat exchanger using a two-phase model: Effects of the shape of the welding points used in the pillow plate. *ZAMM-Journal of Applied Mathematics and Mechanics. Zeitschrift für Angewandte Mathematik und Mechanik.*

Anwar, T., & Kumam, P. (2022). A fractal fractional model for thermal analysis of GO– NaAlg– Gr hybrid nanofluid flow in a channel considering shape effects. *Case Studies in Thermal Engineering*, 31, 101828. DOI: 10.1016/j.csite.2022.101828

Arafa, A. A., & Gorla, R. S. R. (1992). Mixed convection boundary layer flow of a micropolar fluid along vertical cylinders and needles. *International Journal of Engineering Science*, 30(12), 1745–1751. DOI: 10.1016/0020-7225(92)90092-U

Ashraf, H., Siddiqui, A. M., Rana, M. A., & Gawo, G. A. (2021). Flow assessment of the shear rate dependent viscoelastic fluid: Application of biomechanics in growing human embryo transport. *Alexandria Engineering Journal*, 60(6), 5921–5934. DOI: 10.1016/j.aej.2021.04.055

Azar, A. A., Jalili, P., Moziraji, Z. P., Jalili, B., & Ganji, D. D. (2024). Analytical solution for MHD nanofluid flow over a porous wedge with melting heat transfer. *Heliyon*, 10(15). PMID: 39166020

Bahmani, M., Babagoli, M., Jalili, P., Jalili, B., & Ganji, D. D. (2024). Numerical study of the porous cavity with different square size vanes with a high focus on Nusselt number. *International Journal of Thermofluids*, 100860.

Bahmani, M., Babagoli, M., Jalili, P., Jalili, B., & Ganji, D. D. (2024). The numerical study on the MHD natural convection trend of square/circle corrugated porous media. *Journal of Engineering Research*. Advance online publication. DOI: 10.1016/j.jer.2024.05.012

Bahmani, M., Jalili, B., Jalili, P., Mirzaei, A., & Ganji, D. D. (2024). Effect of variations hollow of octagon porous media on heat and mass transfer. *International Journal of Thermofluids*, 21, 100576. DOI: 10.1016/j.ijft.2024.100576

Bai, J., Talebizadehsardari, P., Yekani Motlagh, N., & Goharkhah, M. (2022). Numerical analysis and two-phase modeling of water Graphene Oxide nanofluid flow in the riser condensing tubes of the solar collector heat exchanger. *Sustainable Energy Technologies and Assessments*, 53, 102408. DOI: 10.1016/j.seta.2022.102408

Barnoon, P., Toghraie, D., Mehmandoust, B., Fazilati, M. A., & Eftekhari, S. A. (2022). Natural-forced cooling and Monte-Carlo multi-objective optimization of mechanical and thermal characteristics of a bipolar disk for use in a proton exchange membrane fuel cell. *Energy Reports*, 8, 2747–2761. DOI: 10.1016/j.egyr.2022.01.199

Bazmi, M., Goharkhah, M., & Talebizadehsardari, P. (2019). Nitrogen-doped carbon nanotubes for heat transfer applications: Enhancement of conduction and convection properties of water/N-CNT nanofluid. *Journal of Thermal Analysis and Calorimetry*, 138(1), 69–79. DOI: 10.1007/s10973-019-08024-y

Berkan, S., Hoseini, S., & Ganji, D. (2017). Analytical investigation of steady three-dimensional problem of condensation film on inclined rotating disk by Akbari-Ganji's method. *Propulsion and Power Research*, 6(4), 277–284. DOI: 10.1016/j.jppr.2016.08.001

Berkan, S., Hosseini, S. R., Jalili, B., Ranjbar, A. A., & Ganji, D. D. (2024). A study on peristaltic flow and nanofluid in medication delivery systems considering heat transfer. *Discover Applied Sciences*, 6(4), 173. DOI: 10.1007/s42452-024-05847-9

Berman, A. S. (1953). Laminar flow in channels with porous walls. *Journal of Applied Physics*, 24(9), 1232–1235. DOI: 10.1063/1.1721476

Bessa, W. M., Ali, M. E., & Said, S. A. (2017). Nanofluid flow and heat transfer inside an annular finned cylinder subject to constant wall temperature. *Powder Technology*, 316, 504–514.

Bhatti, M. M., & Rashidi, M. M. (2017). Numerical simulation of entropy generation on MHD nanofluid towards a stagnation point flow over a stretching surface. *International Journal of Applied and Computational Mathematics*, 3(3), 2275–2289. DOI: 10.1007/s40819-016-0193-4

Borra, S. R., & Das, S. (2021). Impact of permeability on heat and mass transfer of a magnetohydrodynamic nanofluid flow over an inclined plate with variable magnetic field. *Journal of Thermal Science*, 25(5), 1231–1242.

Chari, F. N., Jalili, P., Jalili, B., & Ganji, D. D. (2024). MHD Casson non-Newtonian fluid flow in a channel with expanding/contracting porous walls in the presence of thermal radiation. *Proceedings of the Institution of Mechanical Engineers, Part E: Journal of Process Mechanical Engineering*.

Cheng, H.. (2023). The effect of temperature and external force on the thermal behavior of oil-based refrigerant inside a nanochannel using molecular dynamics simulation. *Journal of Molecular Liquids*, 369, 120893. DOI: 10.1016/j.molliq.2022.120893

Çobanoğlu, N., Banisharif, A., Estellé, P., & Karadeniz, Z. H. (2022). The developing flow characteristics of water-ethylene glycol mixture based Fe3O4 nanofluids in eccentric annular ducts in low temperature applications. *International Journal of Thermofluids*, 14, 100149. DOI: 10.1016/j.ijft.2022.100149

Dawar, A., Shah, Z., Tassaddiq, A., Islam, S., & Kumam, P. (2021). Joule heating in magnetohydrodynamic micropolar boundary layer flow past a stretching sheet with chemical reaction and microstructural slip. *Case Studies in Thermal Engineering*, 25, 100870. DOI: 10.1016/j.csite.2021.100870

Deebani, W. (2023). Convective micropolar fluid over inclined surface with thermal radiation and velocity slip condition effects: Duality and stability. *International Journal of Modern Physics B*, 2450114.

Dehghan Afifi, M., Jalili, B., Mirzaei, A., Jalili, P., & Ganji, D. (2024). The effects of thermal radiation, thermal conductivity, and variable viscosity on ferrofluid in porous medium under magnetic field. *World Journal of Engineering*. Advance online publication. DOI: 10.1108/WJE-09-2023-0402

Dey, D., & Borah, R. (n.d.). Stability analysis on dual solutions of second-grade fluid flow with heat and mass transfers over a stretching sheet.

Dey, D. (2018). Viscoelastic fluid flow through an annulus with relaxation, retardation effects, and external heat source/sink. *Alexandria Engineering Journal*, 57(2), 995–1001. DOI: 10.1016/j.aej.2017.01.039

Dib, A., Haiahem, A., & Bou-said, B. (2015). Approximate analytical solution of squeezing unsteady nanofluid flow. *Powder Technology*, 269, 193–199. DOI: 10.1016/j.powtec.2014.08.074

Dogonchi, A. S., Armaghani, T., Chamkha, A. J., & Ganji, D. D. (2019). Natural convection analysis in a cavity with an inclined elliptical heater subject to shape factor of nanoparticles and magnetic field. *Arabian Journal for Science and Engineering*, 44(9), 7919–7931. Advance online publication. DOI: 10.1007/s13369-019-03956-x

Dogonchi, A. S., Chamkha, A. J., Hashemi-Tilehnoee, M., Seyyedi, S. M., Rizwan-Ul-Haq, , & Ganji, D. D. (2019). Effects of homogeneous–heterogeneous reactions and thermal radiation on magneto-hydrodynamic Cu–water nanofluid flow over an expanding flat plate with non-uniform heat source. *Journal of Central South University*, 26(5), 1161–1171. DOI: 10.1007/s11771-019-4078-7

Dogonchi, A. S., Chamkha, A. J., Seyyedi, S. M., & Ganji, D. D. (2018). Radiative nanofluid flow and heat transfer between parallel disks with penetrable and stretchable walls considering the Cattaneo–Christov heat flux model. *Heat Transfer, Asian Research*, 47(5), 735–753. DOI: 10.1002/htj.21339

Dogonchi, A. S., & Ganji, D. D. (2017). Impact of Cattaneo-Christov heat flux on MHD nanofluid flow and heat transfer between parallel plates considering thermal radiation effect. *Journal of the Taiwan Institute of Chemical Engineers*, 80, 52–63. DOI: 10.1016/j.jtice.2017.08.005

Dogonchi, A. S., & Ganji, D. D. (2018). Effects of Cattaneo–Christov heat flux on buoyancy MHD nanofluid flow and heat transfer over a stretching sheet in the presence of Joule heating and thermal radiation impacts. *Indian Journal of Physics and Proceedings of the Indian Association for the Cultivation of Science*, 92(6), 757–766. DOI: 10.1007/s12648-017-1156-2

Ellahi, R., Hassan, M., & Zeeshan, A. (2016). Aggregation effects on water-based Al2O3 nanofluid over permeable wedge in mixed convection. *Asia-Pacific Journal of Chemical Engineering*, 11(2), 179–186. DOI: 10.1002/apj.1954

Ellahi, R., Hassan, M., & Zeeshan, A. (2017). Effects of magnetic field on water-based nanofluid containing Al2O3 ferroparticles: Mixed convection, a rotating permeable wedge. *Applied Thermal Engineering*, 118, 552–563.

Ellahi, R., Tariq, M. H., Hassan, M., & Vafai, K. (2017). On boundary layer nano-ferroliquid flow under the influence of a low oscillating stretchable rotating disk. *Journal of Molecular Liquids*, 229, 339–345. DOI: 10.1016/j.molliq.2016.12.073

Fakour, M., Vahabzadeh, A., Ganji, D. D., & Hatami, M. (2015). Analytical study of micropolar fluid flow and heat transfer in a channel with permeable walls. *Journal of Molecular Liquids*, 204, 198–204. DOI: 10.1016/j.molliq.2015.01.040

Farahani, S. D., Farahani, A. D., Tayebzadeh, F., & Mosavi, A. H. (2021). Melting of non-Newtonian phase change material in a finned triple-tube: Efficacy of non-uniform magnetic field. *Case Studies in Thermal Engineering*, 28, 101543. DOI: 10.1016/j.csite.2021.101543

Fard, M. S., Torabiyan, A., Jalili, P., Jalili, B., & Ganji, D. D. (2024). Investigating the magnetohydrodynamics non-Newtonian fluid movement on a tensile plate affected by variable thickness with dufour and soret effects: Akbari Ganji and finite element methods. *International Journal of Electrochemical Science*, 19(8), 100701. DOI: 10.1016/j.ijoes.2024.100701

Fathollahi, R. (2022). Analyzing the effect of radiation on the unsteady 2D MHD Al2O3-water flow through parallel squeezing disks by AGM and HPM. *Alexandria Engineering Journal.*

Fatunmbi, E., & Adeniyan, A. (2018). MHD stagnation point-flow of micropolar fluids past a permeable stretching plate in porous media with thermal radiation, chemical reaction, and viscous dissipation. *Journal of Advances in Mathematics and Computer Science*, 26(1), 1–19. DOI: 10.9734/JAMCS/2018/38595

Ganji, D. D., Mahboobtosi, M., Jalili, B., & Jalili, P. (2024). Heat transfer analysis of magnetized fluid flow through a vertical channel with thin porous surfaces: Python approach. *Case Studies in Thermal Engineering*, 60, 104643. DOI: 10.1016/j.csite.2024.104643

Ganji, D. D., Peiravi, M. M., & Abbasi, M. (2015). Evaluation of the heat transfer rate increases in retention pools nuclear waste. *International Journal of Nanodimension*, 6(4), 385–398.

Gao, X., Lv, J., Sun, B., & Liu, Y. (2022). Effects of bed permeability on the hydrodynamic characteristics in a channel with a vegetation patch: A modeling study. *Journal of Hydrology (Amsterdam)*, 612, 128046. DOI: 10.1016/j.jhydrol.2022.128046

Ghadikolaei, S. S., Hosseinzadeh, K., Hatami, M., & Ganji, D. D. (2018). MHD boundary layer analysis for micropolar dusty fluid containing Hybrid nanoparticles (Cu-Al_2O_3) over a porous medium. *Journal of Molecular Liquids*, 268, 813–823. DOI: 10.1016/j.molliq.2018.07.105

Ghasemi, S. E., Mohsenian, S., Gouran, S., & Zolfagharian, A. (2022). A novel spectral relaxation approach for nanofluid flow past a stretching surface in presence of magnetic field and nonlinear radiation. *Results in Physics*, 32, 105141. DOI: 10.1016/j.rinp.2021.105141

Ghosh, D., Meena, P. R., & Das, P. K. (2021). Heat transfer from a ferrofluid during generalized Couette flow through parallel plates in the presence of an orthogonal magnetic field. *International Journal of Thermal Sciences*, 164, 106895. DOI: 10.1016/j.ijthermalsci.2021.106895

Gnaneswara Reddy, M., Punith Gowda, R. J., Naveen Kumar, R., Prasannakumara, B. C., & Ganesh Kumar, K. (2021). Analysis of modified Fourier law and melting heat transfer in a flow involving carbon nanotubes. *Proceedings of the Institution of Mechanical Engineers, Part E: Journal of Process Mechanical Engineering*, 09544089211001353. DOI: 10.1177/09544089211001353

Gorla, R. S. R. (1988). Buoyancy effects on the boundary layer flow of a micropolar fluid along a vertical cylinder. *International Journal of Engineering Science*, 26(8), 883–892. DOI: 10.1016/0020-7225(88)90039-0

Goud, B. S. (2020). Thermal radiation influences on MHD stagnation point stream over a stretching sheet with slip boundary conditions. *International Journal of Thermofluids Science and Technology*, 7(2). Advance online publication. DOI: 10.36963/IJTST.2020070201

Hafez, N. M., Abd-Alla, A. M., & Metwaly, T. M. N. (2023). Influences of rotation and mass and heat transfer on MHD peristaltic transport of casson fluid through inclined plane. *Alexandria Engineering Journal*, 68, 665–692. DOI: 10.1016/j.aej.2023.01.038

Hai, T., Kadir, D. H., & Ghanbari, A. (2023). Modeling the emission characteristics of the hydrogen-enriched natural gas engines by multi-output least-squares support vector regression: Comprehensive statistical and operating analyses. *Energy*, 276, 127515. DOI: 10.1016/j.energy.2023.127515

Hajizadeh, S., Jalili, P., Jalili, B., & Ganji, D. D. (2024). The new analytical and numerical analysis of 2D stretching plates in the presence of a magnetic field and dependent viscosity. *Advances in Mechanical Engineering*, 16(4), 16878132231220361. DOI: 10.1177/16878132231220361

Hamarat, I., Kuram, E., & Ozcelik, B. (2018). Investigation of the mechanical, rheological, and morphological properties of acrylonitrile butadiene styrene blends with different recycling number content. *Proceedings of the Institution of Mechanical Engineers. Part E, Journal of Process Mechanical Engineering*, 232(4), 449–458. DOI: 10.1177/0954408917717994

Harwin, D. A. (2007). *Flows in porous channels* (Doctoral dissertation, University of Bath, United Kingdom).

Hashemi Kachapi, S. H., & Domairry Ganji, D. (2011). Analysis of nonlinear equations in fluids. In Domairry Ganji, D. (Ed.), *Progress in Nonlinear Science* (Vol. 2).

Hatami, M., Ganji, D. D., & Jing, D. (2016). Optimization of a circular-wavy cavity filled by nanofluid under the natural convection heat transfer condition. *International Journal of Heat and Mass Transfer*, 98, 758–767. DOI: 10.1016/j.ijheatmasstransfer.2016.03.063

Hayat, T., Ullah, A., Waqas, M., & Khan, M. I. (2020). Nonlinear radiative three-dimensional MHD nanofluid slip flow towards a rotating surface. *International Journal of Numerical Methods for Heat & Fluid Flow*, 30, 2791–2808.

Hossain, M., Acar, M., & Malalasekera, W. (2017). Optimisation of a thermal heat sink for an automotive PCM storage system using porous media. *Applied Thermal Engineering*, 125, 923–931.

Hosseinzadeh, S., & Ganji, D. D. (2022). A novel approach for assessment of MHD mixed fluid around two parallel plates by consideration hybrid nanoparticles and shape factor. *Alexandria Engineering Journal*, 61(12), 9779–9793. DOI: 10.1016/j.aej.2022.03.031

Humane, P. P., Patil, V. S., Patil, A. B., & Shamshuddin, M. D. (2023). Buongiorno modelled nanoliquid consequence of thermal and solutal convection on the magneto-micropolar fluid inside an inclined porous stretching device. *Journal of Nanofluids*, 12(1), 211–222. DOI: 10.1166/jon.2023.1949

Hussain, T., & Xu, H. (2022). Time-dependent squeezing bio-thermal MHD convection flow of a micropolar nanofluid between two parallel disks with multiple slip effects. *Case Studies in Thermal Engineering*, 31, 101850. DOI: 10.1016/j.csite.2022.101850

Hussein, A. K., Qasem, N. A., & Sheremet, M. A. (2016). Free convection of a nanofluid in a porous square cavity filled with a nanofluid and subjected to a magnetic field. *Journal of Heat Transfer*, 138, 101–114.

Hussein, A. K., Qasem, N. A., & Sheremet, M. A. (2017). Natural convection of a nanofluid in a porous enclosure with discrete heating. *Advances in Heat Transfer*, 49, 1–48.

Jafaripournimchahi, A., Shateri, A., Jalili, B., Jalili, P., & Ganji, D. D. (2024). The effects of magnetic field and thermal radiation on the mixed convection of Al2O3-Cu/water hybrid nanofluid over a permeable vertical flat plate. *Modern Physics Letters B*, 38(25), 2450242. DOI: 10.1142/S0217984924502427

Jalili, A., Sadinezhad, B., Fard, M., Khan, Y., Jalili, P., & Ganji, D. D. (2024). A novel technique to analyze the fractional model of Williamson and Casson non-Newtonian boundary layer flow. *Multidiscipline Modeling in Materials and Structures*, 20(2), 279–294. DOI: 10.1108/MMMS-09-2023-0309

Jalili, B., Aghaee, N., Jalili, P., & Ganji, D. D. (2022). Novel usage of the curved rectangular fin on the heat transfer of a double-pipe heat exchanger with a nanofluid. *Case Studies in Thermal Engineering*, 35, 102086. DOI: 10.1016/j.csite.2022.102086

Jalili, B., Alamdari, S. G., Jalili, P., & Gani, D. D. (2023). Analysis of bio-nanofluid flow over a stretching sheet with slip boundaries. *Results in Physics*, 54, 107083. DOI: 10.1016/j.rinp.2023.107083

Jalili, B., Azar, A. A., Jalili, P., & Ganji, D. D. (2023). Analytical approach for micropolar fluid flow in a channel with porous walls. *Alexandria Engineering Journal*, 79, 196–226. DOI: 10.1016/j.aej.2023.08.015

Jalili, B., Azar, A. A., Jalili, P., & Ganji, D. D. (2023). impact of variable viscosity on asymmetric fluid flow through the expanding/contracting porous channel: A thermal analysis. *Case Studies in Thermal Engineering*, 52, 103672. DOI: 10.1016/j.csite.2023.103672

Jalili, B., Azar, A. A., Jalili, P., Liu, D., Abdelmohimen, M. A., & Ganji, D. D. (2024). Investigation of the unsteady MHD fluid flow and heat transfer through the porous medium asymmetric wavy channel. *Case Studies in Thermal Engineering*, 61, 104859. DOI: 10.1016/j.csite.2024.104859

Jalili, B., Emad, M., Jalili, P., Ganji, D. D., Saleem, S., & Tag-eldin, E. M. (2023). Thermal analysis of boundary layer nanofluid flow over the movable plate with internal heat generation, radiation, and viscous dissipation. *Case Studies in Thermal Engineering*, 49, 103203. DOI: 10.1016/j.csite.2023.103203

Jalili, B., Emad, M., Malekshah, E. H., Jalili, P., Akgül, A., & Hassani, M. K. (2024). Investigating double-diffusive natural convection in a sloped dual-layered homogenous porous-fluid square cavity. *Scientific Reports*, 14(1), 7193. DOI: 10.1038/s41598-024-57395-2 PMID: 38531996

Jalili, B., Ganji, A. D., Jalili, P., Salman Nourazar, S., & Ganji, D. D. (2022). Thermal analysis of Williamson fluid flow with Lorentz force on the stretching plate. *Case Studies in Thermal Engineering*, 39, 102374. DOI: 10.1016/j.csite.2022.102374

Jalili, B., Ganji, A. M., Shateri, A., Jalili, P., & Ganji, D. D. (2023). Thermal analysis of non-Newtonian visco-inelastic fluid MHD flow between rotating disks. *Case Studies in Thermal Engineering*, 49, 103333. DOI: 10.1016/j.csite.2023.103333

Jalili, B., Jalili, P., Rostami, E., & Ganji, D. D. (2023). *A Semi-Analytical Approach to Nonlinear Mechanical Engineering Issues*. Cambridge Scholars Publishing.

Jalili, B., Jalili, P., Sadighi, S., & Ganji, D. D. (2021). Effect of magnetic and boundary parameters on flow characteristics analysis of micropolar ferrofluid through the shrinking sheet with effective thermal conductivity. *Zhongguo Wuli Xuekan*, 71, 136–150. DOI: 10.1016/j.cjph.2020.02.034

Jalili, B., Jalili, P., Shateri, A., & Ganji, D. D. (2022). Rigid plate submerged in a Newtonian fluid and fractional differential equation problems via Caputo fractional derivative. *Partial Differential Equations in Applied Mathematics : a Spin-Off of Applied Mathematics Letters*, 6, 100452. DOI: 10.1016/j.padiff.2022.100452

Jalili, B., Rezaeian, A., Jalili, P., Ganji, D. D., & Khan, Y. (2023). Squeezing flow of Casson fluid between two circular plates under the impact of solar radiation. *ZAMM-Journal of Applied Mathematics and Mechanics. Zeitschrift für Angewandte Mathematik und Mechanik*, 103(9), e202200455. DOI: 10.1002/zamm.202200455

Jalili, B., Rezaeian, A., Jalili, P., Ommi, F., & Ganji, D. D. (2023). Numerical modeling of magnetic field impact on the thermal behavior of a microchannel heat sink. *Case Studies in Thermal Engineering*, 45, 102944. DOI: 10.1016/j.csite.2023.102944

Jalili, B., Roshani, H., Jalili, P., & Ganji, D. D. (2023). Numerical study of Newtonian fluid flow characteristics in triangle enclosure considering various obstacle geometries. *International Journal of Thermofluids*, 20, 100515. DOI: 10.1016/j.ijft.2023.100515

Jalili, B., Roshani, H., Jalili, P., Jalili, M., Pasha, P., & Ganji, D. D. (2023). The magnetohydrodynamic flow of viscous fluid and heat transfer examination between permeable disks by AGM and FEM. *Case Studies in Thermal Engineering*, 45, 102961. DOI: 10.1016/j.csite.2023.102961

Jalili, B., Sadighi, S., Jalili, P., & Ganji, D. D. (2019). Characteristics of ferrofluid flow over a stretching sheet with suction and injection. *Case Studies in Thermal Engineering*, 14, 100470. DOI: 10.1016/j.csite.2019.100470

Jalili, B., Shateri, A., Akgül, A., Bariq, A., Asadi, Z., Jalili, P., & Ganji, D. D. (2023). An investigation into a semi-porous channel's forced convection of nano fluid in the presence of a magnetic field as a result of heat radiation. *Scientific Reports*, 13(1), 18505. DOI: 10.1038/s41598-023-44275-4 PMID: 37898603

Jalili, H., Ahmadi Azar, A., Esmaeili, K., Liu, D., Jalili, P., & Domiri Ganji, D. (2024). A novel approach to micropolar fluid flow between a non-porous disk and a porous disk with slip. *Zhongguo Wuli Xuekan*, 87, 118–137. DOI: 10.1016/j.cjph.2023.11.023

Jalili, J., Sadighi, S., Jalili, P., & Ganji, D. D. (2022). Numerical analysis of MHD nanofluid flow and heat transfer in a circular porous medium containing a Cassini oval under the influence of the Lorentz and buoyancy forces. *Heat Transfer*, 51(7), 6122–6138. DOI: 10.1002/htj.22582

Jalili, P., & Afifi, , M., Jalili, B., Mirzaei, A. M., & Ganji, D. (2023). Numerical study and comparison of two-dimensional ferrofluid flow in semi-porous channel under magnetic field. *International Journal of Engineering*, 36(11), 2087–2101. DOI: 10.5829/IJE.2023.36.11B.13

Jalili, P., Ahmadi Azar, A., Jalili, B., Asadi, Z., & Domiri Ganji, D. (2022). Heat transfer analysis in cylindrical polar system with magnetic field: A novel hybrid analytical and numerical technique. *Case Studies in Thermal Engineering*, 40, 102524. DOI: 10.1016/j.csite.2022.102524

Jalili, P., Ahmadi Azar, A., Jalili, B., & Domiri Ganji, D. (2023). A novel technique for solving unsteady three-dimensional brownian motion of a thin film nanofluid flow over a rotating surface. *Scientific Reports*, 13(1), 13241. DOI: 10.1038/s41598-023-40410-3 PMID: 37582817

Jalili, P., Alamdari, S. G., Jalili, B., Shateri, A., & Ganji, D. D. (2024). Analytical and numerical investigation of heat transfer of porous fin in a local thermal non-equilibrium state. *Heliyon*, 10(4). Advance online publication. DOI: 10.1016/j.heliyon.2024.e26424 PMID: 38420422

Jalili, P., Asadi, Z., Shateri, A., Jalili, B., Ahmad, H., Albalwi, M. D., & Ganji, D. D. (2024). Thermal analysis of nanofluid magnetic flow on a rotating disk in the presence of radiation considering response surface method. *Modern Physics Letters B*, 38(24), 2450217. DOI: 10.1142/S0217984924502178

Jalili, P., Azar, A. A., Jalili, B., & Ganji, D. D. (2023). Study of nonlinear radiative heat transfer with magnetic field for non-Newtonian Casson fluid flow in a porous medium. *Results in Physics*, 48, 106371. DOI: 10.1016/j.rinp.2023.106371

Jalili, P., Azar, A. A., Jalili, B., & Ganji, D. D. (2023). The HAN method for a thermal analysis of forced non-Newtonian MHD Reiner-Rivlin viscoelastic fluid motion between two disks. *Heliyon*, 9(6), e17535. DOI: 10.1016/j.heliyon.2023.e17535 PMID: 37408890

Jalili, P., Azar, A. A., Jalili, B., & Ganji, D. D. (2024). A Novel analytical investigation of a swirling fluid flow and a rotating disk in the presence of uniform suction. *Arabian Journal for Science and Engineering*, 49(8), 10453–10469. DOI: 10.1007/s13369-023-08391-7

Jalili, P., Jalili, B., Ahmad, I., Hendy, A. S., Ali, M. R., & Ganji, D. D. (2024). Python approach for using homotopy perturbation method to investigate heat transfer problems. *Case Studies in Thermal Engineering*, 54, 104049. DOI: 10.1016/j.csite.2024.104049

Jalili, P., Mahboob, M., Shateri, A., Jalili, B., & Domiri Ganji, D. (2024). A Comparative Study of Hybrid Analytical and Laplace Transform Approaches for Solving Partial Differential Equations in Python. *International Journal of Engineering*, 37(2), 352–364. DOI: 10.5829/IJE.2024.37.02B.11

Jalili, P., Mirzaei, A., Jalili, B., Shateri, A., Ganji, D. D., Ozsahin, D. U., & Ahmad, H. (2024). Thermal analysis of transverse fluid flow in a gradient porous media with the exponentially boundary conditions. *Modern Physics Letters B*, 38(10), 2350229. DOI: 10.1142/S0217984923502299

Jalili, P., Narimisa, H., Jalili, B., & Ganji, D. D. (2023). Micro-polar nanofluid in the presence of thermophoresis, hall currents, and Brownian motion in a rotating system. *Modern Physics Letters B*, 37(01), 2250197. DOI: 10.1142/S0217984922501974

Jalili, P., Narimisa, H., Jalili, B., Shateri, A., & Ganji, D. D. (2023). A novel analytical approach to micro-polar nanofluid thermal analysis in the presence of thermophoresis, Brownian motion and Hall currents. *Soft Computing*, 27(2), 677–689. DOI: 10.1007/s00500-022-07643-2

Jalili, P., Sadeghi Ghahare, A., Jalili, B., & Domiri Ganji, D. (2023). Analytical and numerical investigation of thermal distribution for hybrid nanofluid through an oblique artery with mild stenosis. *SN Applied Sciences*, 5(4), 95. DOI: 10.1007/s42452-023-05312-z

Jalili, P., Sharif Mousavi, S. M., Jalili, B., Pasha, P., & Ganji, D. D. (2024). Thermal evaluation of MHD Jeffrey fluid flow in the presence of a heat source and chemical reaction. *International Journal of Modern Physics B*, 38(08), 2450113. DOI: 10.1142/S0217979224501133

Jalili, P., Shateri, A., Ganji, A. M., Jalili, B., & Ganji, D. D. (2023). Analytical analyzing mixed convection flow of nanofluid in a vertical channel using python approach. *Results in Physics*, 52, 106908. DOI: 10.1016/j.rinp.2023.106908

Joens, M. A., & Swan, J. W. (2022). Unsteady and lineal translation of a sphere through a viscoelastic fluid. *Physical Review Fluids*, 7(1), 013301. DOI: 10.1103/PhysRevFluids.7.013301

Khan, S. A., Waqas, H., Muhammad, T., Farooq, U., & Alghamdi, M. (2021). Recent progress in melting phenomenon for magnetized hybrid nanofluid flow over a stretching surface with temperature dependent viscosity: A comparative study. *Journal of Materials Research and Technology*, 15, 3965–3973. DOI: 10.1016/j.jmrt.2021.09.099

Kodi, R., & Konduru, V. (2022). Heat and mass transfer on MHD convective unsteady flow of a Jeffrey fluid past an inclined vertical porous plate with thermal diffusion Soret and aligned magnetic field. *Materials Today: Proceedings*, 50, 2128–2134. DOI: 10.1016/j.matpr.2021.09.436

Krishna, M. V., Swarnalathamma, B., & Chamkha, A. J. (2018). Heat and mass transfer on magnetohydrodynamic chemically reacting flow of a micropolar fluid through a porous medium with Hall effects. *Special Topics & Reviews in Porous Media: An International Journal, 9*(4).

Kumar, V., Mukherjee, J., Sinha, S. K., & Ghosh, U. (2022). Combined electromechanically driven pulsating flow of nonlinear viscoelastic fluids in narrow confinements. *Journal of the Royal Society, Interface*, 19(189), 20210876. DOI: 10.1098/rsif.2021.0876 PMID: 35382577

Larestani, A., Aydin, M. J., & Mohamad, A. A. (2022). Predicting the surfactant-polymer flooding performance in chemical enhanced oil recovery: Cascade neural network and gradient boosting decision tree. *Alexandria Engineering Journal*, 61(10), 7715–7731. DOI: 10.1016/j.aej.2022.01.023

Li, H., Chen, H., Fan, H., & Zang, W. (2023). A comprehensive investigation of thermal conductivity in of monolayer graphene, helical graphene with different percentages of hydrogen atom: A molecular dynamics approach. *Colloids and Surfaces. A, Physicochemical and Engineering Aspects*, 656, 130324. DOI: 10.1016/j.colsurfa.2022.130324

Liu, Z., Gao, Z., Li, Y., & Sheikholeslami, M. (2018). Buoyancy-driven MHD nanofluid convection inside a wavy enclosure under magnetic field. *Journal of Thermal Analysis and Calorimetry*, 131, 3085–3095.

Liu, Z., Zhang, T., Zhao, Y., & Bi, S. (2019). Time-varying stiffness model of spur gear considering the effect of surface morphology characteristics. *Proceedings of the Institution of Mechanical Engineers. Part E, Journal of Process Mechanical Engineering*, 233(2), 242–253. DOI: 10.1177/0954408918775955

Li, Y. M., Sedeh, S. N., Toghraie, D., & Alizadeh, A. (2021). Computational hemodynamics and thermal analysis of laminar blood flow for different types of hypertension. *Mathematics and Computers in Simulation*, 188, 330–341. DOI: 10.1016/j.matcom.2021.04.011

Lou, Q., Ali, B., Rehman, S. U., Habib, D., Abdal, S., Shah, N. A., & Chung, J. D. (2022). Micropolar dusty fluid: Coriolis force effects on dynamics of MHD rotating fluid when Lorentz force is significant. *Mathematics*, 10(15), 2630. DOI: 10.3390/math10152630

Mabood, F., Shamshuddin, M. D., & Mishra, S. R. (2022). Characteristics of thermophoresis and Brownian motion on radiative reactive micropolar fluid flow towards continuously moving flat plate: HAM solution. *Mathematics and Computers in Simulation*, 191, 187–202. DOI: 10.1016/j.matcom.2021.08.004

Mahboobtosi, M., Ganji, A. M., Jalili, P., Jalili, B., Ahmad, I., Hendy, A. S., Ali, M. R., & Ganji, D. D. (2024). Investigate the influence of various parameters on MHD flow characteristics in a porous medium. *Case Studies in Thermal Engineering*, 59, 104428. DOI: 10.1016/j.csite.2024.104428

Mahmud, S., & Pop, I. (2016). Thermal and mass transfer characteristics of a porous rectangular cavity saturated with nanofluid under various thermal boundary conditions. *International Journal of Numerical Methods for Heat & Fluid Flow*, 26, 1730–1757.

Md Basir, M. F., Mackolil, J., Mahanthesh, B., Nisar, K. S., Muhammad, T., Anuar, N. S., & Bachok, N. (2021). Stability and statistical analysis on melting heat transfer in a hybrid nanofluid with thermal radiation effect. *Proceedings of the Institution of Mechanical Engineers, Part E: Journal of Process Mechanical Engineering*. DOI: 10.1177/09544089211033161

Mehmood, A., Afsar, K., Zameer, A., Awan, S. E., & Raja, M. A. Z. (2019). Integrated intelligent computing paradigm for the dynamics of micropolar fluid flow with heat transfer in a permeable walled channel. *Applied Soft Computing*, 79, 139–162. DOI: 10.1016/j.asoc.2019.03.026

Mirgolbabaee, H., Ledari, S. T., & Ganji, D. D. (2017). Semi-analytical investigation on micropolar fluid flow and heat transfer in a permeable channel using AGM. *Journal of the Association of Arab Universities for Basic and Applied Sciences*, 24(1), 213–222. DOI: 10.1016/j.jaubas.2017.01.002

Mirzaei, A., Jalili, B., Jalili, P., & Ganji, D. D. (2024). Free convection in a square wavy porous cavity with partly magnetic field: A numerical investigation. *Scientific Reports*, 14(1), 14152. DOI: 10.1038/s41598-024-64850-7 PMID: 38898150

Mirzaei, A., Jalili, P., Afifi, M. D., Jalili, B., & Ganji, D. D. (2023). Convection heat transfer of MHD fluid flow in the circular cavity with various obstacles: Finite element approach. *International Journal of Thermofluids*, 20, 100522. DOI: 10.1016/j.ijft.2023.100522

Mistry, U., & Vadali, M. (2022). A steady-state semi-analytical approximation of melt pool evolution in pulsed laser surface melting. *Journal of Manufacturing Processes*, 74, 123–135. DOI: 10.1016/j.jmapro.2021.11.064

Mittal, A. S. (2022). Study of radiation effects on unsteady 2D MHD Al2O3-water flow through parallel squeezing plates. *International Journal of Ambient Energy*, 43(1), 653–660. DOI: 10.1080/01430750.2019.1662843

Moatimid, G. M., Zekry, M. H., & Ibrahim, D. A. (2023). Nonlinear EHD instability of two viscoelastic fluids under the influence of mass and heat transfer. *Scientific Reports*, 13(1), 357. DOI: 10.1038/s41598-023-27410-z PMID: 36611060

Mohamed, R. A., & Abo-Dahab, S. M. (2009). Influence of chemical reaction and thermal radiation on the heat and mass transfer in MHD micropolar flow over a vertical moving porous plate in a porous medium with heat generation. *International Journal of Thermal Sciences*, 48(9), 1800–1813. DOI: 10.1016/j.ijthermalsci.2009.01.019

Mohammadi, A., Li, Z., Sheikholeslami, M., & Chamkha, A. J. (2018). Numerical simulation of MHD effects on nanofluid flows. *Journal of Molecular Liquids*, 252, 218–232.

Mozafarifard, M., Azimi, A., Sobhani, H., Smaisim, G. F., Toghraie, D., & Rahmani, M. (2022). Numerical study of anomalous heat conduction in absorber disk of a solar collector using time-fractional single-phase-lag model. *Case Studies in Thermal Engineering*, 34, 102071. DOI: 10.1016/j.csite.2022.102071

Muhammad, K., Hayat, T., & Alsaedi, A. (2021). Numerical study for entropy generation and melting heat in flow of modified hybrid nanomaterial (Ag+ MWCNTs+ SWCNTs+ Water). *Proceedings of the Institution of Mechanical Engineers, Part E: Journal of Process Mechanical Engineering*.

Nabi, H., Pourfallah, M., Gholinia, M., & Jahanian, O. (2022). Increasing heat transfer in flat plate solar collectors using various forms of turbulence-inducing elements and CNTs-CuO hybrid nanofluids. *Case Studies in Thermal Engineering*, 33, 101909. DOI: 10.1016/j.csite.2022.101909

Nadeem, S., Fuzhang, W., Alharbi, F. M., Sajid, F., Abbas, N., El-Shafay, A. S., & Al-Mubaddel, F. S. (2022). Numerical computations for Buongiorno nano fluid model on the boundary layer flow of viscoelastic fluid towards a nonlinear stretching sheet. *Alexandria Engineering Journal*, 61(2), 1769–1778. DOI: 10.1016/j.aej.2021.11.013

Nayak, M. K. (2019). HHR impact on 3D radiative stretched flow of Cu-H_2O nanofluid influenced by variable magnetic field and convective boundary condition. *International Journal of Thermofluids Science and Technology*, 6(2), 19060202. DOI: 10.36963/IJTST.19060202

Neuman, S. P., & Paul, A. W. (1970). Finite element method of analyzing steady seepage with a free surface. *Water Resources Research*, 6(3), 889–897. DOI: 10.1029/WR006i003p00889

Niknejadi, M., Goharkhah, M., Talebizadehsardari, P., & Zanganeh, H. (2023). Numerical simulation of the thermal-hydraulic performance of solar collector equipped with vector generators filled with two-phase hybrid nanofluid Cu-TiO2/H2O. *Engineering Analysis with Boundary Elements*, 151, 670–685. DOI: 10.1016/j.enganabound.2023.03.035

Nisar, K. S., Faridi, A. A., Ahmad, S., Khan, N., Ali, K., Jamshed, W., Abdel-Aty, A.-H., & Yahia, I. S. (2022). Cumulative impact of micropolar fluid and porosity on MHD channel flow: A numerical study. *Coatings*, 12(1), 93. DOI: 10.3390/coatings12010093

Noreen, S., Waheed, S., Lu, D., & Hussanan, A. (2020). Heat measures in performance of electro-osmotic flow of Williamson fluid in micro-channel. *Alexandria Engineering Journal*, 59(6), 4081–4100. DOI: 10.1016/j.aej.2020.07.013

Nouar, A., Dib, A., Kezzar, M., Sari, M. R., & Eid, M. R. (2021). Numerical treatment of squeezing unsteady nanofluid flow using optimized stochastic algorithm. *Zeitschrift für Naturforschung. Section A. Physical Sciences*, 76(10), 933–946. DOI: 10.1515/zna-2021-0163

Ojjela, O., Raju, A., & Kambhatla, P. K. (2017). Influence of thermophoresis and induced magnetic field on chemically reacting mixed convective flow of Jeffrey fluid between porous parallel plates. *Journal of Molecular Liquids*, 232, 195–206. DOI: 10.1016/j.molliq.2017.02.061

Paradis, P. Y., & Champliaud, H. (2021). New correlations for heat transfer in parallel-plate ducts at low Peclet number. *International Journal of Heat and Mass Transfer*, 178, 121595. DOI: 10.1016/j.ijheatmasstransfer.2021.121595

Pasha, P., Alibak, A. H., Nabi, H., & Tat Shahdost, F. (2022). Examination of heat transfer on extending sheet by variation iteration method strategy and investigation of arrangements for optimizing liquid properties. *Engineering Reports*, 4(10), 12505. DOI: 10.1002/eng2.12505

Pasha, P., Mirzaei, S., & Zarinfar, M. (2021). Application of numerical methods in micropolar fluid flow and heat transfer in permeable plates. *Alexandria Engineering Journal*.

Pasha, P., Taghinia, F., Nadalinia Chari, F., Jalili, B., Jalili, P., & Domiri Ganji, D. (2023). Chemical reaction-diffusion model around a vessel for studying temperature and concentration of three chemical species by finite element method. *International Journal of Engineering*, 36(1), 171–181. DOI: 10.5829/IJE.2023.36.01A.19

Patil, P. M., Doddagoudar, S. H., & Hiremath, P. S. (2019). Impacts of surface roughness on mixed convection nanofluid flow with liquid hydrogen/nitrogen diffusion. *International Journal of Numerical Methods for Heat & Fluid Flow*, 29(6), 2146–2174. DOI: 10.1108/HFF-11-2018-0703

Patil, P. M., Pop, I., & Roy, S. (2012). Unsteady heat and mass transfer over a vertical stretching sheet in a parallel free stream with variable wall temperature and concentration. *Numerical Methods for Partial Differential Equations*, 28(3), 926–941. DOI: 10.1002/num.20665

Patil, P. M., Roy, S., & Pop, I. (2010). Unsteady mixed convection flow over a vertical stretching sheet in a parallel free stream with variable wall temperature. *International Journal of Heat and Mass Transfer*, 53(21), 4741–4748. DOI: 10.1016/j.ijheatmasstransfer.2010.06.018

Patil, P. M., Roy, S., & Pop, I. (2012). Flow and heat transfer over a moving vertical plate in a parallel free stream: Role of internal heat generation or absorption. *Chemical Engineering Communications*, 199(5), 658–672. DOI: 10.1080/00986445.2011.614978

Peiravi, M. M., & Alinejad, J. (2019). Hybrid conduction, convection and radiation heat transfer simulation in a channel with rectangular cylinder. *Journal of Thermal Analysis and Calorimetry*, 140(6), 2733–2747. DOI: 10.1007/s10973-019-09010-0

Peiravi, M. M., & Alinejad, J. (2021). Nanoparticles distribution characteristics in multi-phase heat transfer between 3D cubical enclosures mounted obstacles. *Alexandria Engineering Journal*, 60(6), 5025–5038. DOI: 10.1016/j.aej.2021.04.013

Peiravi, M. M., Alinejad, J., Domairry Ganji, D., & Maddah, S. (2019). 3D optimization of baffle arrangement in a multi-phase nanofluid natural convection based on numerical simulation. *International Journal of Numerical Methods for Heat & Fluid Flow*, 30(5), 2583–2605. DOI: 10.1108/HFF-01-2019-0012

Peiravi, M. M., Alinejad, J., Domairry Ganji, D., & Maddah, S. (2019). Numerical study of fins arrangement and nanofluids effects on three-dimensional natural convection in the cubical enclosure. *Transport Phenomena in Nano and Micro Scales*, 7(2), 97–112.

Peiravi, M. M., Pasha, P. O. O. Y. A., & Ganji, D. D. (2021). Hydro magnetic effect on thread stretching surface with mounted obstacles in porous medium. *WSEAS Transactions on Heat and Mass Transfer*, 16, 95–105. DOI: 10.37394/232012.2021.16.12

Pop, I., & Ingham, D. B. (2001). *Convective Heat Transfer: Mathematical and Computational Modelling of Viscous Fluids and Porous Media*. Pergamon.

Qasemian, A., Goharkhah, M., & Goharkhah, S. (2021). Hydraulic and thermal analysis of automatic transmission fluid in the presence of nanoparticles and twisted tape: An experimental and numerical study. *Journal of Central South University*, 28(11), 3404–3417. DOI: 10.1007/s11771-021-4864-x

Qiao, L., Chen, H. E., Wang, M., Zhu, Y., Li, X., Tong, Z., Xie, S., Chen, H., & Chen, Z. (2022). Numerical simulation methods for motion-induced eddy current testing signals based on Ar formulation and edge finite elements. *NDT & E International*, 129, 102651. DOI: 10.1016/j.ndteint.2022.102651

Raju, T. L. (2019). MHD heat transfer two-ionized fluids flow between parallel plates with Hall currents. *Results in Engineering*, 4, 100043. DOI: 10.1016/j.rineng.2019.100043

Rashidi, M. M., Ganesh, N. V., Abdul Hakeem, A. K., & Ganga, B. (2014). Buoyancy effect on MHD flow of nanofluid over a stretching sheet in the presence of thermal radiation. *Journal of Molecular Liquids*, 198, 234–238. DOI: 10.1016/j.molliq.2014.06.037

Rashidi, M., Keimanesh, M., Bég, O. A., & Hung, T. K. (2011). Magnetohydrodynamic biorheological transport phenomena in a porous medium: A simulation of magnetic blood flow control and filtration. *International Journal for Numerical Methods in Biomedical Engineering*, 27(6), 805–821. DOI: 10.1002/cnm.1420

Raza, M., & Ahmad, S. (2020). Entropy analysis of MHD and radiation effects on squeezing flow in porous medium with variable fluid properties. *Results in Physics*, 18, 103209.

Reddy Gorla, R. S., Pender, R., & Eppich, J. (1983). Heat transfer in micropolar boundary layer flow over a flat plate. *International Journal of Engineering Science*, 21(7), 791–798. DOI: 10.1016/0020-7225(83)90062-9

Reddy, B. M., Waqas, M., Khan, S. U., & Khan, M. I. (2021). Magnetic field effects in radiative MHD nanofluid flow considering prescribed heat flux and Hall current with cross-diffusion effects. *International Communications in Heat and Mass Transfer*, 125, 105311.

Revnic, C., Ghasemi, E., & Pop, I. (2019). Natural convection in a triangular cavity filled with a nanofluid under the effect of an inclined magnetic field using the Buongiorno model. *Applied Thermal Engineering*, 153, 175–184.

Revnic, C., Sheremet, M. A., & Pop, I. (2016). Double-diffusive natural convection in a square cavity filled with a nanofluid using Buongiorno's model. *Computers & Mathematics with Applications (Oxford, England)*, 71, 362–373.

Reza, A., Houshmand, A., Nazari, F., & Ganji, D. D. (2021). A study of radiative and magnetic fields on squeezing nanofluid flow with heat source and sink using analytical method. *Results in Physics*, 20, 103708.

Riehl, R. R., & Mancin, S. (2022). Estimation of thermophysical properties for accurate numerical simulation of nanofluid heat transfer applied to a loop heat pipe. *International Journal of Thermofluids, 100158.*

Roshani, H., Jalili, P., Jalili, B., Ahmad, I., Hendy, A. S., Ali, M. R., & Ganji, D. D. (2024). The effect of magnetic field on the heat transfer in the porous medium octagonal cavity with Cassini oval barriers. *Case Studies in Thermal Engineering*, 56, 104194. DOI: 10.1016/j.csite.2024.104194

Saeed, A., Shah, R. A., Khan, M. S., Fernandez-Gamiz, U., Bani-Fwaz, M. Z., Noeiaghdam, S., & Galal, A. M. (2022). Theoretical analysis of unsteady squeezing nanofluid flow with physical properties. *Mathematical Biosciences and Engineering*, 19(10), 10176–10191. DOI: 10.3934/mbe.2022477 PMID: 36031990

Salehi, S., Nori, A., Hosseinzadeh, K., & Ganji, D. D. (2020). Hydrothermal analysis of MHD squeezing mixture fluid suspended by hybrid nanoparticles between two parallel plates. *Case Studies in Thermal Engineering*, 21, 100650. DOI: 10.1016/j.csite.2020.100650

Santos, R., Ribeiro, A. P. C., Moreira, A. L. N., & Moita, A. S. (2022). Effect of alumina nanofluids on bubble dynamics and heat transfer under quiescent conditions. *International Journal of Thermofluids, 100168.*

Seyyedi, S. M., Dayyan, M., Soleimani, S., & Ghasemi, E. (2015). Natural convection heat transfer under constant heat flux wall in a nanofluid-filled annulus enclosure. *Ain Shams Engineering Journal*, 6(1), 267–280. DOI: 10.1016/j.asej.2014.09.003

Seyyedi, S. M., Dogonchi, A. S., Hashemi-Tilehnoee, M., Waqas, M., & Ganji, D. D. (2020). Investigation of entropy generation in a square inclined cavity using control volume finite element method with aided quadratic Lagrange interpolation functions. *International Communications in Heat and Mass Transfer*, 110, 104398. DOI: 10.1016/j.icheatmasstransfer.2019.104398

Shadman, P., Parhizi, Z., Fathollahi, R., Zarinfar, M., Yu, A. E., & Pasha, P. (2023). Combined septum and chamfer fins on threaded stretching surface under the influence of nanofluid and the magnetic parameters for rotary seals in computer hardware. *Alexandria Engineering Journal*, 62, 489–507. DOI: 10.1016/j.aej.2022.07.044

Shah, S. A. A., Ahammad, N. A., Din, E. S. M. T. E., Gamaoun, F., Awan, A. U., & Ali, B. (2022). Bio-convection effects on prandtl hybrid nanofluid flow with chemical reaction and motile microorganism over a stretching disk. *Nanomaterials (Basel, Switzerland)*, 12(13), 2174. DOI: 10.3390/nano12132174 PMID: 35808009

Shah, Z., Shutaywi, M., Dawar, A., Kumam, P., Thounthong, P., & Islam, S. (2020). Impact of Cattaneo-Christov heat flux on non-isothermal convective micropolar fluid flow in a hall MHD generator system. *Journal of Materials Research and Technology*, 9(3), 5452–5462. DOI: 10.1016/j.jmrt.2020.03.071

Shamshuddin, M., & Ibrahim, W. (2022). Finite element numerical technique for magneto-micropolar nanofluid flow filled with chemically reactive Casson fluid between parallel plates subjected to rotatory system with electrical and Hall currents. *International Journal of Modelling and Simulation*, 42(6), 985–1004. DOI: 10.1080/02286203.2021.2012634

Shamshuddin, M., Mabood, F., Khan, W. A., & Rajput, G. R. (2023). Exploration of thermal Péclet number, vortex viscosity, and Reynolds number on two-dimensional flow of micropolar fluid through a channel due to mixed convection. *Heat Transfer*, 52(1), 854–873. DOI: 10.1002/htj.22719

Sharma, R. P., & Shaw, S. (2022). MHD Non-Newtonian fluid flow past a stretching sheet under the influence of non-linear radiation and viscous dissipation. *Journal of Applied and Computational Mechanics*, 8(3), 949–961.

Shateri, A., Moghaddam, M. M., Jalili, B., Khan, Y., Jalili, P., & Ganji, D. D. (2023). Heat transfer analysis of unsteady nanofluid flow between moving parallel plates with magnetic field: Analytical approach. *Journal of Central South University*, 30(7), 2313–2323. DOI: 10.1007/s11771-023-5388-3

Shehzad, N., Hayat, T., & Rashidi, M. M. (2016). Convective heat transfer for Maxwell nanofluid with variable conductivity and thermal radiation. *International Journal of Heat and Mass Transfer*, 99, 1037–1046.

Sheremet, M. A., Ghasemi, E., & Ellahi, R. (2020). Double-diffusive natural convection in a nanofluid-saturated porous cavity under the influence of magnetic field. *Microsystem Technologies*, 26, 1415–1432.

Sheremet, M. A., & Pop, I. (2014). MHD mixed convection in a square cavity filled with a nanofluid using the Buongiorno model. *Computers & Mathematics with Applications (Oxford, England)*, 67, 488–503.

Sheremet, M. A., & Pop, I. (2014). Natural convection in a porous cavity filled with a nanofluid under the influence of a magnetic field. *Transport in Porous Media*, 103, 419–436.

Sheremet, M. A., & Pop, I. (2017). Free convection in a trapezoidal porous cavity filled with a nanofluid using Buongiorno's model with heat and mass transfer. *Transport in Porous Media*, 116, 325–344.

Sheremet, M. A., Revnic, C., & Pop, I. (2017). Free convection in a porous wavy cavity filled with a nanofluid using Buongiorno's mathematical model with thermal dispersion effect. *Applied Mathematics and Computation*, 299, 1–15. DOI: 10.1016/j.amc.2016.11.032

Sheykhi, M., Eskandari, A., Ghafari, D., Ahmadi Arpanahi, R., Mohammadi, B., & Hosseini Hashemi, S. (2023). Investigation of fluid viscosity and density on vibration of nano beam submerged in fluid considering nonlocal elasticity theory. *Alexandria Engineering Journal*, 65, 607–614. DOI: 10.1016/j.aej.2022.10.016

Shoukat, Z., Hashir Zubair, M., Farman, M., Akgül, A., Sultan, M., Safarovich Sharipov, S., Botmart, T., Yahia, I. S., & Algarni, H. (2023). Impacts of joule heating with Cattaeno-Christove heat flux model in a MHD flow of Erying-Powell fluid on a Riga plate. *Alexandria Engineering Journal*, 64, 741–748. DOI: 10.1016/j.aej.2022.10.067

Sindhu, S., Gireesha, B. J., & Ganji, D. D. (2020). Simulation of Cu: Γ−ALO(OH)/Water in a microchannel heat sink by dint of porous media approach. *Case Studies in Thermal Engineering*, 21, 100723. DOI: 10.1016/j.csite.2020.100723

Timchenko, V. (2020). Investigation of permeability and tortuosity of gas diffusion layers with respect to the fabrication method. *Journal of Power Sources*, 400, 267–275.

Turkyilmazoglu, M. (2016). Flow of a micropolar fluid due to a porous stretching sheet and heat transfer. *International Journal of Non-linear Mechanics*, 83, 59–64. DOI: 10.1016/j.ijnonlinmec.2016.04.004

Uddin, M. J., Kabir, M. N., Alginahi, Y., & Bég, O. A. (2019). Numerical solution of bio-nano-convection transport from a horizontal plate with blowing and multiple slip effects. *Proceedings of the Institution of Mechanical Engineers. Part C, Journal of Mechanical Engineering Science*, 233(19-20), 6910–6927. DOI: 10.1177/0954406219867985

Uddin, M., Rana, P., Gupta, S., & Uddin, M. N. (2023). Bio-nanoconvective micropolar fluid flow in a Darcy porous medium past a cone with second-order slips and Stefan blowing: FEM solution. *Iranian Journal of Science and Technology. Transaction of Mechanical Engineering*, 47(4), 1–15. DOI: 10.1007/s40997-023-00626-0

Vatani, M., Ghasemi, S. E., & Ganji, D. D. (2016). Investigation of micropolar fluid flow between a porous disk and a nonporous disk using efficient computational technique. *Proceedings of the Institution of Mechanical Engineers. Part E, Journal of Process Mechanical Engineering*, 230(6), 413–424. DOI: 10.1177/0954408914557375

Viswanathan, K. K., Sathiyamoorthy, N., & Ganji, D. D. (2018). Analysis of micropolar fluid flow on a stretching/shrinking surface with multiple slip effects. *Alexandria Engineering Journal*, 57(4), 3047–3055.

Waini, I., Ishak, A., & Pop, I. (2021). Melting heat transfer of a hybrid nanofluid flow towards a stagnation point region with second-order slip. *Proceedings of the Institution of Mechanical Engineers. Part E, Journal of Process Mechanical Engineering*, 235(2), 405–415. DOI: 10.1177/0954408920961213

Wang, K., Li, G., & Lu, Z. (2022). Numerical simulation of dynamic fractures in 2D FGMs using the numerical manifold method. *Engineering Analysis with Boundary Elements*, 140, 32–47. DOI: 10.1016/j.enganabound.2022.04.002

Waqas, H., Farooq, U., Bukhari, F. F., Alghamdi, M., & Muhammad, T. (2021). Chemically reactive transport of magnetized hybrid nanofluids through Darcian porous medium. *Case Studies in Thermal Engineering*, 28, 101431. DOI: 10.1016/j.csite.2021.101431

Waqas, M., Hayat, T., Shehzad, S. A., & Alsaedi, A. (2018). Transport of magnetohydrodynamic nanomaterial in a stratified medium considering gyrotactic microorganisms. *Physica B, Condensed Matter*, 529, 33–40. DOI: 10.1016/j.physb.2017.09.128

Waqas, M., Khan, M. I., Hayat, T., & Alsaedi, A. (2017). Numerical simulation for magneto Carreau nanofluid model with thermal radiation: A revised model. *Computer Methods in Applied Mechanics and Engineering*, 324, 640–653. DOI: 10.1016/j.cma.2017.06.012

Waqas, M., Muhammad, T., Shehzad, N., & Alghamdi, M. (2020). Magnetic dipole and thermal radiation in unsteady radiative nanofluid flow. *Journal of Thermal Analysis and Calorimetry*, 141, 2929–2943. DOI: 10.1007/s10973-020-09570-1

Yazdi, M., Saedodin, S., & Shaterzadeh, P. (2020). Numerical analysis of the thermophysical properties and convective heat transfer enhancement of the two-phase nanofluid in a squeezing flow between two parallel plates. *Results in Physics*, 18, 103160.

Yen, C. Y., Kuo, P.-J., Lin, C.-Y., Nie-Shiuh Chang, N., Hsiao, H.-Y., Chin, Y.-T., Tsai, C.-C., & Lee, S.-Y. (2019). Accuracy of cone beam computed tomography in measuring thicknesses of hard-tissue-mimicking material adjacent to different implant thread surfaces. *Journal of Dental Sciences*, 14(2), 119–125. DOI: 10.1016/j.jds.2019.04.001 PMID: 31210886

Zangooee, M. R., Hosseinzadeh, K., & Ganji, D. D. (2019). Hydrothermal analysis of MHD nanofluid (TiO2-GO) flow between two radiative stretchable rotating disks using AGM. *Case Studies in Thermal Engineering*, 14, 100460. DOI: 10.1016/j.csite.2019.100460

Zar, H., Jalili, B., Jalili, P., & Ganji, D. D. (2024). Thermal study of magnetohydrodynamic nanofluid flow and brownian motion between parallel sheets. *International Journal of Thermofluids*, 23, 100806. DOI: 10.1016/j.ijft.2024.100806

Zar, P. M., Jalili, P., Jalili, B., & Ganji, D. D. (2024). Heat and mass transfer conduct in an unsteady two-dimensional stream between parallel sheets. *South African Journal of Chemical Engineering*, 50, 109–124. DOI: 10.1016/j.sajce.2024.07.011

Zeeshan, A., Shehzad, N., Ellahi, R., & Alamri, S. Z. (2017). Convective Poiseuille flow of Al2O3-EG nanofluid in a porous wavy channel with thermal radiation. *Neural Computing & Applications*. Advance online publication. DOI: 10.1007/s00521-017-2924-9

Zhang, A., Wang, Y., & Wang, H. (2023). Preparation of inorganic-polymer nanoemulsion inhibitor for corrosion resistance of steel reinforcement for concrete. *Alexandria Engineering Journal*, 66, 537–542. DOI: 10.1016/j.aej.2022.11.020

Zhang, G., Ma, Z., Yin, Y., & Wei, X. (2022). Influence of initial permeability on seepage behavior of the lubricant stored in hydrogel surface. *Tribology International*, 173, 107666. DOI: 10.1016/j.triboint.2022.107666

Zhang, Y. X., Gao, Y., Luo, Z., Song, Z., Huang, Q., & Zhou, W. (2023). Investigating the effect of size and number of layers of iron nanochannel on the thermal behavior and phase change process of calcium chloride/sodium sulfate hexa-hydrate with molecular dynamics simulation. *Journal of Energy Storage*, 62, 106762. DOI: 10.1016/j.est.2023.106762

Zhou, Y., Talebizadehsardari, P., & Goharkhah, M. (2023). Computational fluid dynamics and multi-objective response surface methodology optimization of perforated-finned heat sinks. *Journal of the Taiwan Institute of Chemical Engineers*, 145, 104823. DOI: 10.1016/j.jtice.2023.104823

Ziabakhsh, Z., & Domairry, G. (2008). Homotopy analysis solution of micro-polar flow in a porous channel with high mass transfer. *Advances in Theoretical and Applied Mechanics*, 1(2), 79–94.

Zohra, F. T., Uddin, M. J., & Ismail, A. I. M. (2019). Magnetohydrodynamic bio-nanoconvective Naiver slip flow of micropolar fluid in a stretchable horizontal channel. *Heat Transfer, Asian Research*, 48(8), 3636–3656. DOI: 10.1002/htj.21560

About the Authors

Davood Domiri Ganji received his Ph.D. degree from the Department of Mechanical Engineering at Nushirvani Engineering Complex, Mazandaran University, Babol, Mazandaran, Iran. He is currently an Assistant Professor at the same department. His research interests are based on finite difference method, finite volume method, inverse problem, fluid mechanics, aerodynamics, analysis of partial differential equation (PDE), spray modelling (using analytical methods such as homotopy perturbation method (HPM), variational iteration method (VIM), piecewise parabolic method (PPM)), as well as computational fluid dynamic (CFD). He was honored an Appreciation Certificate from the President, Appreciation Certificates from the Minister of Higher Education and Culture, Appreciation Certificates from the Head of Regime Interest Committee. He is the Editor-in-chief of International Journal of Nonlinear Dynamic and Engineering Science (IJNDES), and an Editor of International Journal of Nonlinear Sciences and Numerical Simulation (IJNSNS).

Index

N

P

S

T

W

www.ingramcontent.com/pod-product-compliance
Lightning Source LLC
LaVergne TN
LVHW061544290926
849162LV00004B/901